Hygiene for Management

A text for food safety courses

Ideal for use on CIEH • RIPH RSPH • REHIS Level 4 Courses

**Richard A. Sprenger,
BSc(Hons), DMS, FCIEH, MREHIS, FSOFHT**

Managing Director   Highfield.co.uk Limited

PUBLISHED BY
©HIGHFIELD.CO.UK LIMITED

Highfield House, Sidings Court, Lakeside,
Doncaster, South Yorkshire, DN4 5NL United Kingdom
**Tel: +44 0845 2260350
Facsimile: +44 0845 2260360
E-mail: rsprenger@highfield.co.uk**

**Websites:
www.highfield.co.uk
www.foodsafetytrainers.co.uk**

ISBN 1-904544-79-7

Improving Food Safety

HYGIENE *for* **MANAGEMENT**

TO JAYNE, JASON & SARAH, CHRISTIAN, BECKY, BETHANY, EVIE AND ISAAC

First published 1983
Reprinted 1985, 1988, 1989, 1991, 1993, 1995, 1998, 2002, 2003, 2004, 2005
13th Edition 2007

©HIGHFIELD.CO.UK LIMITED

The information provided in this book should not be taken as an authoritative statement or interpretation of the law. Expert legal advice should be obtained before taking action which could affect health or have financial consequences. Highfield.co.uk ltd does not accept any responsibility for loss or consequential loss by persons or organisations acting on information or guidance provided in Hygiene for Management.

All rights reserved. No part of this publication may be reproduced, stored in a retrieval system, or transmitted in any form or by any means, electronic, photocopying, recording or otherwise, without the prior permission of Highfield.co.uk Limited.

Printed by Trafford Press • Telephone: 01302 367509

Contents

		page
	Acknowledgements	4
	Preface	5
1	An introduction to food hygiene and food poisoning	7
2	Microbiology	20
3	Food poisoning	30
4	Foodborne diseases	52
5	Scientific principles for food safety	63
6	Food contamination and its prevention	70
7	Nutritional safety and food quality	90
8	The storage and temperature control of food	110
9	Food spoilage and preservation	142
10	Personal hygiene	158
11	Training and education of food handlers	171
12	The design and construction of food premises	186
13	The design of equipment	204
14	Cleaning and disinfection	216
15	Pest control	253
16	Control and monitoring of food standards and operations	282
17	Food safety legislation	328

APPENDICES

I	Food processing	358
II	Employee medical questionnaire	376
III	Hygiene courses	377

Glossary	380
Abbreviations	390
Useful websites	392
Bibliography	393
Index	394

Acknowledgements

During the compilation of this book, I have discussed the content with, and received information from, several food hygiene experts. I am grateful for their comments and assistance. I am indebted to Angela Hunter for correcting errors and making the English more readily understandable. I record sincere thanks to the following people and organizations for their contributions and support: Adrian R. Eley (University of Sheffield), Adrian N. Meyer (Acheta), Dr. Chris Barry (Achor Partners), Terry A. Roberts, Andy Bowles, Dr. Slim Dinsdale, Barry S. Michaels, Eunice Taylor (University of Salford), Roger Hart, Vic George, Dennis Farrar, Mark Kieran, Mark Du Val (LACORS), Graham Walker (REHIS), Steve Bagshaw and Peter Littleton (Holchem Laboratories Ltd.), Tony Stevens (Rentokil Initial plc.), Roy Ballam and Dr Sarah Schenker (British Nutrition Foundation), Andy Barnard (Campbell Grocery Products) and Richard North.

I would also like to thank the following companies for providing photographs:

Acheta	Garden Cottage, Horsemoor, Chieveley, Newbury, Berkshire, RG12 8XD. Tel & Fax: +44 (0)1635 248084 www.acheta.co.uk
Anticimex	Thunes vei 2, N-0274 Oslo. Box 472 Skøyen, N-0212 Oslo. Tel: +47 22542880 www.anticimex.se
Biotrace Ltd	The Science Park, Bridgend, Mid Glamorgan, CF31 3NA. Tel: +44 (0)1656 641400 www.biotrace.co.uk
Comark Ltd	Comark House, Gunnels Wood Park, Gunnels Wood Rd, Stevenage, Hertfordshire, SG1 2TA. Tel: 01438 367367 www.comarkltd.com
Drywite	The House of Lee, P.O. Box 1, Park Lane, Halesowen, West Midlands, B63 2RB. Tel: 01384 569556 www.drywite.co.uk
Foster Refrigerator	Oldmedow Rd, King's Lynn, Norfolk, PE30 4JU. Tel: 01553 691122 www.fosterrefrigerator.co.uk
Gemini Data Loggers (UK) Ltd	Scientific House, Terminus Rd, Chichester, West Sussex, PO19 8UJ. Tel: 01243 813000 www.geminidataloggers.com
Holchem Laboratories Ltd	Premier House, 175 Grane Rd, Haslingden, Rossendale, Lancashire, BB4 5ER. Tel: 01706 222288 www.holchem.co.uk
Johnson Diversey	Weston Favell Centre, Northampton, NN3 8PD. Tel: +44(0)1604 405311 www.johnsondiversey.com
Degussa Construction Chemicals (UK)	Albany House, Swinton Hall Road, Swinton, Manchester, MR27 4DT. Tel: 0161 7947411 www.mbtfeb.co.uk
Rentokil Initial plc	Felcourt, East Grinstead, West Sussex RH19 2JY. Tel: Lingfield 01342 833022 www.rentokil-initial.co.uk
Vikan (UK) Ltd.	Unit 23-Ash, Kembrey Park, Swindon, Wiltshire SN2 8UN. Tel: 01793 411130 www.vikan.com

Preface

All organizations work towards achieving objectives. The primary objectives of commercial food businesses are to produce, distribute, store, handle, prepare and sell food at a profit. High standards of hygiene are imperative to prevent food poisoning, food spoilage, loss of productivity, pest infestation and prosecutions for contraventions of legislation. However, these standards must be achieved at a reasonable cost, to ensure that the business remains profitable.

To ensure a profitable operation, managers must implement policies and systems relating to all their activities. One of the most important of these policies is maintaining cost-effective hygiene, by way of a food safety management system based on principles of HACCP (hazard analysis critical control point).

To implement an effective food safety management system, managers will require an understanding of all aspects of food hygiene, including basic microbiology and food science, together with the latest information on causes of foodborne illness, food vehicles and preventive measures.

Managers should be aware of the hazards and controls relating to their food business, especially in relation to temperature and microbiological, physical and chemical contamination. They should be able to identify points that are critical to food safety and implement effective control and monitoring procedures at these points. If monitoring reveals that a problem has occurred at a critical control point, a system must be in place to ensure timely and appropriate corrective action is taken.

High standards of personal hygiene, especially handwashing, are essential to reduce the risk of foodborne illness from low-dose pathogens, such as viruses and *E. coli* O157. Managers, supervisors and food handlers, must be committed to food safety and trained commensurate with their work activities so they are competent to produce safe food. Effective supervision, including the motivation and coaching of staff, is the key to the continuous implementation of good hygiene practices. Managers must lead by example.

The correct storage and handling of food is essential to profitable operation. Food that is stored at incorrect temperatures may result in food poisoning or spoilage, and failure to carry out stock rotation, will result in waste food and pest infestations. Supervision of food throughout processing is essential to avoid contamination, and suitable packaging, storage facilities and distribution vehicles must be provided.

To ensure hygienic and efficient operation, premises must be planned, designed and constructed in accordance with certain basic principles. The correct materials must be selected for surface finishes and equipment. Cleaning, disinfection and maintenance costs must be considered at the planning stage, as a modest increase in capital expenditure may reduce considerably the day-to-day operational costs.

Infestations of insects, rodents or birds are likely to result in food contamination, food wastage, prosecutions and even closure of food premises. Managers must be aware of the basic principles of pest control and the action to take in the event of an infestation.

Control and monitoring are essential functions of management, which ensure the production of safe food and compliance with legal requirements. Managers should provide a food safety policy and ensure the satisfactory implementation of HACCP. The skills to inspect the food operation and, where necessary, an understanding of microbiological sampling, will be advantageous.

Managers must know the legal requirements placed on themselves and the business. A knowledge of relevant food safety acts and regulations, and the role and powers of enforcement agencies and officers should ensure the business complies with legislation and can prioritize its food safety activities to remain profitable.

MANAGEMENT RESPONSIBILITIES

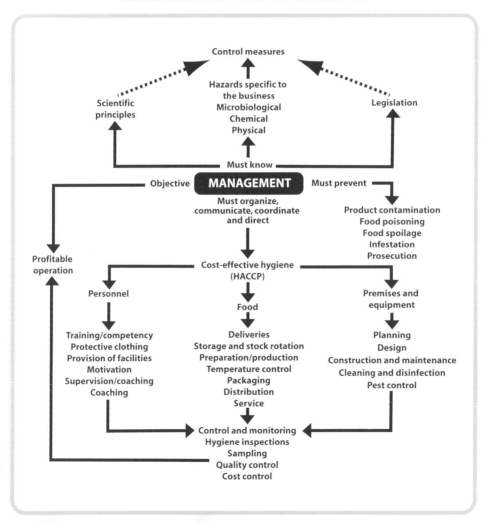

1 An introduction to food hygiene and food poisoning

Hygiene is the science of preserving health. A hygienic operation presents no risk of illness from the operations carried out therein. **Food hygiene** is much more than cleanliness, it involves all measures necessary to ensure the safety and **wholesomeness** of food during preparation, processing, manufacturing, packaging, storage, distribution, handling and offering for sale or supply to the consumer.

This will involve:
- rejecting contaminated food or food from unreliable sources;
- decontaminating food, for example, by washing;
- protecting food from risk of **contamination** of any kind and this includes the effective **cleaning** and **disinfection** of food premises and equipment, and requires high standards of **personal hygiene** and training of staff;
- preventing any organisms multiplying to an extent that would expose consumers to **risk**, or result in premature decomposition of food;
- destroying any harmful **bacteria** in the food by thorough cooking or processing; and
- discarding unfit or contaminated food.

The cost of food poisoning and poor food safety

Persons carrying on a food business have legal, commercial and moral obligations to provide **safe food**. The costs resulting from **food poisoning** can be very high, as are those from poor hygiene. These costs, both financial and social, fall on employers and employees as well as those persons who are ill. Costs for employers include:

- the loss of working days, and productivity, from illness caused by employees eating contaminated food. Even minor infections increase costs for employers through absence of employees;
- the closure of food premises by local authority action;
- a loss of business and reputation, either from bad publicity or from public reaction to poor standards, food poisoning **outbreaks** and even deaths;
- fines and costs of legal action taken because of contraventions of hygiene legislation or because of the sale of unfit or unsatisfactory food. In 2002 buffet caterers were fined £28,500 following an outbreak of salmonella at a wedding reception in East Yorkshire;
- civil action taken by food poisoning sufferers. In the USA compensation for **foodborne illness** is a major business. In 1993 Sizzler International Inc. paid 4.9 million dollars compensation as a result of two outbreaks of E. coli O157 at its restaurants. In May 1998 Ochralla Inc. paid around 15 million dollars to five children who were very ill after consuming apple juice contaminated with E. coli O157;
- food losses due to premature **spoilage** or damage, because of poor stock rotation, incorrect storage temperature or pest infestations;
- higher staff turnover, with attendant costs and inefficiencies from staff unwilling or

unable to tolerate poor standards;
- food complaints and costs of internal investigation;
- loss of production; and
- decontamination, cleaning and replacement of equipment.

Food employees may suffer by:
- losing their jobs because of closure, loss of business or because they become long-term **carriers** of food poisoning organisms, especially salmonella; and
- losing overtime or bonuses.

Cost-effective hygiene

Cost-effective hygiene involves the identification of steps in the operation critical to food safety and the allocation of available resources to control and monitor **hazards** at these **critical points.** To achieve this objective, it is imperative to implement a food safety management system based on the principles of **HACCP** (hazard analysis and critical control point) that is reviewed periodically.

High-risk foods

High-risk foods are those foods that are most likely to be the vehicles of the food poisoning organisms consumed in food poisoning incidents. They are **ready-to-eat foods** that, under favourable conditions, support the multiplication of **pathogenic** bacteria and are intended for consumption without treatment that would destroy such organisms. They are usually high in protein, requiring strict temperature control and protection from contamination and include:
- all cooked meat and poultry;
- cooked meat products including gravy, meat pies, pâté, cook-chill meals and stock;
- milk, cream, artificial cream, custards and dairy produce, especially unpasteurized

Examples of high-risk foods, including doner kebabs, exposed to risk of contamination.

milk and products made from unpasteurized milk, ripened soft and moulded cheeses;
- cooked eggs and egg products, especially those made with raw eggs and not thoroughly cooked, for example, mayonnaise, mousse and home-made ice cream, but excluding pastry, bread and similar baked goods;
- shellfish and other seafoods, for example, raw oysters, mussels and cooked prawns;
- baby foods; and
- cooked rice (not high in protein).

Low-risk foods

These are usually ambient stable foods such as bread, biscuits, cereals, crisps and cakes, but not cream cakes. These foods are rarely implicated in food poisoning and include:
- preserved foods such as sterilized milk or canned food, whilst unopened;
- dried foods or food with little available moisture, such as flour, rice, bread, and biscuits. When adding liquid to powdered food, such as milk, the food becomes high-risk and must be stored under refrigeration;
- acid foods such as fruit, vinegar or products stored in vinegar (pH <4.5);
- fermented products such as salami or pepperoni; and
- foods with high sugar/salt/fat content such as jam or chocolate.

Ready-to-eat raw foods

Raw fruit and salad vegetables would normally be considered low-risk. However, an increasing number of outbreaks of foodborne illness are being associated with ready-to-eat foods which do not support the multiplication of food poisoning bacteria. In the UK in 2001, S. Kedougou was isolated from mushrooms imported from the Republic of Ireland and 14 cases of salmonella resulted from the consumption of contaminated ready-to-eat salads. Between 1990 and 2000 six outbreaks (over 750 cases) of salmonella were reported in the USA and Canada involving cut melon. The salmonellae on the rind are transferred to the flesh when

Several outbreaks of food poisoning have involved ready-to-eat salads.

the melon is cut and will multiply if not refrigerated. Salmonella outbreaks have also involved chocolate and apple juice. Several outbreaks of salmonella and *E. coli* O157, in the USA, have been attributed to the consumption of raw or lightly cooked alfalfa and bean sprouts. In 1996 and 1997 two outbreaks of *E. coli* O157 involved over 6,000 cases. *E. coli* O157 has also been responsible for many outbreaks of illness involving, apple juice, lettuce and raspberries.

It is therefore recommended that all ready-to-eat food, which is intended for consumption without further treatment that would remove or destroy any foodborne pathogens, be treated as high-risk. Some ready-to-eat foods may be stored safely at ambient temperatures but they must always be protected from contamination, handled hygienically and never stored with raw food that may be a source of foodborne pathogens

Bacteria

Bacteria are essential to life. They are minute organisms, often referred to as germs, which are found everywhere, including on and in man and food. Bacteria on the body are usually confined to a particular site, and are known as commensals (part of the normal flora). For example, some species of staphylococcus are found on the skin and in the mouth and nose, other species are transient and may cause skin infections, such as boils. If harmful (pathogenic) species of staphylococcus, especially those which cause skin infections, are transferred to food they may cause illness. Most bacteria are harmless, some being necessary, for example, in yogurt and cheese manufacture. However, two types of bacteria create major problems within the food industry:

- *spoilage bacteria* – responsible for the decomposition of food; and
- *pathogenic bacteria* – responsible for causing illness such as dysentery, typhoid and food poisoning.

Food poisoning

Food poisoning is an acute illness, usually of sudden onset, brought about by eating contaminated or poisonous food. The symptoms normally include one or more of the following; abdominal pain, diarrhoea, vomiting and nausea.

*Food poisoning may be caused by:
- bacteria or their toxins;
- marine toxins produced by dinoflagellates;
- chemicals including metals;
- plants or fish; and
- mycotoxins (moulds).

* As viruses do not multiply in foods and are low-dose organisms that primarily cause illness by routes other than food, they are included with foodborne diseases in chapter 4.

Bacterial food poisoning

Bacterial food poisoning may be defined as "an acute disturbance of the gastrointestinal tract resulting in abdominal pain, with or without diarrhoea and vomiting, due to eating food contaminated by specific pathogenic bacteria or their toxins". The incubation period is normally short (between one and 36 hours). The number of bacteria required to cause illness in the healthy adult is usually large and multiplication of bacteria normally occurs within the food.

Patients usually recover in a few days but where body defences are low, illness may be prolonged and lead to complications. Botulism is usually much more serious, has different symptoms, often results in death, and survivors can take many months to recover. The use of anti-motility agents and **antibiotics** to treat ill people can sometimes be counterproductive. The carrier state can be prolonged and if gut movement is slowed down by using anti-motility agents, toxin from the bacteria stays in the intestines longer and a greater amount can be absorbed into the bloodstream. The destruction of bacteria may also result in more toxin being released and absorbed into the bloodstream.

Carriers

Carriers are people who show no symptoms of illness but excrete food poisoning or foodborne pathogens that may contaminate food, for example, salmonellae or shigellae. Organisms may be excreted intermittently.

Convalescent carriers are people who have recovered from an illness but still harbour and excrete the organism. The convalescent state may be quite prolonged and salmonellae are sometimes excreted for several months.

Healthy carriers are people who have displayed no symptoms but harbour and excrete the causal organism. Healthy carriers may have become infected with pathogenic bacteria from contact with raw food with which they work, particularly poultry or meat.

Symptomless carriers can only be confirmed by bacteriological or, in some cases, serological screening, i.e. examination of faecal specimens or blood. However, routine screening of **food handlers** to detect carriers is not cost-effective. Intermittent excretors may be missed and a person may become a carrier the week after screening.

Medical questionnaires/interviews of new starters (important to identify persons with a history of gastrointestinal illness), induction training, effective communication and supervision of company rules, counselling, bacteriological testing of persons returning to work after illness involving diarrhoea or vomiting or illness whilst on holidays abroad and contacts of persons suffering from food poisoning are all useful to assist in the detection of carriers.

Allergy

An identifiable immunological response to food or food additives may be described as an **allergy**. The allergen is usually a protein and several systems within the body may be affected, for example, the respiratory system, the gastrointestinal tract, the skin and the central nervous system. Symptoms vary considerably and may include bronchitis, vomiting, diarrhoea, urticaria (rash) and migraine. Reactions may be mild or extremely severe and may occur immediately the food is consumed, or up to 48 hours later. An allergic response should not be confused with food poisoning. The first exposure to the specific food, for example, peanuts or sesame seeds, does not produce symptoms, however; subsequent exposure results in a typical allergic response.

Common allergens include:
- peanuts;
- milk (lactose);
- eggs;
- fish;
- shellfish;
- wheat/gluten;
- celery;
- mustard;
- sesame seeds;
- sulphur dioxide and sulphites
- soy;
- tree nuts (walnut, almond, cashew, pecan etc.); and
- fruits.

THE INCIDENCE OF FOOD POISONING

Because high standards of food hygiene should prevent the contamination and multiplication of food poisoning bacteria in food, trends of food poisoning statistics are often used as a measure of hygiene standards or as an indication of the success or failure of control systems or enforcement procedures. In the late 1980s the increase in the reported number of salmonella cases, and the consequential public concern, was one of the main reasons for the enactment of the Food Safety Act, 1990.

Unfortunately, national statistics on the reported number of confirmed food poisoning cases are unavailable. Furthermore, trends in reported food poisoning statistics can be seriously distorted by factors that may have nothing to do with hygiene standards, such as the inclusion of new "infectious intestinal pathogens", which may or may not be transmitted by food, for example, viruses. The number of ill persons seeking medical help and the number of faecal specimens submitted will also affect the statistics, for example, because of new legislation or increased publicity.

Notifications of food poisoning

Doctors attending a case of suspected food poisoning (any disease of an infectious or toxic nature caused, or thought to be caused, by the consumption of food or water) are legally required to notify the Proper Officer of the local authority. Each week the Proper Officer sends details of notified infections to the Communicable Disease Surveillance Centre, which collates this data for England and Wales. This notification scheme facilitates the rapid identification and investigation of suspected cases of foodborne illness by authorized officers and enables implicated food premises to be identified as quickly as possible.

These notifications include cases that are bacteriologically unconfirmed and many cases of persons who are suffering from diarrhoea and/or vomiting that may not be foodborne. They could more accurately be described as "notified cases of gastroenteritis," and should not be used to indicate trends in food poisoning or hygiene standards. In fact, trends of these statistics are just as likely to be affected by such things as non-foodborne viral gastroenteritis and campylobacter enteritis, and the willingness of doctors to notify suspect cases of food poisoning.

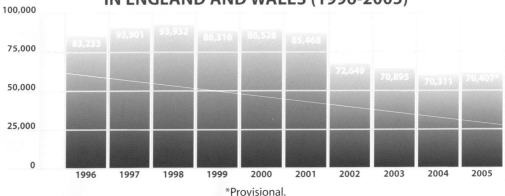

ANNUAL CORRECTED NOTIFICATIONS OF FOOD POISONING IN ENGLAND AND WALES (1996-2005)

1996: 83,233; 1997: 93,901; 1998: 93,932; 1999: 86,316; 2000: 86,528; 2001: 85,468; 2002: 72,649; 2003: 70,895; 2004: 70,311; 2005: 70,407*

*Provisional.

All statistics relating to England and Wales are provided courtesy of the HPA Communicable Disease Surveillance Centre. www.hpa.org.uk

In the absence of information on the number of confirmed food poisoning cases, alternative statistics should be used by persons interested in trends of food poisoning. Arguably, the most representative statistics of actual trends of food poisoning are laboratory isolations relating to specific food poisoning organisms such as salmonella.

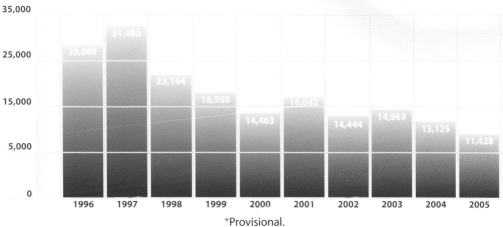

*Provisional.

Salmonella statistics

A case of bacterial food poisoning is a person with symptoms, usually diarrhoea and/or vomiting, who has become ill as a result of eating contaminated food. This can only be confirmed if the person is part of a confirmed food poisoning outbreak or by isolating food poisoning organisms or toxin from faecal or vomitus specimens. Statistics relating to isolates and suspect cases should not be confused with confirmed cases.

The bar chart showing the annual totals of salmonellosis is based on laboratory isolations from faecal specimens. Although the vast majority of isolates will be cases of salmonella food poisoning, these statistics also include symptomless excreters, persons who may have non-foodborne salmonellosis and persons who acquired the illness abroad. Nevertheless, at the present time, these statistics are the most accurate representation of salmonella food poisoning trends over the last ten years.

A survey in Yorkshire and Humberside between June and August 1986 showed that 27.4% of reported cases of salmonella food poisoning in that period originated abroad, although the national percentage is considered to be around 20%.

The Food Standards Agency (FSA)

The Food Standards Agency made a commitment to reduce the level of foodborne disease by 20% by the year 2006 (from a base of 2000). They did not use "notifications of food poisoning statistics", probably for the reasons outlined above. They have acknowledged that the most accurate statistics available are laboratory isolations of specific organisms. The number of isolates of the following organisms, which exclude illness acquired abroad, were used to determine the success of their food safety strategies:

STATISTICS FROM 2000, USED BY THE FSA TO DETERMINE IF FOOD POISONING HAD BEEN REDUCED BY THE YEAR 2006

Bacteria	UK	ENGLAND & WALES	SCOTLAND	N. IRELAND
Salmonella	13,122	11,456	1,338	328
Campylobacter	50,773	43,415	6,359	999
E. coli O157	1,035	790	196	49
Clostridium perfringens	166	124	32	10
Listeria monocytogenes	113	98	11	4
TOTAL	65,209	55,883	7,936	1,390

OUTBREAKS OF FOODBORNE INFECTIOUS INTESTINAL DISEASE (EXCLUDING PRIVATE RESIDENCES) IN ENGLAND AND WALES (1996-2005)

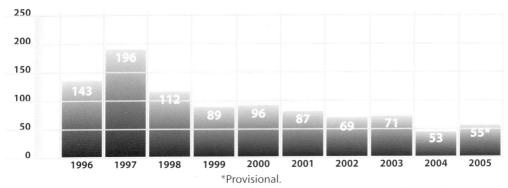

*Provisional.

These figures are a further indication that food poisoning from commercial premises may be much lower than some organizations and the media suggest.

Actual levels of foodborne illness

The number of cases of foodborne illness is significantly under reported. However, claims that millions of people in the UK suffer from foodborne illness as a result of eating food from commercial catering premises are difficult to justify. The Food Standards Agency has indicated that salmonella and campylobacter are the most common organisms implicated in foodborne illness. In 2000 there were 11,456 laboratory reports of salmonella and 43,415 of campylobacter that have been reduced by around 20% to account for illness acquired abroad.

In 1995 the Department of Health "Study of Infectious Intestinal Diseases" suggested that the actual number of salmonella cases was three times greater than reported and eight times greater for campylobacter. This would mean that there would have been a total of 34,368 cases of salmonella and 347,320 campylobacter in England and Wales. Once these figures have been adjusted to take account of non-foodborne cases and those originating from the home, it would appear that the number of actual cases resulting from all commercial premises could be a small percentage of the millions alleged in other books and the media. The Food Standards Agency have suggested that there are approximately 850,000 cases of foodborne illness per year (2005).

CAUSAL FACTORS RELATING TO 1479 OUTBREAKS OF FOOD POISONING IN ENGLAND AND WALES (1970-1982)*

	Causal factor	Total number of outbreaks in which factor recorded (%)**
1	Preparation too far in advance	844 (57)
2	Storage at ambient temperature	566 (38)
3	Inadequate cooling	468 (30)
4	Inadequate reheating	391 (26)
5	Contaminated processed food	246 (17)
6	Undercooking	223 (15)
7	Contaminated canned food	104 (7)
8	Inadequate thawing	95 (6)
9	Cross-contamination	94 (6)
10	Raw food consumed	93 (6)
11	Improper warm holding	77 (5)
12	Infected food handlers	65 (4)
13	Use of leftovers	62 (4)
14	Extra large quantities prepared	48 (3)

*Roberts D. 6th Edition of Food Poisoning and Food Hygiene.
**In some outbreaks more than one factor was involved.

Management failures resulting in food poisoning (Dr Richard North)
- A failure to carry out a risk-assessment when there is a change of menu, ingredients or recipes, for example, increasing the pH of mayonnaise.
- Lack of contingency planning, for example, when the refrigerator breaks down.
- Communication - a failure of management, or head office to provide the front line staff with the correct information. Communication breakdown may also involve the government and government bodies, for example, the failure to give timely warning to pregnant women on the dangers of eating pâté and soft cheeses (listeria).
- Management disincentives, for example, bonus paid in relation to the amount of cleaning chemicals used.
- Commercially driven misuse or abuse of equipment or premises, for example, overloading refrigerators or catering for numbers beyond the capacity.
- A failure to recognize potentially hazardous procedures of the operation, for example, colour-coded equipment all being washed in the same sink with no disinfection.
- Failure to learn lessons or implement recommendations following an earlier outbreak.
- Failure to replace complex or time-consuming operations, for example, refrigerators positioned a considerable distance from workstations. This militates against small amounts of food being prepared, which results in temperature abuse.
- Unrealistic demands placed on junior management or untrained staff.
- The absence of routine planning and consistent procedures.

LABORATORY REPORTS OF FOODBORNE DISEASE IN ENGLAND AND WALES (1996-2005)

ORGANISM	1996	1997	1998	1999	2000	2001	2002	2003	2004	2005*
Campylobacter spp.	43,978	51,360	56,852	56,254	57,674	54,918	47,848	46,181	44,294	46,298
E. coli O157	660	1,087	890	1,084	896	768	595	675	699	950
Giardia lamblia	5,379	5,340	4,660	4,387	4,008	3,529	3,267	3,434	3,179	2,924
Hepatitis A	1,086	1,310	1,104	1,391	1,047	801	1,356	996	669*	
Listeria monocytogenes	117	124	107	105	100	146	136	237	213	192
Shigella spp.	1,858	1,946	1,286	1,268	1,095	1,187	1,052	1,049	1,269	1,382
Cryptosporidium	3,587	4,393	3,670	5,045	5,774	3,625	3,011	5,860	3,614	4,527
Norovirus	2,450	2,182	1,874	1,901	1,922	1,745	4,308	2,327	3,133	2,922
Typhoid	178	145	133	155	172	180	147	203	206	231
Paratyphoid	146	172	189	189	158	253	161	204	217	229
Yersinia enterocolitica	150	148	76	70	26	29	30	34	19	26

*Provisional.

STATUTORY NOTIFICATIONS OF FOOD POISONING AND LABORATORY ISOLATIONS OF SALMONELLA IN SCOTLAND 1996-2005

The reported cases of food poisoning in Scotland include laboratory isolations of campylobacter.

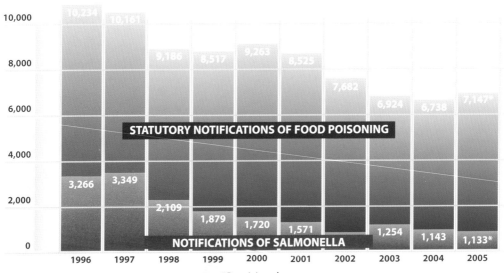

*Provisional.

LABORATORY REPORTS OF SPECIFIC FOODBORNE DISEASE IN SCOTLAND (1996-2005)

ORGANISM	1996	1997	1998	1999	2000	2001	2002	2003	2004	2005*
Campylobacter spp.	5,107	5,533	6,381	5,865	6,482	5,435	5,121	4,445	4,365	4,581
E. coli O157	506	423	217	294	197	235	229	148	209	172
Giardia lamblia	357	336	360	296	281	251	207	192	188	197
Listeria spp.	14	6	14	7	11	15	20	15	15	31
Shigella spp.	182	119	114	88	90	90	77	76	104	116
Cryptosporidium	618	690	879	598	867	569	646	822	465	708
Norovirus	87	81	108	163	213	348	1,477	1,442	1,320	1,552
Yersinia enterocolitica	52	25	30	8	9	15	12	47	51	28

*Provisional.
All statistics relating to Scotland courtesy of Health Protection Scotland.
www.hps.scot.nhs.uk

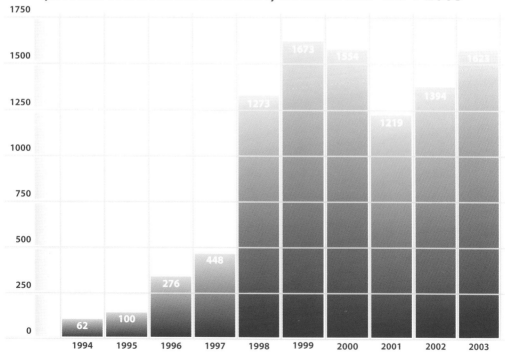

FOOD POISONING NOTIFICATIONS OF BACTERIA (OTHER THAN SALMONELLA) IN IRELAND 1994-2003

Values by year: 1994: 62; 1995: 100; 1996: 276; 1997: 448; 1998: 1273; 1999: 1673; 2000: 1554; 2001: 1219; 2002: 1394; 2003: 1623.

All statistics relating to the Republic of Ireland are provided courtesy of Health Protection Surveillance Centre. www.ndsc.ie

HYGIENE for MANAGEMENT *An introduction to food hygiene & food poisoning*

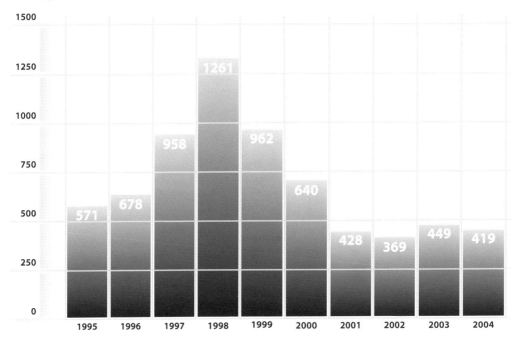

NOTIFICATIONS OF SALMONELLA IN IRELAND 1995-2004

Year	Cases
1995	571
1996	678
1997	958
1998	1261
1999	962
2000	640
2001	428
2002	369
2003	449
2004	419

CASES OF E. COLI O157 IN IRELAND 1998-2004

Year	Cases
1998	76
1999	51
2000	42*
2001	52†
2002	70
2003	88
2004	52

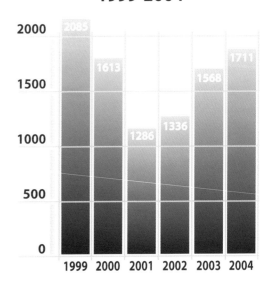

CASES OF CAMPYLOBACTER IN IRELAND 1999-2004

Year	Cases
1999	2085
2000	1613
2001	1286
2002	1336
2003	1568
2004	1711

*Including five non-residents.
†Including two non-residents.

NOTIFICATIONS OF INFECTIOUS INTESTINAL DISEASES IN NORTHERN IRELAND 1996-2005

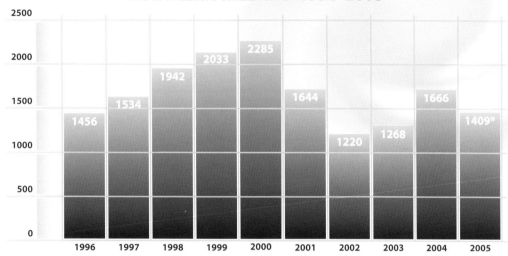

*Provisional

LABORATORY REPORTS OF FOODBORNE DISEASE IN NI (1996-2005)

ORGANISM	1996	1997	1998	1999	2000	2001	2002	2003	2004	2005
Campylobacter spp.	653	778	775	862	1001	885	817	743	849	890
F. coli O157	14	30	29	54	54	52	27	53	19	49
Giardiu lamblia	45	24	21	37	30	16	12	18	19	18
Hepatitis A	40	37	70	67	18	5	11	3	5	7
Listeria spp.	2	4	6	1	4	5	2	3	4	3
Salmonella spp.	413	432	534	689	425	367	240	214	451	180
Shigella spp.	154	24	14	12	11	16	9	13	8	7
Cryptosporidium spp.	98	82	180	181	417	360	126	140	137	165
Norovirus	7	11	35	90	68	131	396	115	276	209

Courtesy of the Communicable Disease Surveillance Centre Northern Ireland. www.cdscni.org.uk

2 Microbiology edited by Adrian Eley

Bacteria are single-celled organisms found everywhere; in soil, air, water, on people, animals and food. Bacteria are measured in micrometres (μm) and 1μm = 1/1000mm. Staphylococci have a diameter of around 0.75μm and salmonellae are about 3μm in length. The naked eye can see to approximately 75μm and consequently bacteria are only visible in large numbers when they form colonies or occasionally as a slime on the surface of food. Bacteria may be examined at a magnification of 1000 times using a powerful microscope. (A housefly similarly magnified would be nine metres long.)

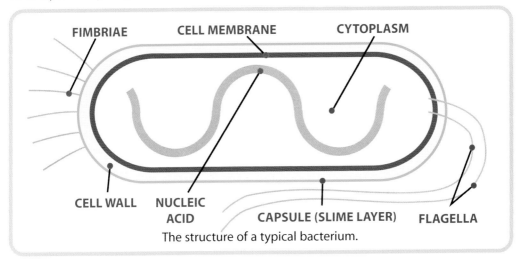

The structure of a typical bacterium.

Capsule	a gel-like secretion, which surrounds many bacteria and protects them against desiccation and harmful chemicals.
Cell wall	a rigid, permeable structure, which surrounds the bacterium and gives the cell its shape and strength.
Cell membrane	a selectively permeable membrane, which controls the passage of nutrients and waste products both into and out of the cell.
Cytoplasm	the major part of the cell and the medium within which the metabolic reactions occur.
Flagella	present in some bacteria for locomotion.
Fimbriae (or pili)	often involved in adherence.
Nucleic acid (such as DNA)	determines the genetic characteristics of the bacterium.

Appearance of bacteria

Bacteria vary considerably in shape:

- *Cocci* are spherical; some form chains, for example, streptococci. Others form clusters, for example, staphylococci.
- *Rods* are sausage-shaped, for example, salmonellae and *Escherichia coli*.
- *Spirochaetes* are spiral.
- *Vibrios* are comma-shaped.

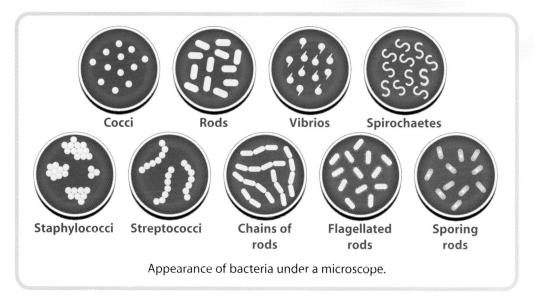

Appearance of bacteria under a microscope.

Spore formation

With minor exceptions only two bacterial genera, *Bacillus* and *Clostridium*, are able to form spores that are capable of surviving adverse conditions, such as lack of nutrients. Temperatures achieved in normal cooking may destroy the vegetative bacteria but any spores present will probably survive. Furthermore, these high temperatures trigger the spore into germination and as the food cools down, providing the environment is suitable, a new mature vegetative cell capable of reproduction is formed.

Spores may be resistant to desiccation, disinfectants and heat. Very high temperatures in excess of boiling (100°C) are often required, for long periods, to ensure their destruction. The time and temperature used in the canning of low-acid foods is based on the destruction of *Clostridium botulinum* spores (121°C for three minutes). The survival of spores is a hazard.

Toxin production

Certain bacteria release poisons known as toxins. Some toxins, such as those produced by *Cl. botulinum,* are often the cause of death of persons consuming food contaminated by this organism.

Exotoxins

These are highly toxic proteins usually produced during the multiplication, or sporulation, of some Gram-positive bacteria, for example, *Staphylococcus aureus* and *Bacillus cereus*. Quite often these toxins are produced in food and occasionally are heat-resistant so that, although

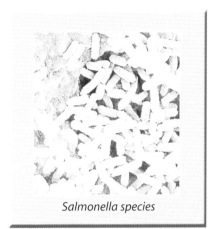
Salmonella species

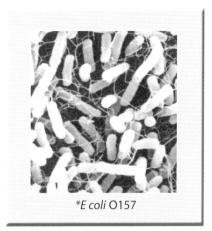
*E coli O157

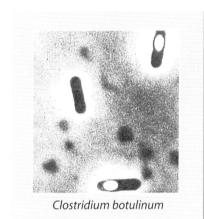
Clostridium botulinum

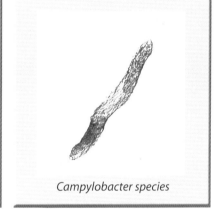
Campylobacter species

Bacillus cereus

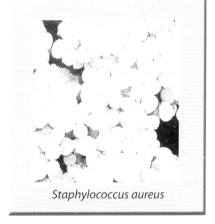
Staphylococcus aureus

Examples of pathogenic bacteria as viewed under an electron microscope. (Not to scale) (Courtesy of Unilever PLC.)

*(Courtesy of Professor T H Pennington, University of Aberdeen.)

cooking may destroy the bacteria, the toxin is unaffected and can still cause problems if the food is eaten. Toxins which are not destroyed are a hazard. Botulinum **exotoxin** is not heat-resistant. **Enterotoxins** are exotoxins that affect the gastrointestinal tract.

Endotoxins
These reside in the outer membrane and are released on the death of many Gram-negative bacteria. **Endotoxins** are commonly produced in the intestines of persons consuming food contaminated by such organisms as salmonella.

The identification of bacteria

As individual bacteria are so minute it is not practical to work with a single cell and the first stage in identification is to obtain a pure **culture** containing millions of organisms. This involves spreading the bacterial mixture on to the surface of a solid culture medium in a petri dish, using a sterile wire loop. The usual medium is nutrient agar, which consists of beef extract, peptone water and agar to solidify the medium.

Pre-enrichment involves the use of a liquid medium, such as peptone water, to allow damaged cells to recover. Selective enrichment involves using a medium to suppress unwanted competitors and encourage the multiplication of the desired species to a detectable level. For example, the addition of crystal violet inhibits the growth of Gram-positive bacteria. MacConkey agar, which contains bile salts to inhibit the growth of non-intestinal bacteria, is used to grow some enteric bacteria such as salmonella. Selective atmospheres are used in a similar manner, for example, strict **anaerobes** such as *Clostridium perfringens* will not grow in air.

The inoculated medium is then incubated at a constant temperature for an appropriate length of time, for example, 24 hours. The temperature is often critical in promoting or precluding growth. Most of the common **food poisoning**/borne disease organisms prefer a temperature of 37°C whereas 42°C is used to isolate *Campylobacter* species. During incubation each bacterium can become a pure culture of millions of bacteria, which appear as small, discrete characteristic spots on the agar and are known as colonies.

The shape, size, colour and consistency of a colony can be of great value to identification. However, other methods will probably need to be used, including staining reactions, microscopical examination and biochemical reactions. In order to identify individual members of bacterial groups it may be necessary to use more complex tests such as **serotyping, phage typing** and plasmid typing.

The Gram stain
Bacteria can be divided into two major groups, the Gram-positive and Gram-negative types, depending on whether or not they retain a crystal violet/iodine dye complex after treatment with alcohol and safranin. Gram-positive bacteria retain the dye and are blue-purple whereas Gram-negative are pink. Staining is carried out after a suspension of the bacteria is applied to a microscope slide, using a sterile wire loop. The slide must be heated to fix the bacteria before the application of the stain.

Microscopical examination
Light microscopes are normally used to observe staining reactions, cell shape, the presence of spores and the motility of a particular bacterium.

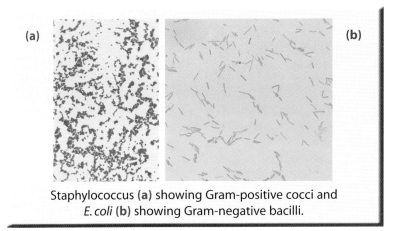

Staphylococcus (a) showing Gram-positive cocci and
E. coli (b) showing Gram-negative bacilli.

Biochemical reactions

Bacteria that cannot be identified by morphology (shape and size) and cultural characteristics may be distinguished by their biochemical reactions. For example, some bacteria will ferment sugars to produce an acid, which changes the colour of a liquid medium, and gas which can be collected in a small inverted test tube immersed in the medium. Other biochemical properties of bacteria include the production of coagulase by *S. aureus*; the production of indole by *E. coli*; the ability to digest gelatin, egg and serum; to split fats; reduce dyes such as methylene blue; and produce ammonia and hydrogen sulphide.

Sub-typing

Molecular and genetic methods are beginning to revolutionize "sub-typing", for example, if *Listeria monocytogenes* is isolated from a specimen from a sick person and from a food vehicle. If they are the same "sub-type" the food vehicle was probably implicated, if they are different sub-types, a different vehicle or mode of transmission was probably involved.

Serological typing

This test is used to distinguish different sub-types of the same species of bacteria. It involves observing the agglutination (clumping together) reaction between the antigens of the test bacteria and the antibodies produced in the blood serum (antisera) of human beings or animals infected with the same bacteria. Examples include *Salmonella* Enteritidis and *Salmonella* Typhi.

Phage typing

Phage typing enables even finer differentiation of sub-types. Bacteriophages are viruses parasitic on bacteria. Each virus is host specific and will only attack one particular type of bacteria. The bacteriophages are inoculated on to the bacteria culture. Susceptibility of the bacteria to the virus enables positive identification of the phage type. Phage typing is particularly useful for identifying salmonellae and staphylococci. Examples include *S.* Enteritidis PT4 and *S.* Enteritidis PT2.

Immunoassay

The use of specific monoclonal antibodies to detect bacterial antigens or toxins, for example, ELISA (enzyme linked immunosorbant assay).

THE MULTIPLICATION OF BACTERIA

Bacteria reproduce vegetatively (non-sexually) by the process of **binary fission** (dividing into two). After cell division each daughter cell grows to maturity and itself divides. The time between each division (known as the generation time) varies between species but, given the right conditions, is commonly around 20 minutes. However, under **optimum** conditions *Cl. perfringens* and *Vibrio parahaemolyticus* are both capable of division in ten minutes or less.

Given a generation time of ten minutes 1,000 bacteria could become 1,000,000 in only one hour 40 minutes (1,000,000 bacteria per gram of food may cause food poisoning and 1,000 bacteria per gram of food is not an uncommon level of **contamination**). The multiplication of pathogens in food is a hazard.

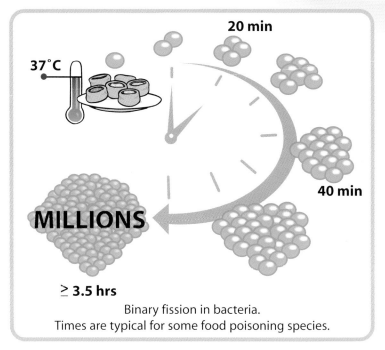

Binary fission in bacteria.
Times are typical for some food poisoning species.

Multiplication

When bacteria contaminate food and conditions are favourable, they begin to multiply. The increase in numbers does not occur at a uniform rate but passes through a succession of phases. These phases are described as follows:

1 to 2 the **lag phase** –	no multiplication as the bacteria "acclimatize".
2 to 3 the logarithmic phase –	rapid multiplication.
3 to 4 the stationary phase –	numbers of bacteria remain constant as the number produced is equal to the number dying.
4 to 5 the decline phase –	numbers decrease as numbers dying exceeds those produced.

The lag phase is short if there are large numbers of young bacteria in optimum growth conditions.

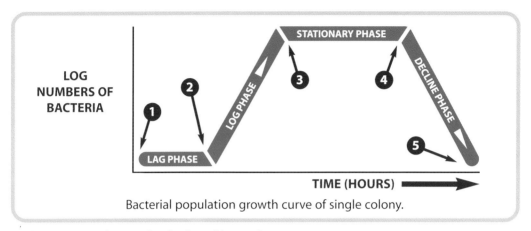
Bacterial population growth curve of single colony.

Multiplication of several colonies of bacteria

Although the population growth curve may follow the standard pattern in the laboratory, this is unlikely in food in practice. If the food is mixed or the colony disturbed in some way, a new supply of nutrients will be provided and levels of toxins reduced. It is extremely unlikely that a single colony or species would exist in isolation. Therefore, there will be many colonies at different stages of growth as they compete and as the substrate is changed by the preceding culture. The number of bacteria in food in favourable conditions will probably remain in a series of logarithmic phases interspersed by stationary and lag phases. The numbers of bacteria overall will therefore continue to increase until unfavourable growth conditions occur.

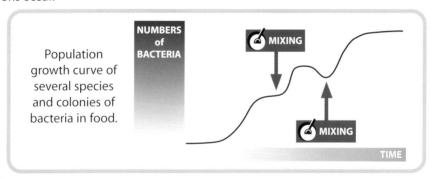
Population growth curve of several species and colonies of bacteria in food.

Factors influencing bacterial multiplication
Nutrients

Bacteria require carbon, hydrogen, oxygen and nitrogen with smaller amounts of sulphur and phosphorus, together with trace elements such as sodium, potassium, magnesium, iron and manganese. All nutrients must be in solution (dissolved in water) before they can enter the bacterium. The elements are converted into the constituents of cytoplasm by a process described as a **metabolic reaction**. Certain vitamins may be required to facilitate these reactions.

Most **pathogenic** and **spoilage bacteria** obtain the essential basic elements from sugar, amino acids, fats and minerals. The types of food favoured include high protein food such as milk, eggs, meat and fish. Foods with a high sugar or salt content are unsuitable for the growth of most bacteria and are therefore potentially safe. *S. aureus* and *Listeria spp.* are exceptions.

HYGIENE for MANAGEMENT Microbiology

Hydrogen-ion concentration (pH)

The **pH** of a solution is measured on a scale of 1 to 14 (hydrogen-ion concentration). Acid foods have pH values below 7 and alkaline foods above 7; a pH value of 7 being neutral.

All bacteria have their own optimal, minimal and maximal pH for growth with most preferring a pH near neutrality. Most bacteria will not grow in food with a pH below 4·0 and at this pH it allows lower processing temperatures to be used in preservation techniques, such as canning, to render food safe and free of spoilage organisms. However, a large number of pathogens introduced into an acid food will not die off immediately and may still cause illness. S. Typhimurium has survived in fruit juices (pH 3.2 at 22°C) for up to five days. A food poisoning outbreak occurred in New Jersey, USA, following the consumption of apple juice (pH 3.6) contaminated with S. Typhimurium.

Generally, **moulds** and **yeasts** are less sensitive to the pH of food than bacteria.

Moisture

All life requires water, which is used to transport nutrients into the cell and take away waste products. Dry products are poor media for the multiplication of bacteria, for example, flour, biscuits and bread, and may be described as low-risk foods. However, some bacteria, especially spore formers, are able to survive **dehydration** and when dried egg or milk powder is reconstituted by the addition of liquid, the food once again becomes **high-risk**. Most other foods contain sufficient moisture to promote bacterial growth.

Water activity (a_w)

The amount of moisture in any food available to bacteria is normally considered in terms of water activity. The a_w of pure water is 1.00 (a saturated salt solution 0.75). Bacteria prefer an a_w around 0.99 and many will not grow below 0.95. However, S. aureus will grow at an a_w of 0.89 and some bacteria can exist below 0.75. Yeasts and moulds tolerate lower levels of a_w, some as low as 0.62. Fresh meat has an a_w of 0.95 to 1.00, bread 0.94 to 0.97, cured meat 0.87 to 0.95, jam 0.75 to 0.80, flour 0.67 to 0.87 and sugar 0.19.

Temperature

Bacteria have a maximum and minimum temperature for growth between which there is an optimum temperature when multiplication is the most rapid. The range of temperature for the fastest multiplication of most pathogens is 20°C to 50°C. Some pathogens grow between 0°C and 20°C but they multiply more slowly at the lower temperatures. The lowest recorded temperature for the growth of pathogenic bacteria is −2°C. Usually bacteria are separated into four groups:

	Optimum	Range	Importance
Psychrophiles	< 20°C	−8° to 25°C	Include bacteria which cause spoilage in refrigerators and cold stores.
Psychrotrophs	>20°C	−5° to 40°C	Include *Cl. botulinum* type E, *Listeria monocytogenes*, *Yersinia enterocolitica*, and *Aeromonas hydrophila*.
Mesophiles	20° to 45°C	10° to 56°C	Include most common pathogens that cause food poisoning.
Thermophiles	> 45°C	35° to 80°C	Important in canning – some are very heat-resistant and if not destroyed will cause spoilage if cans are stored at high temperatures.

Presence of oxygen

Some micro-organisms will only grow in the presence of oxygen and these are known as obligate **aerobes**. Others flourish in the absence of oxygen (obligate anaerobes). Micro-organisms that grow either with or without free oxygen are known as facultative. Some food poisoning bacteria are anaerobic, such as clostridia, others are facultative anaerobes, such as salmonellae and staphylococci. However, they usually prefer aerobic conditions. **Microaerophilic** bacteria grow in the presence of reduced quantities of free oxygen, for example, campylobacter.

Free oxygen is normally present in food, except those foods with a high liquid content that have been thoroughly boiled, roast joints and vacuum-packed foods.

Competition

When there are many different bacteria present, they will compete for the same food. Fortunately, most food poisoning bacteria are not as competitive as the normal flora found on food and, unless present in high numbers, will usually die. Modified atmospheres will favour anaerobes.

Moulds

Moulds are aerobic, or facultative anaerobes, multinucleate, chlorophyll-free **fungi** that produce thread-like filaments (hyphae) and form a branched network of mycelium. Moulds, which may be black, white or of various colours, will grow on most foods, whether moist or dry, acid or alkaline and high in salt or sugar concentrations. The optimum growth temperature is usually 20°C to 30°C, although they will grow well over a wide range of temperatures and may cause problems in refrigerators. Growth has been recorded as low as −10°C. High humidities and fluctuating temperatures accelerate mould growth.

Microscopic **mould spores** are inseparable from atmospheric dust and large numbers may be present in food premises. Moulds commonly affect bread and other bakery products, and although spores are usually destroyed in baking, subsequent contamination is difficult to avoid. Rapid cooling of bread and wrapping below 27°C will reduce mould problems. Also important is the removal of dampness and mould growth in factories and the thorough cleaning and disinfection of plant and equipment.

Food must always be stored in accordance with the manufacturer's instructions and never sold outside its **use-by** date. Mishandling of vacuum packs of cheese may result in punctures and consequential mould growth. As the mycelium grows over the food, hyphae penetrate the substance and consequently mould soon returns if scraped off the surface. Regular checking of stock is imperative to avoid customer complaints. The presence of mould on food is usually considered to render it unfit for human consumption. Cheeses produced with specific moulds are an exception.

Certain species of mould, for example, *Aspergillus flavus* and *Fusarium spp.*, are capable of producing poisonous metabolites, known as **mycotoxins**, which may affect animals and humans. Even toxins produced by some species of *Penicillium* have induced **carcinogenic** effects in animals including marked damage to the liver and kidneys and may, in large doses, constitute a hazard to man. Both aspergilli and penicillia are common in nuts and cereals, especially rice. Incidents of poisoning by mycotoxins in the UK over the last 50 years have been extremely rare: a few cases of ergotism were recorded in Manchester in 1927. However, constant vigilance is required by authorized officers employed at ports to prevent the import of mouldy food or animal feed. In 1984 several tons of peanuts intended for human consumption were found to be contaminated by small amounts of aspergilli and were condemned.

Yeasts

Yeasts are microscopic unicellular fungi that reproduce by budding. They vary in size from around 5μm diameter to 100μm in length. Most yeasts grow best in the presence of oxygen, although fermentative types grow slowly anaerobically. The majority of yeasts prefer acid foods (pH 4 to 4.5) with a reasonable level of available moisture (a_w above 0.88). However, many yeasts will grow in high concentrations of sugar and salt with an a_w as low as 0.62. The optimum growth temperature for yeast is around 25°C to 30°C with a maximum of around 47°C. Some yeasts can grow at 0°C and below.

Yeasts are used in the manufacture of foods such as bread, beer and vinegar. They are not usually considered to be responsible for food poisoning, although large numbers of yeast in unpasteurized, home-made fermented beer has resulted in mild illness. However, yeasts do cause spoilage of many foods including jam, fruit juice, honey, meats and wines.

Mouldy food complaints may result in prosecutions.

3 Food poisoning *edited by Adrian Eley*

Classically, **food poisoning** is commonly associated with symptoms such as diarrhoea and vomiting and may also include headache, stomach cramps and fever. **Bacteria** are responsible for most food poisoning **cases**, with **mycotoxins**, poisonous plants, chemicals or metals occasionally causing problems. Viral **gastroenteritis** is caused by **low-dose viruses**, which do not multiply in food, and is therefore dealt with in chapter 4 "Foodborne diseases".

Because food poisoning organisms are commonly found in **foods**, it is impossible to operate a food business without them being present in either small or large numbers. This is why good hygiene practices are so important to minimize **risk** of illness from consuming **contaminated** food.

BACTERIAL FOOD POISONING

There are at least eight types of bacteria known to be responsible for causing most food poisoning **outbreaks** in the United Kingdom:

Salmonella spp.
Clostridium perfringens
Staphylococcus aureus
Bacillus cereus and other *Bacillus* spp.
Clostridium botulinum
Vibrio parahaemolyticus
Escherichia coli
Yersinia enterocolitica

TYPES OF FOOD POISONING

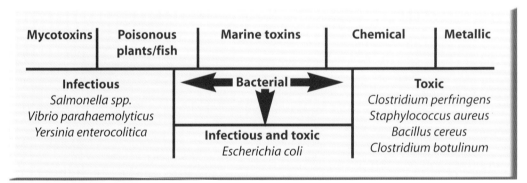

Salmonella

Salmonellae are Gram-negative, facultative **anaerobic** rods. There are approximately 2,500 **serotypes** of salmonella, the commonest being *S.* Enteritidis and *S.* Typhimurium.

Nomenclature of the salmonellae does not follow the conventional rules in that the classification systems usually refer to serotypes rather than to species. More recently, it has been proposed that all serotypes of salmonella be considered as members of a single species, *S.* Enterica. In this scheme, individual serotypes are referred to as, for example, *S.* Enterica subspecies *enterica* serotype Typhimurium. However, it is likely that for routine use *S.* Enterica serotype Typhimurium, *S.* serotype Typhimurium or *S.* Typhimurium will be acceptable.

CLASSIFICATION OF SALMONELLA

	Old classification	New classification
Family	Enterobacteriaceae	Enterobacteriaceae
Genus	Salmonella	Salmonella
Species/serotype	*enteritidis*	
Species		*enterica*
Subspecies		*enterica*
Serotype (Serovar)		Enteritidis
Type	*enteritidis* PT4	Enteritidis PT4

The growth range of these bacteria is from 7°C to 47°C with a generation time of ten hours at 10°C. Some salmonella can grow at a pH as low as 3.8. If present in sufficient numbers in consumed food (normally 100,000+ per gram), the digestion process may not destroy them. The bacteria multiply in the intestine and diarrhoea results after penetration of the intestinal wall. The **endotoxin** released on the death of the cell probably produces the fever associated with salmonellosis. Salmonellae are able to survive in soil for several months.

The primary **sources** of salmonella are the intestinal tracts of animals and birds. 90% of reptiles carry salmonella. Animals may become infected from the consumption of contaminated feed or contact with **carriers** or infected animals. Ill animals should always be isolated and animal feed treated to ensure it is free of salmonellae. Raw meat, especially poultry, is the main source of salmonellae in food premises. In 2001, the Food Standards Agency published the results of a survey of salmonella contamination of fresh and frozen chicken on retail sale in the UK. Findings showed an average contamination level of 5.8% (4.1% in fresh chicken and 10.8% in frozen chicken), which is significantly lower than in previous surveys. The contamination levels in imported chickens were slightly lower.

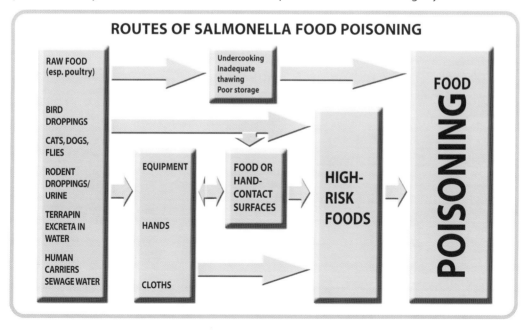

Outbreaks of salmonella have involved a variety of foods. In 1984 (UK) aspic glaze contaminated with S. Enteritidis PT4 resulted in 766 cases and two deaths. In 1984 in Canada, S. Typhimurium in cheddar cheese (<10 organisms per gram) resulted in 2,700+ cases. An outbreak of S. Typhimurium at Stanley Royd Hospital, Wakefield in 1984 involved 455 cases and 19 deaths. In 1985 a UK outbreak involving dried baby milk contaminated with S. Ealing resulted in 60 cases and one death. The adverse publicity resulted in the closure of the factory responsible. In 1985 pasteurized milk with S. Typhimurium in the USA, resulted in 16,284 cases and seven deaths. The dairy was permanently closed. An egg sandwich shared by two women at Manchester Airport in 1988 resulted in compensation of £183,500 for food poisoning, irritable bowel syndrome, constipation and heartburn. In 1991 a sandwich bar in Colchester paid £25,000 in legal fees and £100,000 in compensation to 76 victims. The sandwiches, which contained home-made mayonnaise (raw egg), had been stored in a car boot at temperatures exceeding 30°C. In 1988 (Japan) cooked eggs contaminated with *Salmonella spp.* resulted in 10,476 cases. In 1994 (USA) ice-cream contaminated with S. Enteritidis resulted in 740 cases.

Although most outbreaks of salmonella involve high-risk foods such as cooked poultry and ready-to-eat products containing raw or lightly cooked egg and cheese, usually made with unpasteurized milk, several outbreaks have involved low-risk foods. In 1982, S. Napoli in imported Italian chocolate bars resulted in an unusual outbreak affecting 272 people. The chocolate was at least seven months old, the incubation period was up to ten days and the infective dose was commonly less than 100 organisms. The low moisture and high sugar content of chocolate increase the temperature resistance of the salmonellae and enable some organisms to survive the process. The low infective dose of salmonella in chocolate is thought to be due to the high fat content, which protects against the stomach acid. Chocolate was also responsible for outbreaks of salmonella in the USA (1973, S. Eastbourne), Finland (1987, S. Typhimurium), and Canada (1985-6, S. Nima). Towards the end of 2001 and into early 2002, an international outbreak of S. Oranienburg infection was traced to chocolate. In June 2006, Cadbury Schwepps recalled bars of chocolate following allegations of links to S. Montevideo food poisoning cases. Between 1990 and 2000 at least six outbreaks of salmonella occurred in the USA or Canada due to the consumption of contaminated, cut melons left at ambient temperatures. Bean sprouts have been implicated in several outbreaks including mung bean sprouts in England (1988, S. Saint-paul) and alfalfa sprouts in Finland and Sweden (1994, S.

A SEQUENCE OF THREE PHOTOGRAPHS FROM AN ELECTRON MICROSCOPE SHOWING INCREASING MAGNIFICATION OF SALMONELLAE ON RAW BEEF

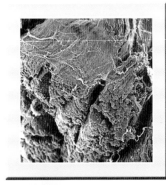
(Magnification x68).

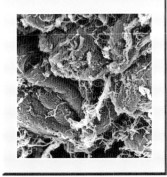
(Magnification x1000).

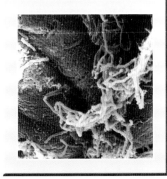
(Magnification x3400).

Bovismorbificans) USA and Finland (1995, S. Stanley) and Denmark, Canada and USA (1995-6, S. Newport). Outbreaks have also involved mustard cress in England (1989, S. Gold-coast). Unpasteurized orange juice in the USA (1995, S. Hartford, S. Gaminara and S. Rubislaw) and two outbreaks involving tomatoes in the USA (1990 & 1993). In Germany in 1993 paprika crisps resulted in 1,000 cases of salmonella and in 1982 black pepper was the vehicle in Norway.

Since 1988 S. Enteritidis has been the most common serotype and is found in chicken muscle and in the albumen of raw eggs. In 2001, about 70% of salmonellae implicated in human infections were S. Enteritidis.

The increase in multi-antibiotic resistant S. Typhimurium DT104 is of major concern (259 in 1990, 4,006 in 1996 and 1,142 in 2000), possibly due to misuse of antibiotics in animal husbandry.

On a positive note, the incidence of human salmonellosis in England and Wales has fallen dramatically since 1997 (>30,000) to only 12,725 in 2004, which is attributable in part to vaccination of poultry flocks.

Clostridium perfringens

These bacteria are Gram-positive, anaerobic rods, which form spores under adverse conditions such as exposure to free oxygen. The normal growth range is between 15°C and 52°C and there is no division at 10°C. *Cl. perfringens* can reproduce every ten minutes at its optimum growth temperature of 43°C to 47°C. Growth in meat was shown to be preceded by a lag of two to four hours at 35°C but no lag at 46°C*. Spores may survive boiling for several hours.

Illness is caused when large numbers of ingested organisms sporulate in the intestine and in so doing release an exotoxin, also described as an enterotoxin. The infective dose for adults is around 1,000,000 per gram of food.

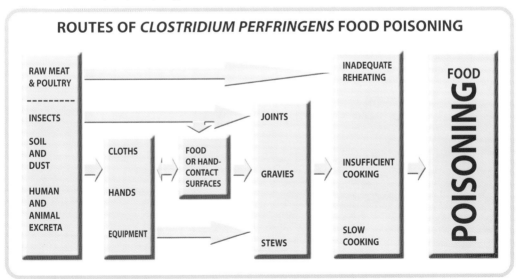

Cl. perfringens food poisoning usually occurs as a result of food handling malpractices in large kitchens, especially institutional catering where large quantities of food are prepared

*Hall H. and Angelotti R. (1965). "*Clostridium perfringens* in meat and meat products."
Appl. Microbiol., 13:352-357.

hours before service. On cooking, oxygen is driven from the food creating anaerobic conditions ideal for the growth of the heat activated spores. Stews, gravies and large joints of meat allowed to cool slowly in warm kitchens are most commonly associated with outbreaks of *Cl. perfringens* food poisoning. Because *Cl. perfringens* is commonly found in the intestine of healthy individuals, sporadic cases are difficult to identify. Statistics therefore relate to numbers of cases involved in outbreaks and may be seriously underestimated.

Cl. perfringens has been shown to sporulate in meat products and produce enterotoxin in the food. Although not common, this explains why the incubation period for this type of food poisoning has been reduced to as little as two hours.

It is estimated that for every 343 cases of *Cl. perfringens* in the community only one is reported and that there are around 144,000 cases per year in the UK.

Cooking and simmering bulk fluids

Temperature distribution within bulk liquids during cooking is not uniform. Variations can be extreme and under certain conditions cool spots can form. Because of the nature of heat transference in liquids, stable convection currents are set up and liquid outside these currents may remain stationary at a low temperature.

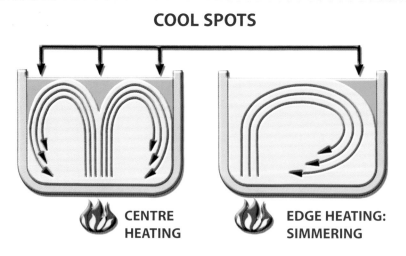

Circulation currents and the formation of cool spots.

Where a number of adverse factors combine, parts of a bulk liquid may be visibly boiling, whilst other areas of the liquid may be at a temperature that allows the multiplication of *Cl. perfringens*. The most dangerous combination is when a large volume of a viscous liquid, such as gravy or stock, is boiled and then simmered in an open pan on a solid type cooker, especially when the pan has no lid, the kitchen is cool and the pan contents are left unstirred. The initial boiling activates *Cl. perfringens* spores from which the vegetative bacteria grow. Furthermore, the boiling destroys other organisms, thereby reducing competition. Clostridia within the cool spot survive and multiply under virtually ideal conditions.*

*North R. A. E.: Unpublished information.

THE FORMATION AND PREVENTION OF COOL SPOTS

Problem	Remedy
Volume of liquid, especially greater than 25 litres	Cook in smaller volumes: use several small pans in preference to one large pan
Intensity of heat, especially low edge heating	Use heat source equal in diameter to pan base
Large tall pans	Use wide low pans
Failure to stir	Stir frequently, at least every ten minutes using a clean utensil
Cool draughts and absence of lids	Keep excessive cold air draughts away from cooking area. Keep lids on pans between stirring
It is preferable to use alternative equipment such as jacketed kettles or bratt pans.	

Clostridium botulinum

Clostridium botulinum is a Gram-positive, anaerobic, spore-forming, gas-producing rod. There are several types of *Cl. botulinum* that produce **neurotoxins** in food, which, if consumed, often cause fatalities. Types A, B, E and F are associated with food poisoning. Types A and B are commonly found in the soil and are usually associated with vegetables and meats. They produce objectionable smells when they grow in foods. **Psychrotrophic** type E is found in mud, streams, oceans and fish. The optimum growth temperature is 20ºC to 30ºC but toxin production is possible down to a temperature of 3ºC. Spores of type E are much more sensitive to heat and are killed at around 85ºC. Non-proteolytic strains producing types E and F toxins can produce lethal amounts of toxin with no detectable changes in the food, i.e. no smell. The toxin of all strains is **heat labile** (sensitive to heat) and is destroyed by boiling.

Unfortunately, the spores of proteolytic types A and B *Cl. botulinum* are very heat-resistant, for example, a temperature of 121ºC for three minutes is used as a basis for ensuring their destruction in canned **low-acid foods**. Strict canning controls are essential. Toxin production does not occur below a pH of 4.5. If consumed by adults, spores are usually harmless as they are unable to grow and produce toxin in the human intestinal tract.

Outbreaks of botulism are most frequently associated with improperly processed or handled low-acid canned foods, fermented products, traditional and smoked fish products, vegetables in oil, and garlic in oil (neutral pH, long shelf-life, grown in soil, ambient storage, anaerobic conditions). Vegetables should be rendered safe prior to placing them in oil, for example, by heat, acidification or high salt. If contaminated, fish products may become toxic before they become unacceptable through spoilage. It is preferable, therefore, to store such products below 5ºC to ensure their safety. Nitrates and nitrites are often added to cooked meats, bacon and ham to prevent the **germination** of spores in vacuum packs. Refrigerated processed foods, especially fish, anaerobically packaged with long **shelf-lives** are of particular concern.

Although gas may be produced, which causes cans contaminated with *Cl. botulinum* to blow, this does not always occur and cannot be relied on as a method to detect suspect cans. In 1978, four elderly people in Birmingham developed botulism after eating a contaminated can of Alaskan salmon. Two of them died. In 1989, 26 cases of botulism type B occurred following the consumption of hazelnut yogurt and only one person died, primarily due to the administration of polyvalent botulinum antitoxin. Underprocessing of canned hazelnut purée, used in the yogurt production, was stated to be the cause, and a change from sugar to saccharin increased the a_w and allowed germination of the spores. In 1996, eight cases of botulism in Italy resulted from the consumption of contaminated acidified dairy cream (mascarpone cheese). In 1998 two people were ill and one died following the consumption of home bottled mushrooms, brought back from Italy.

To July 2002, there have been six cases of infant botulism reported in the UK, where spores have apparently germinated in the infant gut and toxin has subsequently been produced. In the UK in 2001, a baby food manufacturer recalled cans of powdered baby milk after a five-month-old girl contracted infant botulism. Infant botulism was first described in the USA and since 1976 over 900 cases have been recorded. Some cases have implicated honey contaminated with *Cl. botulinum* spores. In 1991 in Egypt 91 people were ill and 18 died after eating salted fish that was spoilt before salting. In 1999 in Morocco 80 people were ill and 15 died following the consumption of a temperature-abused meat and chicken dish contaminated with *Cl. botulinum*.

Staphylococcus aureus

This organism is a Gram-positive, facultative anaerobe that is salt-tolerant and can, therefore, survive in some cured products. Approximately 40% of adults carry *S. aureus* in the nose and throat and 15% on their skin, especially the hands. Around half of these strains are enterotoxigenic.

If present in food, *S. aureus* will, under ideal conditions, produce an exotoxin, which may survive boiling for 30 minutes or more. The growth range for *S. aureus* is between 7°C and 48°C and the generation time at 10°C is 30 hours. However, it requires at least 100,000 organisms per gram and a temperature of at least 10°C to form toxin. (No toxin formation above 45°C).

The majority of outbreaks are caused by the direct contamination of cooked foodstuffs by hands soiled with secretions from the nose, mouth and skin lesions. Frequently, ready-to-eat food has been handled while warm and storage conditions have encouraged the organism to multiply and produce its toxin. *S. aureus* is commonly added to cooked meat during slicing or

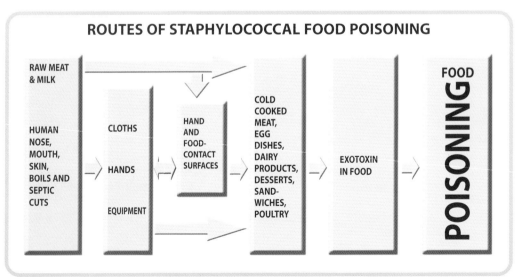

deboning at levels of around ten to 100 per gram. Such levels are harmless but storage at ambient temperatures for a few hours would allow significant levels of toxin to be produced. It is not very competitive unless the food has a significant salt content.

In 1965 an outbreak of *S. aureus* involving Cheddar cheese resulted in 42 cases. A typical outbreak in Texas, USA involved 1,364 children out of 5,824 attending 16 schools. They had eaten chicken salad that had been prepared the day before and had not been cooled correctly after boning. After transport on the day of the outbreak it was left at **ambient temperature**. Outbreaks have also involved fermented sausage. In 1985, 47 cases of *S. aureus,* in the UK, were caused by the consumption of imported Italian lasagne.

Bacillus cereus

Bacillus cereus is a facultative anaerobic, Gram-positive, spore-forming bacterium that grows predominantly **aerobically**. It is capable of causing two distinct forms of food poisoning. The more common type in this country is characterized by nausea and vomiting one to five hours after ingestion of contaminated food, including rice dishes, vegetables, cornflour, custards, soups, vanilla slices and raw vegetable sprouts. The spores survive normal cooking, and rapid growth and exotoxin production will occur if the food is not cooled quickly and refrigerated. One of the exotoxins produced is very heat-resistant and can survive a temperature of 126°C for 90 minutes. Consequently, once the toxin is produced subsequent reheating is of little value. The second type of *B. cereus* food poisoning is rare in the UK and has an **onset** period of eight to 16 hours. Toxins are produced in the intestine and symptoms are usually diarrhoea and abdominal pain.

B. cereus grows between 4·4°C and 48°C with an optimum between 28°C and 35°C. It is distributed throughout the environment but is particularly common in soil, vegetation, cereals and spices. It also causes spoilage of cream – referred to as bitty cream.

B. cereus is often isolated in low numbers in many types of food. However, large numbers (100,000+ per gram) must be isolated to prove that a contaminated food has caused food poisoning. It is recommended that faecal specimens be sent to the laboratory under refrigeration to avoid other organisms multiplying.

Bacillus subtilis and *Bacillus licheniformis*

Bacillus subtilis and *Bacillus licheniformis* are spore-forming bacteria, which have occasionally been implicated in food poisoning incidents. The main symptoms are nausea and vomiting, usually within one hour of consuming contaminated food. Diarrhoea is less common but may occur after approximately 12 hours. In 1994 an outbreak of *B. subtilis* involved meat pasties seasoned with contaminated black pepper. The spores had survived the cooking process. Main vehicles include meat or vegetables and pastry products.

Vibrio parahaemolyticus

Vibrio parahaemolyticus is a Gram-negative, marine facultative anaerobe that grows between 8°C and 44°C, with an optimum of 35°C to 37°C. Given ideal conditions the generation time may be as short as five to seven minutes. Rapid growth may occur in seafoods or the human intestine.

Outbreaks of *V. parahaemolyticus* food poisoning in the UK are infrequent and usually caused by the consumption of imported seafood, such as cooked prawns and dressed crab from the Far East. However, it is occasionally isolated from the UK coastal waters, especially in the South. In 1998, 28 faecal **isolates** were reported to the Communicable Disease

Surveillance Centre (CDSC). The organism is sensitive to heat and outbreaks are normally caused by contamination after cooking and multiplication due to unrefrigerated storage. *V. parahaemolyticus* is one of the commonest causes of food poisoning in Japan as it prefers the warmer coastal waters, and seafoods form a high percentage of the staple diet. In 2002 an outbreak of *V. parahaemolyticus* in Hong Kong involved 56 suspect cases. The suspect foods were escargots and squid which had been stored at ambient temperatures for too long.

Escherichia coli

Escherichia coli is a Gram-negative, facultative anaerobic rod that is a normal part of the intestinal flora of man and animals. It is usually used as an **indicator** of faecal contamination of food or water. However, certain strains are known to be pathogenic and some produce a toxin in the intestine that results in symptoms of abdominal pain and diarrhoea. One group of pathogenic *E. coli* is responsible for severe infantile diarrhoea and another group causes travellers' diarrhoea. A major outbreak in the USA was traced to the consumption of soft, fermented cheese. Growth of *E. coli* has been recorded as low as 4°C. *E. coli* O157 is a particularly serious foodborne pathogen and is dealt with in chapter 4.

Yersinia enterocolitica

Yersinia enterocolitica is a psychrotrophic, Gram-negative, facultative anaerobic rod that is quite widespread in nature, particularly in wild and domestic animals. Not all strains of *Y. enterocolitica* are pathogenic for humans. Faecal–oral transmission involving infected animals or persons or the consumption of contaminated food and water are the usual ways of spread and it is not heat-resistant. *Yersinia* has been incriminated in several major outbreaks of food poisoning in the USA and is commonly isolated as a cause of diarrhoea in Sweden, Norway and Germany. Foods involved are usually of animal origin and outbreaks have involved raw pork and pork products, chocolate milk, reconstituted powdered milk, turkey and chow mein. In recent years, there has been a marked decline in the reporting of confirmed cases of *Y. enterocolitica* infection from laboratories in England and Wales. In 1989, there were 571 laboratory reports, which had fallen to 31 in 2003.

Symptoms include diarrhoea, abdominal pain, fever, headache, pharyngitis, occasional vomiting, acute mesenteric lymphadenitis, pseudoappendicitis (especially in children) and arthritis. The incubation period is usually one to five days (occasionally longer) and bacteria may persist in the stools for several months. *Yersinia* has been isolated from raw milk, shellfish, raw and cooked meat, particularly pork, poultry, liquid egg, salami, dairy products, salads and vegetables. Of particular concern is the ability of *Y. enterocolitica* to grow on food during refrigerated storage, despite the normal microflora. This is in contrast to the majority of pathogens, which are usually not able to grow competitively at low temperatures.

Other bacteria

Vibrio vulnificus is a rare cause of food poisoning but has a high death rate. In the USA it is responsible for around 47 cases and 18 deaths each year. The onset period is around 16 hours. Symptoms include fever, chills and nausea. It has been associated with the consumption of raw or recontaminated oysters, clams and crabs.

Very occasionally, *Enterococcus faecalis*, organisms belonging to the *Streptococcus viridans* group and *Aeromonas* species, have been implicated in food poisoning outbreaks. However, there is still controversy with the role of these and other bacteria as a cause of food poisoning and further studies are warranted.

THE PREVENTION OF BACTERIAL FOOD POISONING

Businesses must implement a food safety management system based on the principles of **HACCP** (hazard analysis critical control point). This will include a satisfactory environment, competent and properly trained personnel, suitable equipment and facilities to prevent contamination and adequate temperature control to ensure that food at all stages of production, storage, distribution and sale is as safe as reasonably practicable.

The food poisoning chain

The chain of events associated with an outbreak of food poisoning usually consists of:
- contamination of ready-to-eat food with food poisoning bacteria;
- multiplication of these bacteria; and
- consumption of the food.

In order to prevent illness this chain must be broken and it is essential that care be taken to ensure that:

- **The contamination of food is kept to an absolute minimum. This may be achieved by:**
    - separating raw and ready-to-eat food at all stages of preparation, storage and distribution. The same equipment or working surface must not be used to handle raw and high-risk foods, or ready-to-eat foods such as lettuce, unless **disinfected** between uses. The liquid from thawed, frozen meat and poultry must not come into contact with high-risk or ready-to-eat foods;
    - maintaining the highest standards of **personal hygiene** at all times. Hand washing, particularly after using the toilet, is essential;
    - the exclusion of ill persons and, where appropriate, carriers;
    - high standards of **pest** exclusion and control (**integrated pest management**); and
    - using the correct **cleaning** procedures. **Hand-contact** and **food-contact** surfaces must be kept clean and, where necessary, disinfected at appropriate frequencies. Care should be taken with wiping cloths, disposable paper cloths are preferable.

Failure to separate raw and cooked food during storage.

- **Bacteria already in food are prevented from multiplying by:**
    - keeping food out of the **danger zone**. Food should be kept below 5°C, for example, in a refrigerator, or kept above 63°C, for example, in a bain marie;
    - ensuring that refrigerators and cold stores are maintained at the correct temperature, usually below 5ºC;

- ensuring that when food is removed from chilled storage for preparation, this work is carried out as quickly as practicable. High-risk food must not be left in the ambient temperatures of kitchens, serving areas or food rooms;
- cooling food as rapidly as possible, prior to refrigeration, or serve and consume immediately;
- precooling all salad ingredients, especially for sandwiches, overnight;
- not allowing dried foods to absorb moisture; and
- using suitable preservatives.

- **Those bacteria within food are destroyed by:**
  - thorough cooking, usually to a core temperature of 75°C;
  - heat processing such as pasteurization, sterilization or canning; and
  - a suitable preservation technique, for example, fermentation.

A combination of a suitable temperature and sufficient time is always required to destroy bacteria. The time and the temperature required will depend on the particular organism. For example, spores of *Cl. perfringens* are much more heat-resistant than salmonellae. Furthermore, a time/temperature process only destroys a percentage of vegetative bacteria and the greater the numbers that are present the more likely it is that some will survive. Practical work carried out in Hull indicated that under certain circumstances large joints of ham had to be cooked to a centre temperature of 70°C for one hour to ensure the destruction of *Salmonella* Give.

- **The consumption of potentially contaminated food is avoided by:**
  - obtaining food from reputable suppliers;
  - washing all raw fruit and vegetables. Blanching at around 71°C for 15 seconds may be preferable for ready-to-eat products to be consumed raw;
  - never adding leftovers/old food to fresh food; and
  - avoiding consuming raw foods known to be a common source or vehicle of pathogens, for example, raw oysters, raw milk or raw egg dishes.

FOOD POISONING BACTERIA

SOURCE	FOOD VEHICLES	ONSET (HOURS)	SYMPTOMS & DURATION	ADDITIONAL SPECIFIC CONTROLS
Salmonella (Infection)				
Raw meat, milk, eggs, poultry, birds, carriers, pets, rodents, terrapins, sewage/water	Poultry, "raw egg products", meat, dairy products, cheese, mayonnaise, sauces and salad dressings, bean sprouts and coconut	6 to 72 usually 12 to 36

10 days in the case of low-dose outbreaks | Severe abdominal pain, diarrhoea, dehydration, nausea, vomiting and fever (1 to 7 days) | Sterilization and strict control of animal foodstuff. Slaughterhouse hygiene. Safe sewage disposal and chlorination of water. Screening of carriers or suspects. Avoid products made with raw eggs and not fully cooked. Thorough thawing and cooking of frozen poultry. Avoid raw milk, use heat treated |

HYGIENE for MANAGEMENT *Food poisoning*

SOURCE	FOOD VEHICLES	ONSET (HOURS)	SYMPTOMS & DURATION	ADDITIONAL SPECIFIC CONTROLS
***Clostridium perfringens* (Toxin in intestine)**				
Animal and human excreta, soil, dust, insects and raw meat	Meat, meat products, poultry, gravy, rolled joints and stews	4 to 24 usually 8 to 12	Acute abdominal pain and diarrhoea. Vomiting is rare (12 to 48 hours)	Eat immediately after cooking, store above 63°C or rapid cooling and refrigeration within 1.5 hours of reaching 63°C. Joints should not exceed a size of 2.5kg. Thorough reheating of foods
ced*Staphylococcus aureus* (Toxin in food)				
Human nose, mouth, skin, boils and cuts. Raw milk from cows or goats with mastitis	Cooked meat, meat products, poultry, egg products, salads, cream products, milk and dairy products, and dried foods (toxins are heat stable)	1 to 7 usually 2 to 4	Abdominal cramps, nausea, vomiting, some diarrhoea, subnormal temperatures and collapse (6 to 24 hours)	Avoid handling food, use utensils. Good personal hygiene, especially regarding handwashing. Exclude handlers with respiratory infections involving coughing or sneezing. Cover cuts with waterproof dressings. Exclude persons with boils or septic cuts. Avoid the use of raw milk
***Vibrio parahaemolyticus* (Infection)**				
Seafoods	Raw, improperly cooked or recontaminated shellfish and fish	2 to 96 usually 12 to 18	Diarrhoea, nausea, abdominal cramps, some vomiting, fever, **dehydration**, blood or mucus in stools (1 to 7 days)	Cook fish well. Take care to avoid cross-contamination between raw and cooked shellfish and fish. Avoid raw shellfish
***Clostridium botulinum* (Toxin in food)**				
Soil and aquatic sediments, fish, meat and vegetables	Improperly canned low-acid or fermented foods. Smoked fish in vacuum packs. Vegetables in oil	2 hours, to 8 days usually 12 to 36 hours	Difficulties in swallowing, talking and breathing, weakness, double vision and vertigo. Fatalities common, recovery of survivors may take months	Strict control over processing low-acid canned foods. Discard blown cans or those with holes or defective seams. Strict control over smoking and handling of smoked fish. Store smoked fish in a freezer. Care in gutting and preparing raw fish. Strict attention to the shelf life of vacuum packed food

The Health Protection Agency (preventing person-to-person spread following gastrointestinal infections:) guidelines for public health physicians and environmental health practitioners/officers (2004) were used as the main reference for onset times in this table.

SOURCE	FOOD VEHICLES	ONSET (HOURS)	SYMPTOMS & DURATION	ADDITIONAL SPECIFIC CONTROLS
Bacillus cereus (i) (Toxin in food)				
Cereals, especially rice, herbs, spices, dried foods, milk and dairy products, meats, dust and soil	Rice products, starchy food such as pasta and potato, cornflour, vanilla slices, custards, soups and vegetables	1 to 6	Acute nausea, vomiting, abdominal pains and some diarrhoea (12 to 24 hours)	Thorough cooking and rapid cooling. Storing at correct temperatures. Avoid rewarming
Bacillus cereus (ii) (Toxin in intestine)				
Cereals, especially rice, herbs, spices, dried foods, milk and dairy products, meats, dust and soil	Rice products, starchy food such as pasta and potato, cornflour, vanilla slices, custards, soups and vegetables	6 to 24	Abdominal pain, fever, watery, diarrhoea and some vomiting (1 to 2 days)	Thorough cooking and rapid cooling. Storing at correct temperatures. Avoid rewarming
Escherichia coli (Infection and toxin)				
Human sewage, water and raw meat	Cooked foods especially buffets, water, milk, cheese, seafoods and salads	10 to 72 usually 12 to 24	Abdominal pain, fever, diarrhoea and vomiting (1 to 5 days)	High standards of hygiene. Thorough cooking. Avoid cross-contamination. Storage at correct temperature. Safe sewage disposal and chlorination of water used for drinking or food production
Yersinia enterocolitica (Infection)				
Milk and raw pork	Meat, raw milk, fish, shellfish, oysters, poultry, dairy products and salads	3 to 7 days	Severe abdominal pain, fever and diarrhoea, rarely nausea and vomiting (1 to 7 days). Complete recovery in most cases	Thorough cooking. Care with shelf life of foods stored under refrigeration (psychrotrophic)

CHEMICAL AND METALLIC POISONING

Some chemicals are extremely poisonous and if ingested may result in severe vomiting within a few minutes, and in some cases fatalities. Other chemicals may cause cancer or other serious **chronic disease**. Environmental contaminants such as dioxins may, for example, end up in cow's milk as a result of contaminating grazing land. In 2001 the Food Standards Agency (FSA) asked two companies to remove cod liver oil from sale because it contained high levels of dioxin and polychlorinated biphenyls (PCBs). In the same year the FSA ordered the withdrawal of Greek, Spanish, and Italian olive-pomace oil contaminated with high levels of potentially **carcinogenic** polycyclic aromatic hydrocarbons. In 2002 one of Britain's main suppliers of freshwater prawns withdrew several brands when tests demonstrated that 16 out of 84 samples contained nitrofurans, a potential cancer-causing drug.

Chemicals can enter foodstuffs by leakage, spillage, or other accidents during processing or preparation. Unacceptable levels of benzene migration from plastic packaging has been known to occur. However, acute chemical poisoning from food premises is rare and is usually caused by negligence, for example, storing weedkiller, pesticide or cleaning chemicals in unlabelled food containers. Problems may occasionally be caused by naturally occurring toxicants, for example, the processing contaminant ethyl carbamate formed during the fermentation of alcohol and other fermented food. In 2001 the FSA advised that certain soy sauce brands should be avoided as they contained high levels of cancer-causing chemicals.

Chemical **additives** of food have to undergo rigorous tests before they are allowed and are usually harmless. However, some may cause problems if ingested in large amounts, for example, monosodium glutamate. The onset of symptoms usually occurs in less than one hour and may include burning sensations in the chest, neck and abdomen. The use of antibiotics in food production is usually prohibited because of the dangers of developing resistant bacteria, which could prove harmful to the populace. Nitrates and nitrites are added to ham and bacon to inhibit the growth of *Cl. botulinum* and *Cl. perfringens* and to produce a desirable red colour. However, the maximum amounts allowable have been reduced because of the carcinogenic effect on animals. In 1997, three persons became ill after consuming sausages containing sodium nitrate and sodium nitrite at levels of 200 times that permitted for cured meats (not permitted in sausages). Symptoms included drowsiness, dizziness and a greying colour of the skin.

Residues of drugs, **pesticides** and fertilizers may be present in deliveries of raw materials. Pesticides sprayed on to fruit and vegetables just prior to harvesting may result in cumulative toxic effects and should be strictly controlled. FSA warnings have been issued regarding levels of dieldrin residues in eels and in 1995 the need to peel carrots. Two outbreaks of chemical food poisoning in 1991 resulted from the contamination of watermelons (USA) and cucumbers (UK) with the pesticide aldicarb. In August 2001, over 500 people in the Assam state of India became ill following the consumption of food snacks contaminated by **insecticides**. In 2002 the FSA ordered the removal of several brands of Chinese honey which were contaminated with chloramphenicol. This antibiotic has been linked with blood disorders and cancer.

Chemical poisoning may also occur because of waste, such as mercury compounds, polluting river water used for drinking or food production. Poisons may be taken up by fish, which may cause illness in humans if consumed. In 2002 the FSA issued a warning for pregnant women and children to stop eating shark, swordfish and marlin because of fears of high levels of mercury, in the form of methylmercury. Symptoms include blurred vision and a prickling sensation of the skin. In 1984 a large outbreak of gastrointestinal illness in the north-west of England and North Wales was caused by drinking mains water contaminated by phenol. From May 1981 to December 1982 approximately 20,000 Spanish people became ill after using olive oil sold by street vendors, which allegedly contained industrial waste oil. At least 350 people died.

Several metals are toxic and if ingested in sufficient quantities can give rise to food poisoning. The symptoms, mainly vomiting and abdominal pain, usually develop within an hour. Diarrhoea may also occur. Metals may be absorbed by growing crops or contaminate food during processing. Acid foods (pH <7) should not be cooked or stored in equipment containing any of the following metals.

Aluminium

Although concern has been expressed about aluminium being a factor in some cases of pre-senile dementia, contaminated food is unlikely to play a significant part.

Antimony

Antimony is used in the enamel coating of equipment. Under certain conditions antimony poisoning can occur. Vessels with chipped enamel should not be used.

Cadmium

Cadmium is used extensively for plating utensils and fittings for electric cookers and refrigeration apparatus. It is readily attacked by some acids including fruit and wines. Foods, such as meat, placed directly on refrigerator shelves containing cadmium may become poisonous. Earthenware from some countries may contain acid-extractable cadmium, which may be hazardous if used with acid food.

Copper

Copper poisoning has been caused by a worn copper dispensing outlet on a soft-drink machine. Copper is also an initiator of rancidity development in fats and oils. Copper, and bronze fittings, must be avoided when processing foods such as milk.

Lead

Lead is a very poisonous metal and if ingested may cause acute poisoning. Consumption of food and drink is the main exposure route of people to lead, especially fruit, leafy vegetables, cereals (airborne lead from petrol and incinerators), kidneys, wine and shellfish (industrial effluent or geological sources). Lead is also released from containers made of ceramics and lead crystal. The glazing of earthenware vessels is compounded with lead oxide and the storage of acid products in such vessels should be avoided. Chronic poisoning from lead absorbed by soft water passing through lead pipes is not classed as food poisoning. The Lead in Food Regulations provide a maximum general limit for lead of 1ppm in most foods. Some imported cans may have lead seams and must be carefully examined.

Tin and iron

Most cans used for the storage of food are constructed of tin-plated iron sheet. Occasionally, due to prolonged storage, certain acid foods, such as pineapple, rhubarb, strawberries, citrus fruits and tomatoes, react with the tin-plate and hydrogen gas is produced. Iron and tin are absorbed by the food which may become unfit for human consumption. The Tin in Food Regulations, 1992 impose a statutory limit of 200ppm for the amount of tin permitted in food. Hundreds of students in Florida suffered from tin poisoning following the consumption of pineapple. The corrosion of the poorly lacquered imported cans from South Africa was responsible. In 2001 the EU Scientific Committee on Food concluded that levels of 150mg/kg in canned beverages or 250mg/kg of tin in other canned foods could cause illness in some people. In July 2002, cans of organic spaghetti were recalled, by a major UK company, because of high tin levels.

Zinc

Zinc is used in the galvanizing of metals. Galvanized equipment should not be used in direct contact with food, particularly acid foods.

POISONOUS PLANTS AND FISH

Poisonous plants are rarely the cause of food poisoning in food premises. Plants responsible for causing acute poisoning include deadly nightshade, death cap (which may be mistaken for edible mushroom) and rhubarb leaves. In 1981 apricot kernels containing enzymatically produced hydrogen cyanide caused the poisoning of 24 children and four deaths in Israel. In 1983 several people became ill after drinking herbal tea contaminated with deadly nightshade. DEFRA advised consumers to avoid drinking comfrey preparations containing a high level of pyrrolizidine alkaloids. Natural toxicants in plants include: glycoalkaloids (sprouts and peel of potatoes, especially if green); furocoumarins (parsnips, celery and limes); phytoestrogens (soya-based products, cereals, peas, beans, cabbage, cucumbers, coffee and plums) and hydrazine derivatives (mushrooms). In 1999 a 34-year-old man went into a coma following suspected mushroom poisoning.

The gonads, liver and intestines of some fish are highly toxic. For example, the puffer fish, which is considered a delicacy by the Japanese, has caused fatalities because of the failure to remove the toxic organs during preparation.

Several incidents of red whelk poisoning have been recorded in the UK since 1970. Symptoms include tingling of the fingers, disturbance of vision, paralysis, nausea, vomiting, diarrhoea and prostration. Recovery is usually within 24 hours. Intoxication is due to poisoning by tetramine, which is present in the salivary glands of these whelks.

Red kidney beans

Between 1976 and 1982 there were 26 reported incidents of food poisoning, affecting at least 118 persons, associated with the consumption of raw or undercooked red kidney beans *(Phaseolus vulgaris)*.

The onset period is short, one to six hours, and recovery normally rapid. Symptoms may develop after eating only four raw beans and include nausea, vomiting and abdominal pain followed by diarrhoea. Recovery is usually around three hours but may be up to two days.

A naturally occurring haemagglutin, which can be destroyed by boiling for ten minutes, is responsible for the illness. The temperatures achieved by "slow cookers" are usually insufficient to destroy the haemagglutin.

In 1986, a number of people became ill after eating vegetable burgers containing haricot beans. The symptoms resembled those caused by red kidney beans and it is likely that undercooking resulted in a failure to destroy the high levels of haemagglutin in the raw beans.

Scombrotoxic fish poisoning

Scombrotoxic fish poisoning is caused by toxins that accumulate in the body of some fish, mainly the Scombridae family, during storage, especially above 4ºC. It is responsible for about 65% of the outbreaks of food poisoning associated with the consumption of fish in England and Wales. The onset period is between ten minutes and three hours. Symptoms last up to eight hours and include headache, nausea, vomiting, abdominal pain, a rash on the face and neck, a burning or peppery sensation in the mouth, sweating and diarrhoea. Scombrotoxic poisoning results from the conversion of histidine, an amino acid found in dark-fleshed fish, to histamine by the action of spoilage bacteria such as *Morganella spp.*, *Hafnia spp.* and *Klebsiella spp*. Histamine is normally found at levels in excess of 50ppm in fish responsible for illness and these levels may be reached before the fish appears spoiled. Problems arise in canning fish as, once formed, the toxin is very heat-resistant and will not be destroyed during processing.

Refrigerated storage of fish should prevent toxin formation.

From 1992 to 2000 there were 78 outbreaks and 40 sporadic cases of scombrotoxic fish poisoning involving 341 people. No outbreaks were reported in 2001/2002. Incidents were confirmed by histamine analysis (levels greater than 5mg per 100g of flesh). Fish involved included raw and canned tuna, smoked, canned and soused mackerel, canned sardines, pilchards, herring, anchovies and salmon. Around 50 cases a year are investigated.

Ciguatera poisoning

Ciguatoxin enters the food chain through fish eating the toxic dinoflagellate, *Gambierdiscus toxicus*, which is associated with dead coral reefs and marine algae. Carnivorous fish accumulate toxins when consuming the smaller herbivorous fish. Species of fish involved include the moray eel, red bass, coral trout, Spanish mackerel, reef cod, grouper and surgeonfish. Fish affected look, smell and taste normal. The head, liver, gonads and roe are usually the most toxic parts of the fish. The toxin is unaffected by cooking, freezing or gastric juices.

The onset period is usually one to six hours and symptoms include malaise, headaches, disturbed vision, alternating feelings of hot and cold, respiratory paralysis, numbness or burning of the mouth, throat and tongue, weakness, numbness of the extremities, abdominal pain, vomiting and diarrhoea. Occasional deaths are recorded. In 1990, three people suffered ciguatera poisoning after eating a home-cooked red snapper imported from Oman. Ciguatera poisoning is the most common fish poisoning problem in the Pacific and Indian oceans and the Caribbean. In June 2001 an outbreak of over 50 cases occurred in the Philippines following the consumption of ciguatoxic barracuda.

Paralytic shellfish poisoning (PSP) and diarrhetic shellfish poisoning (DSP)

PSP and DSP may result from the consumption of mussels and other bivalves that have fed on poisonous plankton (various species of dinoflagellates). The aquatic biotoxins causing PSP and DSP may withstand cooking. Symptoms for PSP include a tingling or numbness of the mouth almost immediately and this spreads to the neck, arms and legs within four to six hours. Death, when it does occur, is usually caused by respiratory paralysis within two to 12 hours. DSP symptoms include nausea, vomiting, abdominal pain, diarrhoea and chills with an onset time of 30 minutes to 12 hours.

In July 1997, an outbreak of DSP involving 30 cases resulted from the consumption of contaminated mussels served in two restaurants. Action is taken to prevent the consumption of bivalves if tests are positive for the presence of DSP toxin or if levels of PSP toxin exceed 80g/100g shellfish flesh. In 2002, cockle fishing was banned on the Burry inlet in South Wales and a prohibition notice stopped the fishing of mussels, cockles, oysters and clams from Foulness to Canvey Island because of fears of DSP.

MYCOTOXINS

Mycotoxins are metabolites of some **moulds** that cause illness, and sometimes death, for example, aflatoxin and ochratoxin A. Symptoms vary depending on the particular mould. They are often carcinogenic but some may cause gastrointestinal disturbances, for example, *Aspergillus ustus*.

Aflatoxin M1 has been detected in milk and milk powder and there is increasing interest in the presence of patulin, a mutagenic mycotoxin produced by certain species of *Penicillium* and

Aspergillus moulds, which infect apples and pears. Because of migration, cutting out the visible mould may not remove the risk. Patulin is not destroyed by pasteurization. The FSA issued guidance on patulin, following concern over high levels in apple juice sold in the UK. Juice containing levels of patulin in excess of 50ppb should not be used. In 1992, 26% of samples of juices from pressed apples (cloudy juices) exceeded 50ppb, however, in 1996 only 2% of samples of juices from pressed apples exceeded this limit. In 2002, the FSA ordered the recall of James White apple juice because of unacceptable levels of patulin.

Aflatoxicosis

Aflatoxicosis is a **foodborne illness** caused by the consumption of aflatoxin produced by the common moulds, *Aspergillus flavus* and *Aspergillus parasiticus*. The illness may be either acute or chronic but the effects include liver cirrhosis and the induction of tumours. Foods usually implicated include cereals and nuts. Most illnesses caused by aflatoxins occur in underdeveloped countries where considerable amounts of mouldy cereals are consumed over prolonged periods. The Aflatoxins in Nuts, Nut Products, Dried Figs and Dried Fig Products Regulations, 1992 (SI. 3236) prohibit sales of the above products if levels of aflatoxin exceed 4μg/kg. In 1994, two samples of peanut butter and one sample of peanuts from a major retailer had a mean aflatoxin concentration of 20μg/kg for the peanut butter and 23μg/kg for the peanuts. Voluntary withdrawal was undertaken. A UK survey of herbs and spices in 1994 found 5% had levels of aflatoxin above 10μg/kg, but none above the action level of 50μg/kg. Domestic cooking will not destroy aflatoxin. In 1997 the EC Commission banned the import of pistachio nuts from Iran because of high levels of aflatoxin B1. In 2001 a survey by Suffolk C.C. Trading Standards found that five out of 20 peanut butter samples contained unacceptably high levels of aflatoxin.

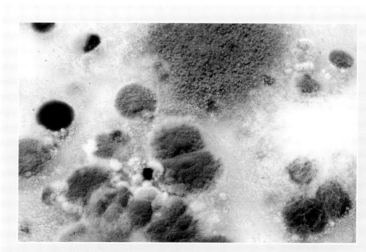

Mycotoxin producing mould (Courtesy of Anticimex).

THE INVESTIGATION OF FOOD POISONING OUTBREAKS

The definition of an outbreak is an incident in which two or more people, thought to have a common exposure, experience a similar illness or proven infection (at least one of them having been ill). A general outbreak is an outbreak affecting members of more than one private residence or residents of an institution. Some general outbreaks are, therefore, from private residences and they are not all from commercial food premises. They may also include waterborne outbreaks.

The effective investigation of food poisoning outbreaks, by all disciplines involved, is essential to limit the spread and to obtain accurate information for use in surveillance. This information can then be disseminated to food businesses and enforcement officers to assist in a risk-based approach to food production and control and make a major contribution to preventing foodborne illness. The sale of food that results in food poisoning is a criminal offence under the Food Safety Act, 1990, and may also result in successful claims for civil damages by sufferers.

The objectives of the investigation of an outbreak are to:

- contain the spread of the outbreak;
- identify the **outbreak location** (place where food vehicle was prepared or served);
- identify the food vehicle(s) involved (the food eaten which gave rise to illness);
- identify the **causative agent** (the organism, toxin or poison associated with the illness that is recovered from sufferers and/or food and/or the environment under investigation, for example, salmonella, scombrotoxin or mercury);
- trace cases and carriers, especially food handlers;
- trace the source of the causative agent (the vehicle that brought the causative agent into the outbreak location, for example, raw food or a food handler, and/or the origin of the causative agent, for example, a cow in the case of raw milk);
- determine **causal factors** (how the food vehicle became contaminated and what stage of food preparation allowed bacterial multiplication);
- recommend how food should be prepared in the future to prevent recurrence;
- provide data for use in surveillance; and
- provide evidence for legal action.

The isolation of the causative agent in food and/or from environmental swabs of surfaces and equipment provides the best evidence to confirm the food vehicle and the outbreak location in the event of a food poisoning outbreak. Consequently, the suspect outbreak location should be inspected, and food remains and environmental swabs taken, at the earliest possible opportunity. Causal factors may be grouped as:

- pre-process contamination (includes eating contaminated raw food);
- survival of pathogens (inadequate cooking/processing/reheating/thawing);
- post-process contamination;
- multiplication of pathogens (abuse of time/temperature, inadequate cooling);
- infected food handlers; and
- management failures.

The role of management
A manager will become involved in food poisoning investigations when:

- a food handler reports that he or she is suffering from diarrhoea and/or vomiting;
- persons purchasing food from, or eating at, the premises complain that they are ill.

The complainants may contact the local Environmental Health Department, who will inform the manager;
- persons attending a function begin to vomit whilst on the premises. If a manager is notified of a suspected case of food poisoning, he or she should immediately contact the Environmental Health Department (EHD) and then make enquiries to find out:
  - whether any staff have been ill;
  - which food or meals were implicated and how they were handled or prepared; and
  - whether or not any other persons were affected.

It is essential to cooperate fully with the investigating officer and to provide required information as soon as possible. Immediately upon becoming aware of a foodborne outbreak associated with the premises, the person in charge should:
- stop serving any more food and stop any more food being eaten;
- telephone a doctor or ambulance if necessary (if people are ill on the premises);
- retain any remaining food, cans, cartons or other forms of food packaging that may be needed to trace suspect food consignments;
- telephone the EHD to arrange for an investigation by an Environmental Health Practitioners/Officers (an-out-of hours emergency number should be available);
- telephone the owner of the business;
- not clean up until bacteriological investigations have been carried out;
- prepare a relevant menu list and, as far as possible, details of food preparation (when purchased, storage temperature, method of cooking, cooling or reheating and monitoring records);
- if possible, make a complete list of persons consuming the suspect meal, including their addresses;
- prepare a complete list of all food handlers and other staff, especially those involved with the suspect meal. Absent staff should be included, the reason for the absence should be given;
- prepare a list of suppliers;
- remain available on the premises to give the necessary information to the investigating officer, for example, when the meal was served, the number of people known to be affected, etc; and
- when the investigation has been completed, thoroughly clean and disinfect the premises. Advice should be obtained from the Environmental Health Practitioner/Officer.

The HACCP system will need to be examined and any necessary amendments implemented.

The role of the authorized officers (environmental health practitioners/officers) in food poisoning investigations

The circumstances of an outbreak investigation may vary considerably. Notifications require different levels of response, however, all activities must be carefully coordinated and structured. Investigations should be undertaken in a way that protects public health without prejudicing future legal action (including formal cautions). There are six key stages:
- *Preliminaries*: reviewing notifications; formulating an initial hypothesis of the nature of the disease; determination of case definition, population at risk and general case search area; deciding on resources needed to investigate; defining responsibilities and reporting systems; setting up communication channels; briefing investigators;

and issuing documents/forms etc. Officers should keep accurate and comprehensive personal daily logs of events.
- *Case finding*: searching for cases, carriers and contacts; recording detailed case histories of sufferers, food consumption details, symptoms, times of onset, personal and occupational details, obtaining specimens; assembling data to formulate a hypothesis as to outbreak location, meals/foods eaten, and approximate time of food preparation. Possible **food vehicles** should not be suggested. Telephone interviews can be useful and quick. Questionnaires, if used, should also be completed by persons attending the same function who were not ill. Appropriate detailed information must accompany all specimens and samples sent to the laboratory to maintain the integrity of evidence.
- *Site investigation:* secure assistance of managers, collect menus, details of food and suppliers; identification of all food production staff and their functions, staff records and sickness returns; locating production areas, food and waste stores and meal remains; seizing or detaining suspect food; postponing cleaning and **disinfection**; terminating or modifying food production, if necessary, to prevent spread of disease; collecting food and other relevant samples and swabbing surfaces (in accordance with the Food Safety [Sampling and Qualifications] Regs, 1990); inspecting and recording site conditions; interviewing managers and staff (contemporaneous notes of the interviews should be made); arranging for collection of specimens from staff. (Food handlers with symptoms must be excluded from food handling duties.) Officers should not allow the sale of food from the premises to be resumed until they have, as far as is reasonably practicable, satisfied themselves that food can be produced safely with no identifiable risk to the consumer. Site investigations are the most important part of the investigation to ascertain the causal factors and require the special skills of an officer experienced in inspecting food premises and interviewing techniques.
- *Intermediate review:* assemble all available data; assess control measures needed; formulate a provisional hypothesis as to suspect food vehicle(s), causal factors and sources; review investigation progress and determine whether the outbreak is foodborne, as opposed to person-to-person spread, and whether a **due diligence** defence is likely to succeed. If a prosecution is likely, all witness statements will need to be dealt with in accord with the Criminal Justice Act, 1967, as amended, and interviews of persons likely to face prosecution should be carried out under the Police and Criminal Evidence Act, 1984. Determine what further investigations are needed, prepare an interim report and if appropriate issue public warnings.
- *Source tracing:* sample ingredients used, to identify contamination levels; assess possibility of organisms surviving processing; analysis of food sample results, swabs and specimens; trace the distribution chain of suspect ingredients or other foods; investigate primary produce/animals or birds.
- *Consolidation:* collating and processing all data to test the hypothesis; carry out, as necessary, further detailed investigations of critical factors in the outbreak; discussion of a draft final report with investigators, agreeing final hypothesis and formulating remedial and control measures; publishing the final report.

The Food Law Code of Practice provides advice in relation to food **hazards** and foodborne illness. In the case of a serious localized incident or a wider problem the relevant government department should be notified. Responsibility at local level remains with the Food Authority. Where a public health hazard has been identified and **risk assessment** indicates consumers could be exposed to the hazard, it is essential to provide information to assist in protecting public health. Consultation with the relevant food industry, government departments and the business affected should occur prior to the release of information to the media to avoid contradictory statements. However, delays should not be allowed to expose the public to risk.

4 Foodborne diseases edited by Adrian Eley

A **foodborne disease** may be considered to differ from **food poisoning** in that:
- a relatively small number of organisms is capable of causing the illness; and
- the **food** acts purely as a **vehicle** and the multiplication of the organism within the food is not an important feature of the illness.

There are potentially a very large number of organisms that could **contaminate** food and lead to foodborne disease. In such a text as this it is not possible to give a complete review. Instead, important examples are given, some of which may lead to symptoms unrelated to **gastroenteritis**.

Typhoid and paratyphoid fever

Sometimes known as enteric fever, typhoid is caused by the **bacterium** *Salmonella* Typhi and paratyphoid by the bacterium *Salmonella* Paratyphi. Only a few *S*. Typhi are needed to cause illness, whereas 100,000 *S*. Paratyphi may be needed to cause illness. The **incubation period** is 3 days to 1 month (usually 8 to 14 days). Symptoms include sustained fever, severe headache, nausea, loss of appetite, slow pulse, spleen enlargement, rose spots on the trunk and constipation or sometimes diarrhoea. The fatality rate for typhoid is between 2% and 10% depending on use of **antibiotics**. Paratyphoid is generally much less severe and symptoms may be similar to salmonella food poisoning.

The organism is excreted in the faeces and urine of patients and **carriers**. Up to 5% of persons affected become permanent carriers. Typhoid fever affects 17 million people worldwide each year with approximately 600,000 deaths. There are a few hundred cases of enteric fever in England and Wales each year, the majority of cases being acquired from **endemic** areas abroad. Enteric fever may be waterborne, due to contamination by urine or sewage, or foodborne, for example, shellfish taken from sewage-contaminated beds, vegetables fertilized by human manure and eaten raw, milk or cooked meat contaminated by polluted water or by carriers who are **food handlers**. Laboratory confirmation is by bacteriological examination of blood, faeces or urine. In 1993 an outbreak of *S*. Paratyphi B in France resulted from eating cheese made from raw goats' milk. In July 2001 an outbreak of typhoid fever occurred in the Newport area of Wales. The only link between the five cases (four children and one young adult) was the consumption of food from a local kebab restaurant.

Prevention
- Ensure the safety of all water supplies. Water used for food preparation or drinking should be chlorinated.
- Ensure the satisfactory disposal of sewage.
- Ensure the heat treatment of milk and milk products, including ice cream.
- Prevent the sale of raw shellfish from sewage-polluted waters.
- Identify carriers and ensure that they are not employed within the food industry. Medical questionnaires should be used as an aid to recruitment.
- Maintain high standards of **personal hygiene** amongst food handlers, especially with regard to thorough handwashing after visiting the toilet.
- Ensure high standards of **hygiene** in food production and distribution.
- Double-wash ready-to-eat fruit and vegetables.

Dysentery
Bacillary dysentery

In the UK bacillary dysentery is usually caused by the Gram-negative bacterium *Shigella sonnei* and less frequently by *Shigella flexneri* and *Shigella boydii*. It is an acute disease of the intestine characterized by diarrhoea, fever, stomach cramps and often vomiting. Stools may contain blood, mucus and pus. Fatality is normally less than 1% and the severity of illness depends on the type of *Shigella* causing the infection. The incubation period is normally around 1 to 3 days, although it varies between 12 hours and 7 days. Dysentery is almost always spread through faecal–oral transmission from an infected person or more rarely by the consumption of contaminated foods, including water or milk. In 1994 imported iceberg lettuce was implicated in several outbreaks of *Sh. sonnei*. In December 2001 Japan banned oysters from S. Korea after 159 people became ill with dysentery and the organism was isolated from the oysters. Outbreaks often involve children. Parents caring for ill children commonly become infected. Consequently some food businesses exclude food handlers with household contacts suffering from dysentery. Preventive measures are similar to those used for typhoid, with the emphasis on personal hygiene.

Amoebic dysentery

This form of dysentery is rare in the UK apart from persons becoming infected whilst visiting tropical and sub-tropical countries where it is endemic. There were 161 reported cases in England and Wales in 2002. The causative organism is *Entamoeba histolytica*, a microscopic single-celled protozoan. Symptoms include abdominal pain and mild or severe diarrhoea alternating with constipation. The incubation period is usually 2 to 4 weeks.

Campylobacter enteritis

Campylobacter has been implicated as a cause of acute enteritis since 1977. High-risk food is rarely proven to be a vehicle in incidents and the main vehicle for transmission has yet to be identified. However, it is thought that up to 70% of cases may be foodborne. Even though the epidemiology of campylobacter infection remains poorly understood, it is the commonest cause of acute bacterial gastroenteritis in England and Wales and this must be due in part to its low-infective dose (up to 500 organisms).

Campylobacters are Gram-negative spirally curved rods and most species are microaerophilic; oxygen at normal atmospheric pressure prevents growth and numbers decline. They multiply quickly between 37ºC and 43ºC but not below 28ºC. At 48ºC they are inactivated.

The causal species of human enteritis are *Campylobacter jejuni* (approximately 90%) and *Campylobacter coli* (approximately 10%). Symptoms vary in severity and include headache, fever, diarrhoea (often blood-stained), persistent colicky abdominal pain (may mimic acute appendicitis) and nausea (vomiting is rare). The incubation period is from one to ten days but usually between two and five days. The normal duration of illness is one to seven days. A second "dose" of symptoms may occur three weeks after the first.

Most reported cases are sporadic and peak in the late spring. Mortality is minimal and complications are rare. Campylobacters disappear from the stools within a few weeks of illness and long-term carriers have not been detected. Because of the low transmissibility and inability of campylobacter to multiply at room temperatures, infected food handlers should usually be allowed to resume normal duties when their stools are firm.

Animals and wild birds are the main reservoirs of infection and as campylobacters can survive in water for several weeks at low temperatures, untreated natural water is a potential **source**. Furthermore, water storage tanks could be contaminated by birds or small animals. Campylobacters are commonly found on raw poultry, in raw milk and sewage and on carcase meat and offal. A survey undertaken by the Food Standards Agency in August 2001 indicated that 63% of fresh and 33% of frozen poultry was contaminated by campylobacter. Further investigation is necessary to determine why fresh poultry have almost twice the level of contamination of frozen poultry. **Cross-contamination** from raw poultry is extremely likely and hands can carry campylobacters for up to an hour.

Transmission is thought to be from raw and undercooked poultry and meat, raw milk, bottled milk pecked by birds, especially magpies, contaminated water (private supplies) and infected dogs and cats. Person-to-person spread and secondary cases are rare. Freezing reduces the number of campylobacters but once frozen they can survive for several months. The organisms may be destroyed by heating food to 75°C and are sensitive to drying, chlorination and salt in excess of 2%.

Control measures include reducing the numbers of campylobacters in raw meat and the food chain, hygiene training of food handlers, especially on the dangers of cross-contamination and the importance of thorough cooking, and raising the hygiene awareness of consumers.

Listeriosis

Although food is not the only means of transmission, and in the majority of cases a vehicle of infection is not identified, listeriosis may be considered as a foodborne disease and is caused by *Listeria monocytogenes*, a Gram-positive, **psychrotrophic** bacillus that is widely distributed in the environment (soil, water and drains). It is commonly found in effluents and sewage sludge and survives many weeks after spraying. In one outbreak involving coleslaw, the cabbage, stored for several months, had been grown in a field fertilized by sheep manure. Human or animal carriers may excrete the bacteria, and many cases of cross-infection have been recorded. The incubation period is one day to three months. Symptoms include influenza-like illness, fever, diarrhoea, **septicaemia**, meningitis and abortion; **neonates**, pregnant women, **immunocompromized** persons and the elderly are most at **risk**. It causes little illness in healthy people and not all subtypes of *L. monocytogenes* have been proven to be pathogenic.

The organism, which is described as a low temperature pathogen, is salt tolerant, multiplies at a **pH** as low as 4.4 and between −1.5°C and 42°C, although growth at 3°C is very slow. It has the lowest temperature growth of any pathogenic bacteria. Under refrigeration, especially above 3°C, small numbers of listeria can soon dominate competing organisms and may also become more infective due to the production of a **toxin** (listeriolysin O). The trend to longer **shelf life** makes this of particular concern. Hypochlorite dips may be ineffective at removing the bacteria from contaminated vegetables.

Thermal death time is widely disputed: the bacteria may have a greater ability than normal to recover from heat damage. However, most reports state listeria is not especially heat-resistant, and is rapidly destroyed at 75°C.

Numbers increased from the 1970s with a peak of 291 cases and 63 deaths in 1988, probably due to imported Belgium pâté. In the mid 1980s an outbreak in Los Angeles associated with soft cheese resulted in 148 cases and 48 deaths among Mexicans. In 1987 Switzerland banned the production of Vacherin Mont d'Or cheese following 60+ deaths over a period of around four years. In 1989 the UK government issued a belated warning to risk-groups to avoid soft cheeses, reheated cook-chill meals and pâté. An outbreak in France in 1992, probably due to pork tongue in aspic, resulted in 279 cases, 63 deaths (including seven neonates) and 22 abortions. In 2001, Fresh Products Northwest, USA, recalled Crunch Pak Fresh Sliced Apple Packages due to possible contamination by *L. monocytogenes*. The death rate for listeriosis is usually around 30% but mainly involves persons with other serious illness or with immune deficiencies. Surveys in the UK have shown that soft cheese and pâté now contain lower levels of listeria, which is also isolated in raw meat/poultry/fish/vegetables/milk, fried rice, salads, dairy products and retail processed-chilled food. Sporadic cases have implicated turkey frankfurters, dried mushrooms, salami, vegetable rennet and undercooked chicken.

Streptococcus zooepidemicus

This bacterium may cause a serious flu-like illness together with septicaemia, meningitis and bacterial endocarditis. *Streptococcus zooepidemicus* is found in a variety of animals including horses and cows, when it may cause mastitis.

In 1984, 12 persons in West Yorkshire, ten over 70 years of age, were hospitalized after being infected with *S. zooepidemicus*. Eight of these subsequently died, including a day-old premature baby. All eleven adults drank raw milk. The organism was isolated from a bottle of milk and three cows from the supplier's herd. In 1994 two persons drinking raw milk were infected by *S. zooepidemicus* and one of them subsequently died.

Brucellosis (undulant fever)

A systemic disease caused by the Gram-negative bacteria *Brucella abortus* and *Brucella melitensis*. Symptoms include intermittent fever, extended depression, headache, weakness and generalized aching. The illness may last from a few days to several months. Fatality is less than 2%. The incubation period is usually five to 21 days. Transmission is normally by contact with an infected cow or by the consumption of raw milk or milk products. The brucellosis eradication scheme, which involved slaughtering all infected animals, and the heat treatment of milk have largely eliminated this disease. In 2001, six cases of *Brucella melitensis* were reported in England and Wales (probably contracted abroad).

Tuberculosis

A chronic bacterial disease caused by *Mycobacterium tuberculosis*, the human tubercle bacillus, and *Mycobacterium bovis* in cattle. The bovine type also affects man and can result in extra-pulmonary tuberculosis involving the bones, lymph nodes, kidneys, intestines and skin. The incubation period is four to six weeks but development of the later stages may take years. Airborne transmission and contact with infected sputum occur, although bovine tuberculosis is primarily spread by the consumption of raw milk or dairy products. Tuberculosis used to be a very common disease of bovines but is now rare, due to the slaughter of all infected animals. Heat treatment of milk was introduced primarily to combat this disease.

Vero cytotoxin-producing *Escherichia coli* O157:H7 (O157 VTEC)

Most *E. coli* that are found in the intestine are harmless. However, those capable of producing Vero cytotoxins often cause serious illness, which is sometimes fatal, particularly in young children and the elderly. Although there are many strains of VTEC, the main one associated with human disease in the UK is O157. The first documented outbreak of haemorrhagic colitis caused by *E. coli* O157 was in 1982, when it was linked to consuming hamburgers from a fast-food restaurant in the USA.

Symptoms vary from a watery diarrhoea, nausea and abdominal pain to bright red bloody diarrhoea (haemorrhagic colitis) and severe abdominal cramps, usually without fever. Up to 30% of patients develop haemolytic uraemic syndrome (HUS), although it is usually between 2% and 7%. This generally involves young children and *E. coli* O157 is the major cause of acute renal failure in children in the UK. Some adults develop HUS together with neurological complications. Fatality rates range from 1% to 5%, but in some outbreaks, for example, involving the elderly, these may be much higher. The onset period is one to eight days, usually three to four days. The duration of illness is approximately two weeks, unless complications, such as HUS, develop. *E. coli* O157 disappears from adult faeces within a few days and specimens should be examined within six days of the onset of symptoms. Young children may excrete *E. coli* O157 intermittently for three or more weeks.

Although *E. coli* O157 can multiply in food, it has a very low infective dose involving fewer than 100 bacteria. Infection results from eating contaminated foods, person-to-person spread and direct contact with animals, especially farm animals and their faeces. There have been many outbreaks from children "petting" farm animals. The main food vehicles are undercooked meat products, especially burgers and mince. Other foods implicated include raw milk, cheese made with unpasteurized milk, and apple juice. (*E. coli* O157 can grow at a pH of 4.4.) However, because of the low infective dose, cross-contamination of many **ready-to-eat foods** from raw meat is likely to result in illness. There have been several outbreaks associated with cooked meat, especially from butchers. Bathing in contaminated water has occasionally caused infection.

The main reservoir of *E. coli* O157 is the stomach and intestines of cattle and, possibly, sheep. Manured vegetable and fruit crops may therefore be a **source** of the organism. Cattle excrete *E. coli* O157 intermittently and the infected animals show no signs of illness. The first isolation of *E. coli* O157 in food was in Sheffield in 1993, when it was discovered in raw milk, although it had been isolated on cattle carcasses in Sheffield in 1992. Surveys by the Public Health Laboratory Service and local authorities have demonstrated 1.5% raw beef and 5.9% raw lamb samples from butchers' shops have been contaminated with *E. coli* O157.

The bacterium survives freezing and is relatively tolerant to **acid** conditions. It survives and multiplies in some foods between 3°C and 46°C although numbers may decline below 3°C. It can double every four hours at 10°C. It is destroyed by normal, effective cooking processes. Most outbreaks occur in summer and secondary cases from person-to-person spread are common. Most investigations fail to identify a food vehicle.

In 1996 over 10,000 people, mainly young children in Japan, were affected in a major *E. coli* O157 outbreak(s) with over 15 deaths. Also, in 1996, several linked outbreaks of *E. coli* O157 in Central Scotland were traced back to a butcher's shop in Wishaw. The food vehicles involved included cold cooked meats and cooked steak/gravy. The outbreak(s) resulted in over 500 cases and 21 deaths, most victims being over 69 years old. As a result the government set up an expert group, chaired by Professor Hugh Pennington, to examine the Wishaw outbreak(s) and recommend control measures.

The largest outbreaks of *E.coli* O157 recorded in England and Wales occurred in Cumbria (over 100 cases) during the spring of 1999 and in Wales during October 2005 (over 150 cases and one death). In Cumbria the food vehicle was raw milk as a result of **pasteurization** failure. In Wales cooked meat was involved.

A task force set up in Scotland in September 2000 confirmed that the majority of infection with *E.coli* O157 was originating from environmental sources and human contact with animals and animal **faecal material**, although contaminated food remained a significant vehicle.

The 1997 Pennington Group recommendations for the control of *E. coli* O157

- An education/awareness programme for farm workers to be repeated and updated periodically.
- Consistent and rigorous enforcement of hygiene standards at slaughterhouses to avoid carcase faecal contamination with particular reference to dirty animals, the implementation of HACCP and the training of abattoir workers.
- The accelerated implementation of HACCP, including documentation and licensing of butchers' shops, with particular reference to hygiene training, temperature control and monitoring and product recall.
- The physical separation of raw meat and unwrapped cooked meat/meat products and other ready-to-eat foods.
- Improved food hygiene training of all food handlers, especially those working with vulnerable groups. Supervisory staff and one-person operations to be trained to at least intermediate level.
- Further targeted and prioritized research.
- Food hygiene training to be provided within the primary and secondary school curriculum.
- More resources to address enforcement and education/awareness issues, especially relating to the dangers of cross-contamination and eating undercooked products.
- Improved surveillance, research and control of outbreaks, including full written reports and consideration of their publication.

Norovirus

Incidents of gastroenteritis are being increasingly attributed to viruses and undoubtedly contribute to a large number of notified cases of "food poisoning". However, rotaviruses, adenoviruses, astroviruses and calciviruses are predominantly faecal-oral, paediatric infections and are rarely incriminated in foodborne transmission. Outbreaks in hospitals, nursing homes and other "closed" communities are common and are often considered to be person-to-person spread. However, airborne spread from vomit may be of greater importance. Inhalation of particles from desiccated vomit and faeces may be a significant transmission vehicle in "closed settings".

Outbreaks have been associated with contaminated ice and sewage polluted water, although standard water treatment should prevent problems.

Noroviruses, formerly known as small round structured viruses (SRSVs), are the major cause of viral foodborne cases and outbreaks in the UK. Symptoms include vomiting, the predominant symptom and usually projectile, some diarrhoea, abdominal pain, fever, nausea and dehydration. Symptoms usually last around 12 to 60 hours and patients are infectious for a further two days. Recent studies have suggested symptoms may last up to two weeks. Virus particles can be shed for up to ten or more days. The incubation period is 10 to 50 hours,

Raw oysters grown in polluted waters are often implicated in viral gastroenteritis.

usually 24 to 48 hours, and is dose dependent. Specimens are required within 48 hours of the first symptoms. The infective dose is very low, between ten and 100 particles. Humans were thought to be the only reservoir for Noroviruses but pigs and cattle are now suspected. Astroviruses are occasionally foodborne and have a four day onset period.

Viruses are around three hundredths the size of bacteria and can only be seen through an electron microscope. They multiply rapidly in living cells not in food. They survive well on inanimate surfaces and in the environment. Transmission depends on contamination of food by food handlers or sewage. Filter feeding bivalves, such as oysters, harvested from sewage-polluted waters, are a major problem because of their ability to concentrate viruses from contaminated water. Foods that are handled the most present the greatest risk, ice, desserts, fruit, cold meats and salads are common vehicles, but the vehicle is often unidentified. Fresh produce can be contaminated before, during or after harvesting.

The virus is present in vomit and the mode of transmission is person to person, through **fomites** or environmental contamination with up to 10% being foodborne or waterborne. Outbreaks peak in the winter and are characterized by a high secondary attack rate. Re-infection is common. Outbreaks of Norovirus can be suspected if stool specimens are negative for bacteria, average duration of the illness is 12 to 60 hours, vomiting is present in the majority of cases and the incubation period is 10 to 50 hours. The average attack rate is usually 45% or more. **Asymptomatic** infections are also common (up to 5%). Viruses thrive in cold conditions and are destroyed at temperatures above 60°C. A heat treatment of 90°C for 90 seconds is recommended for destruction in shellfish, but is disliked as it makes the shellfish rubbery.

Depuration of shellfish is ineffective against viral contamination. Freezing does not destroy the virus and 15 outbreaks in Finland were traced to imported frozen raspberries. Illness has also been caused by eating shellfish pickled in brine and vinegar. They also survive high levels of chlorine.

Control measures include:
- the exclusion of **symptomatic** food handlers until 48 hours after they are symptom-free. This may be extended to ten days if they cannot be relied on to wash their hands properly and/or handle high-risk food. (Disposable gloves may be considered);
- the use of reputable suppliers, especially for bivalves, salad vegetables and fruit;
- the washing or **blanching** of ready-to-eat raw fruits such as melons and berries;
- staff training, particularly in relation to handwashing;
- the implementation of HACCP;
- environmental decontamination of public areas. Steam cleaning of carpets, furniture and soft furnishings;
- ensuring that toilets always have adequate accessible supplies of liquid soap, disposable paper towels, nailbrushes and warm running water;
- the destruction of potentially contaminated foods;
- the separation of potential sources of Noroviruses, such as bivalves, from ready-to-eat foods;
- ensuring that any contaminated food preparation surfaces, hand-contact surfaces and toilets are thoroughly cleaned and disinfected using 500ppm hypochlorite; and
- the detection and elimination of Noroviruses from the food chain.

The first recorded outbreak of foodborne viral gastroenteritis was identified in an American school in the town of Norwalk, Ohio in 1968. The first time that similar virus particles were detected in both oysters and the faecal specimens of ill persons who had consumed the oysters was in Kingston-upon-Hull in 1980. The Communicable Disease Surveillance Centre received reports of 1,877 laboratory confirmed outbreaks of Norovirus infections between 1992 and 2000 in England and Wales. These affected 57,000 people, 24 of whom died. There were 754 outbreaks in hospitals, 724 outbreaks in residential care homes, 147 outbreaks in hotels, 105 in restaurants and 73 in schools. Noroviruses are the most common microbial cause of infectious intestinal disease in the USA and of major concern on cruise ships.

Hepatitis

Hepatitis is caused by a virus and the two main types are hepatitis A and hepatitis B. Type A is the more common. Some statutory notifications of Hepatitis A are foodborne and associated with shellfish. Hepatitis type B is transmitted sexually and by blood. It is often spread by the use of contaminated needles.

The onset of hepatitis A is abrupt and symptoms include fever, **malaise**, nausea, abdominal pain and later jaundice. The duration of the illness varies from one week to several months. Man is the reservoir for the virus and transmission is often by the faecal–oral route. Faeces, blood and urine may be infected and can contaminate food, especially water, shellfish and milk. The incubation period is from 15 to 50 days. The fatality rate is less than 1%. Preventive measures are similar to those used for typhoid. A temperature of 90°C for 90 seconds will inactivate hepatitis A virus.

An outbreak in Japan involved 30,000 cases and was traced to infected clams. In 1994, 79 cases of hepatitis A resulted from the consumption of cinnamon buns in New York. The baker, who was positive for hepatitis, had applied cinnamon glaze to the buns. Handwashing was unsatisfactory. Another outbreak in the USA involved 71 cases who had eaten iced pastries.

Transmissible spongiform encephalopathy

Spongiform encephalopathy is thought to be caused by prions (proteinaceous infectious particles) and is responsible for scrapie in sheep, bovine spongiform encephalopathy (BSE) and Creutzfeldt-Jacob disease (CJD) including variant-CJD (vCJD) in humans.

vCJD is a rare and fatal neurodegenerative disease that is considered to be related in part to the consumption of BSE-contaminated food. Between 1970 and December 2001 there were 944 cases of CJD in the UK. 730 cases were classified as definite, the remainder being probable. Risk factors for the development of vCJD include age, host genotype and residence in the UK.

Proposals agreed by the European Commission in February 2001 on restrictions to limit infected material entering the food chain were:

- the removal of the vertebral column from all cattle over 12 months and prohibiting mechanically recovered meat from bones of ruminants;
- requirements to pressure-cook rendered animal fats from ruminants for food and feed; and
- authorizing certain hydrolyzed proteins from fish and feathers.

Unfortunately, much remains unknown about BSE and vCJD. For example, assuming a causal relationship between vCJD and oral consumption of BSE-infected food, we do not know the infective dose and whether it is a single dose or cumulative. It is obvious that there is an urgent need to carry out research in a number of areas to answer basic but important questions. There is currently no treatment or cure for BSE or vCJD.

Information on CJD is available from www.cjd.ed.ac.uk

PARASITES AFFECTING FOOD AND/OR MAN

A parasite is a plant or animal that has the ability to live on another plant or animal known as the host. The parasite obtains its nourishment from the host. Parasites may pass through complicated life cycles that depend on intermediate hosts for the cystic stage of their life.

Giardiasis

Giardiasis is a type of gastroenteritis caused by the parasitic protozoan *Giardia duodenalis* (*Giardia lamblia or Giardia intestinalis*), which is a common inhabitant of the intestinal tract. The incubation period is 3 to 25 days and symptoms include watery diarrhoea, flatulence, abdominal pain and distension. Alternatively, there may be less acute diarrhoea with fatty stools (steatorrhoea), lassitude and weight loss. The infective dose is very low and may only require one viable cyst. The main sources are human cases and carriers and animals. Transmission is by the faecal–oral route. Cases involving food are rarely proven. Waterborne outbreaks occasionally occur and chlorination without filtration is ineffective. Most laboratory reports of giardiasis in England and Wales affect children, and many of these probably originated abroad.

Cryptosporidiosis

Cryptosporidiosis is a gastrointestinal infection caused by parasitic protozoa, *Cryptosporidium spp*. Symptoms include diarrhoea, vomiting, abdominal pain and anorexia. The incubation period is probably around ten days. The sources are cattle and other domestic and wild animals. The protozoa cling to the surface of the intestine and produce oocysts that pass out in the faeces. The oocysts (4 to 6mm in diameter) are unaffected by the chlorine levels in tap water. Transmission is by the faecal–oral route, although it is believed that close contact with infected animals is more significant than contaminated food. Most large outbreaks involve water, although other vehicles have included raw milk and unpasteurized apple juice.

Cyclosporiasis

Cyclosporiasis is a gastrointestinal infection caused by the parasitic protozoan *Cyclospora cayetanensis*. Symptoms include a prolonged relapsing watery diarrhoea with a duration of one to eight weeks; there may also be vomiting, abdominal pain, fever, flatulence, bloating, anorexia and weight loss. The incubation period is often seven to eleven days. Oocysts are excreted unsporulated in human faeces but need to sporulate to be infective. The main vehicles identified in outbreaks include untreated drinking water, soft fruit especially raspberries, lettuce, basil and vegetables. In 1996 and 1997 outbreaks of cyclosporiasis in the USA and Canada were linked to raspberries imported from Guatemala. Transmission is unknown although it is considered that the infective dose is low. Around 50 laboratory reports of *C. cayetanensis* are made in England and Wales each year and the biggest risk seems to be from imported food. However, current laboratory procedures in the UK might fail to identify the organism, and hence the true incidence could be underestimated.

Toxoplasma gondii

Toxoplasma gondii is a protozoan which may increase in significance as it is alleged that there are millions of people infected in the USA each year. Most infected people don't have symptoms but it can be serious for developing foetuses and people with compromised immune systems.

Infections mainly result from contact with the faeces of infected cats containing the oocysts and less frequently by the handling or consumption of undercooked pork, lamb, beef, or venison. Cooking to 75°C should destroy the parasite.

Taenia saginata

The adult tapeworm attaches itself to the intestine of man (adult host), eggs are passed out in faeces and if they reach grazing land may be eaten by cattle (intermediate host). The eggs hatch in the gut and the resultant cysts *(Cysticercus bovis)* migrate to the muscle via the blood. If infected, undercooked meat is eaten, the cysts become adult tapeworms. In 2001, 97 cases were reported in England and Wales.

Trichina spiralis

Adult worms live in the gut of the host, for example, rats. Larvae are produced and are carried to the muscles, via the blood, and encyst. If infected meat is eaten the life cycle begins again. In 1996, 23 cases in France were linked to eating imported horse meat, which had probably been undercooked. In late 2001 an outbreak of trichinosis in the former Yugoslavia affected over 247 cases following the consumption of contaminated smoked sausages.

Fasciola hepatica
The adult fluke lives in the bile duct of the host, for example, cattle. Eggs are passed out in faeces and hatch on wetland. If the cysts find a mud snail (intermediate host), further development occurs before they end up on vegetation waiting to be eaten by sheep or cattle. They infect persons who, for example, eat watercress grown in contaminated water. Cases are rare in the UK.

Echinococcus granulosus (hydatid cyst)
The adult host for this small tapeworm is usually the dog but also foxes. The worms attach themselves to the wall of the small intestine. Ova are passed out in faeces after around seven weeks and may be ingested by cattle, sheep, horses and pigs. The embryos are released and may end up in the liver, lungs, heart, kidneys or spleen. Man may become infected by eating food contaminated by ova. Infection is more likely from a dog licking the hands or face after licking its anus. Ten cases were reported in England and Wales in 2001.

5 Scientific principles for food safety by Dr Chris Barry

Introduction

Microbiological food safety is dependent on a range of interactions whose foundations are based on the ecology and physiology of **micro-organisms**. The achievement of **safe food** then lies with the actions and beliefs of those involved in its handling. Thus, the cultural, social and behavioural normalities experienced by **food handlers** and their managers is the defining issue in the final safety of the food. This includes the consumer behaviour as **HACCP** will not protect food that is abused by the purchaser.

Until we recognize, as Chadwick did, that public health has many facets and recognize that food safety is simply one aspect of infection control, **food poisoning** incidents will continue at unacceptable levels. All infection control is based on the understanding of basic facts about the biology, chemistry and physics of the **pathogens** and the food that supports them. Any law or action not based in these realities will be ineffective. However, the scientific connections made in **food hygiene** training or advice are often tenuous and obscure. Trainers try to remember laws and statements rather than universally applicable principles, thus consistent behaviours leading to safe food are sparsely distributed in the food industry and in the community at large. Change and effective response to the challenge of foodborne illness will only be achieved when the common understandings and legislation on which food safety decisions are made is based on the biological facts and biochemical realities of living things.

> A few years ago, I was working with a church in Togo, West Africa. My African colleague and I went to visit the holder of a subsistence farm financed by the church. In normal African fashion, as soon as we arrived we were offered a cup of water each. My colleague took his but insisted that I drank a bottle of 'pop' we had brought with us, fearful I would get a dysenteric infection or parasite from the water because it had not been boiled. Our hosts understood at once - they knew no microbiology but knew only too well the link between unboiled water and illness.
>
> **A simple principle always applied gave me protection from a potentially fatal disease.**

Biological principles for safe food

All living things on earth are constructed from and driven by complex associations of proteins, carbohydrates, fats and nucleic acids. To these are added a range of generally smaller chemical compounds and ions playing key roles in the maintenance of life. Life as we know it is founded on a range of complex chemical interactions occurring in a medium of liquid water or at the surface between liquid water and fatty membranes. Living things exist in a state of higher energy than their normal surroundings and their chemical processes and biological activities are geared to maintaining them in a "steady state" of energy. Part of this activity concentrates on maintaining their cells and structures in a highly organized state, which allows the biochemical processes to proceed in an orderly manner and so contribute to the steady state.

So, destroying micro-organisms involves preventing them from maintaining their steady state either by disrupting the chemical reactions that sustain it or by disrupting the structures that give order to the chemical reactions. Controlling the increase of micro-organisms consists of slowing down or partially disabling those processes that maintain the steady state, but not to the point where they are entirely disrupted and, so, destroyed.

Such disruption is usually achieved through heat, chemical treatment or denial of moisture. The remainder of this chapter explores what is actually going on in the organism and using some basic principles to make decisions about the value of different approaches to the control of micro-organisms.

What is certain is that prevention is better than cure and all good practice starts by attempting to keep the numbers of micro-organisms as low as possible. No method of control can substitute for hygienic preparation, storage and practices throughout the food chain that aim to prevent and reduce contamination.

Some months ago, I was asked to prepare a HACCP plan for a hospital catering department. In the kitchens, the staff were impressive in their attention to good hygienic practice – food was moved from delivery to storage and storage to cook with the minimum of handling and meticulous attention to control of temperature, be it hot or cold.

I followed the food service to the ward. On the ward, service of this excellently prepared food was the responsibility of nursing assistants and under the direction of the duty staff nurse. Their care of the acute needs and cleanliness of the patients and the ward was as meticulous as the care of the kitchen staff for the food.

Unfortunately, the delightful nursing assistants charged with food service moved directly from handling patients, bedding and other matter associated with those confined to bed, directly to food service without a look at the wash-hand basin. Their uniform was unchanged and no apron was taken to give a barrier between their potentially soiled uniform and the food.

In this one service of care, the food was subject to the high possibility of contamination with a variety of pathogenic bacteria acquired on the ward, many of them potentially resistant to standard antibiotics.

In this one action, the nursing assistants placed a ward of sick people at risk and set at nought the best efforts of an entire catering department.

Heat and cold

Heat is a form of energy, its use causes chemical bonds to bend, extend and contract, sometimes so much that they shake themselves apart or collide in unpredictable ways. Cooling, by contrast, causes the slowing down in chemical movements and therefore of chemical reactions. A rule of thumb is that a reduction of 10°C will result in a halving of the rate of a given chemical reaction. A converse rise can be expected with a rise in temperature, though this has limits in living things, for reasons that will become obvious later.

Cooling below freezing results in the removal of water in a usable form; ice is water interacting with other molecules of water. In this form, the water molecules are denied to the molecules that make up cells and that rely on water for their activity.

Cooling and heating food (or bacteria) is demanding on time since the crucial factor is the heating or cooling of water. All living things are between 50% and 80% water and water demands 4.2 kJ of energy to raise 1kg through one degree Celsius. Cooling food demands that this energy be lost and hence the difficulty of bringing food through the 5°C to 63°C **danger zone**, in which the majority of pathogens survive and especially that middle range of 20°C to 50°C to which they, as we, are adapted. We attempt to accelerate the process in either direction by creating a bigger difference between the surrounding temperature and the temperature of the food, thus creating a steeper temperature gradient and consequently steeper energy gradient. The maintenance of the desired temperature is dependent on how long the gradient is applied, so the effectiveness of heating or cooling is equally dependent on the time of cooking or cooling.

However, care must be exercised since as the surface of the food alternatively heats up (cooking) or cools down, changes will occur that will alter the effect at deeper layers in the food. It is this that leads to the production of warm spots and cool spots in food and consequent points for continued or accelerated growth of micro-organisms in the food. Each food type must be treated on its own merits such that, highly structured foods, such as meat, will exchange heat in either direction more slowly than liquid or amorphous foods (potato or custard). Their transition cannot be hurried, as this will bring about differential changes within the food, leading to an appearance of cooking or cooling that belies the mosaic of temperatures within the food. It is necessary then to **probe** foods in several places to be sure that it is cooked through or chilled through, as required.

> The dilemma of the pie is a familiar one – to brown and set pastry we need to cook hot and fast while the meat needs to be cooked long and medium why?
>
> Well – the pastry is strands of starch and protein with lots of air – heat moves in and out quickly, the strands are directly heated without much water to cushion the blow. So the strands effectively start to burn if they are heated for too long, like charcoal from wood sticks.
>
> In contrast, the meat and all the molecules that make it up are loaded with water, so the meat is effectively boiled in its own juice! Water takes a lot of heat to get the temperature to change and takes a long time to lose it and cool.
>
> Give it a thought next time you take a bite and get a blister on the roof of your mouth from the gravy!

In discussing cooking and cooling a special consideration of the myths and realities of microwave cooking seems appropriate. Microwave ovens work by emitting waves that interact with the chemical bonds in the water molecule, causing them to vibrate and the molecules to move around more rapidly. This energy of movement translates into heat and, so, the internal water of cells and other structures in the food becomes hot, cooking the proteins and other contents of the cells and the food as a whole. The penetration of microwaves into a food is no more than a few centimetres from the surface at best, and the bulk of the cooking is done by the transmission of heat down a temperature gradient from the outside of the food inwards, the newly heated layer imparting heat to the next layer in and so on. Hence, it is true that microwave cooking utilizes the internal water of the food to cook it, but it is wholly untrue to say that the microwave cooks "from the inside out", as is all too frequently suggested.

The cooking of food has its most profound effects in three ways:
- disruption of cell membranes;
- denaturation of proteins and nucleic acids (DNA and RNA); and
- creation of toxic substances in the cell.

When I was a teenager, my mother and stepfather ran a small hotel on the South Coast. At Christmas time, they would invite all the family, together with various people who would be on their own for Christmas to come for Christmas Dinner. Well on this occasion, we got to about mid-day and the guests were arriving, when my mother rushed up from the kitchen to say that 'lunch would be some time'. The reason? Although the turkey looked nicely brown, she had checked the inside of the carcass to find it quite cold and the remains of ice crystals.

She turned the oven down and cooked on. Suffice it to say that the hotel bar declined in stock and an exceptionally 'merry' Christmas was had by all an hour later than expected!

It is the length of time and the temperature that does the cooking right, since the structure of food demands time for the heat to move through the layers and raise the temperature of all.

Cell membranes

The successful development of life on earth has relied on the encapsulation of water and the concentration of certain molecules within fatty bags to form cells. The membranes, which comprise the fatty bags, have protein channels and gateways built in them, allowing the control of the movement of other substances into and out of the cells. This ability to control the contents of the cell is central to life, allowing, as it does, the cell to develop enormously high concentrations of some very specific molecules inside and maintain concentrations of other substances at low levels, by comparison to its surroundings.

Usually molecules will separate and diffuse into the surrounding water until the concentration achieves equilibrium. Damaging the outer membrane of any cell will cause the cell to leak and lose the control of its internal environment on which life depends.

Cooking causes the membrane to shake itself apart, and also causes the denaturation (disruption) of the proteins embedded in the membrane that control the movement of charged molecules and atoms (ions) in and out of the cell. Similarly, if the cell is frozen, the water in the cell may crystallize as ice, expanding and forming sharp edged crystals that may physically puncture the cell membrane, though freezing does not destroy all micro-organisms and is not effective at reducing micro-organisms to safe levels.

Denaturation of proteins and nucleic acids

Proteins are very large molecules composed of long chains of smaller molecules – amino acids. The amino acids follow the same basic plan, but there are 21 distinct amino acids and the sequence of the particular amino acids making up the chains in the protein gives the protein its essential properties. This sequence is referred to as the "primary structure"; the amino acids in the given protein chain are joined to each other by a special type of strong chemical bond, the "peptide bond". All amino acids have a hydrogen H^+ end and an oxygen O^- end as well as a number of other attached chemical functional groups, which give the particular amino acid its characteristic properties.

Interactions between the H^+ and O^- ends of the amino acids in the protein chain cause the chain to form a helix. This helix is referred to as the "secondary structure" of the protein. The interactions between the functional groups attached to the amino acids in the protein chain causes the helix to fold over on itself. The folding of the helical, amino acid chain gives rise to a third level of organization, or "tertiary structure". Finally, several protein chains may interact with each other to give a fourth level of organization, the "quaternary structure". Some proteins do not have a tertiary structure; instead, the helices interact with others without further folding on themselves. Each of these levels of structure are largely held in shape by weak bonds caused by the interaction between the H^+ and O^- as before, these bonds are referred to as "hydrogen bonds". As well as hydrogen bonds, the other parts of the functional groups, which are slightly charged, interact. These weak attractions are referred to as "Van der Waal's forces" after the chemist who first described them.

It is the three dimensional shape given by the higher orders of organization that give the protein its biologically important properties. The disruption of this shape is described as **denaturing**, since it takes away the protein's "nature", in biological terms.

Addition of heat brings about denaturation of proteins leading to loss of activity of the **enzymes** that catalyze and control the orderly chemical changes in the cell that sustain life – leading to biochemical chaos and death. It denatures fibres that support the structure of the cell. Denaturation for many proteins starts at around 63°C, but some proteins will survive intact to temperatures as high as 80°C. This may be extended further, where the **bacterium** forms a **spore** or is protected by the food that forms its environment. **Viruses** too are protein based, consisting of a protein case wrapped around a strand of nucleic acid and are susceptible to heat treatment due to the denaturation of their proteins and nucleic acid.

A further point to remember is that all organisms utilize proteins in the same ways and are therefore changed by the effect of heat. The increased tenderness and palatability of cooked meat, for example, is largely due to the denaturation and break up of the long fibrous proteins that give structure and resistance to the muscle fibres. The white of an egg becomes white due to the opening up of the ball-like (globular) albumen proteins, their fibres forming new links with each other and so giving an opaque protein network in place of the watery suspension of individual protein balls. These changes open up the tightly structured proteins and render them more easily digested by ourselves and by microbes.

Furthermore, effective cooking will drastically reduce the numbers of living microbes present, removing not only the pathogens, but also the harmless **commensal** organisms whose competition puts a significant brake on the numbers of harmful ones. This means that if cooked food is contaminated or is insufficiently cooked, leaving pathogenic survivors, and then left in warm conditions, it is potentially more dangerous than if it had never been cooked at all!

Nucleic acids are the molecules that carry the information to make another organism. **Deoxyribo nucleic acid** (DNA) is constructed from sequences of three subunits (nucleotides) from a choice of four. Each three-nucleotide sequence codes for either an amino acid, or another guidance signal. Genes are strings of three nucleotide sequences with a start and a stop sequence. The majority of genes code for sequences of amino acids and their translation leads to the formation of the proteins of the cell.

The wonder of DNA is that it conserves its structure whilst being read for the gene sequence and in the replication of the cell. It is able to do this by being constructed from two sequenced molecules, mirroring each other in an intertwined (double) helix, the helices being held together by hydrogen bonds. Enzymes enable the double helix to partially separate and allow the sequences on the opposing molecules to be copied by a similar molecule, ribo nucleic acid (RNA), which then carries the information into the cell and leads to the production of precise copies of the cell's proteins. This amazing molecule – DNA – is highly susceptible to denaturing by heat. It denatures at around 63ºC. It is this fact that probably gave rise to that mythical upper point of the "danger zone" beloved of food hygiene trainers and regulators.

Water, micro molecules and microbial death

Most essential biochemical reactions occur in aqueous solution or at an interface with an aqueous solution and denial of adequate liquid water will result in disruption of cell chemistry and death. Added to this is that cells maintain a carefully balanced relationship between the concentration of ions on the outside of their cell membranes, and the inside, even slight distortion of these concentrations through a reduction in available water will lead to widespread biochemical damage and death.

Finally, water and some other small molecules found in the cell will respond to heat or to ionizing radiation by breaking down into highly charged particles, especially OH^-. These "free radicals" play havoc with other chemical processes and interfere with the replication of DNA and RNA and hence cell replication and repair, leading to death. Free radicals are the basis of the effect of irradiation and there are health concerns expressed in relation to using ionizing radiation as a means of microbial decontamination, since our cells are also susceptible to damage by free radicals, including those ingested in our diet. The question is "how susceptible?", given the very small doses carried in treated foods.

Conclusions

- There is no substitute for keeping microbial numbers as low as possible through high standards of hygiene and good practice at every stage in the food production chain.
- Food must be cooked sufficiently to obtain a minimum temperature of 75ºC to bring about death of micro-organisms by the denaturation of their enzymes and other proteins. Even this will be ineffective where the microbe is able to form spores.
- Cooked food must be kept free from contamination and should be consumed as soon as possible after cooking. If it must be kept, then the rate of chemical activity, and therefore reproduction of pathogens and production of toxic products, may be retarded by chilling. The effectiveness of the chilling will depend on the air velocity and temperature. Slight differences in temperature of chilling have large effects. In general, therefore the lower the better.
- The effects of both cooking and cooling will depend on the structure of the food and care must be taken to ensure that a false appearance of attainment of the required internal temperature is not obtained. The time and temperature for cooking and cooling must be realistic and be sure not to distort the process by producing surface effects that alter the rate of change. A *disinfected* probe thermometer should be used to check the temperature at several points and depths in the food in order to verify the required temperature.

THE ROUGH ENERGETICS OF THAWING A CHICKEN
- or why some advice that seems sensible can be dangerous !

CHICKEN FROZEN
Core temperature of meat = −5°C
Cavity containing ice crystals

Minimum energy input: > 3.78 kJ
(In fact the effect of latent heat of crystallization and the influence of Raoult's law phenomena make this a substantial under estimate).

CHICKEN TEMPERATURE RISES TO BETWEEN −1°C AND +1°C
Ice crystals begin to melt. Energy is used to melt ice.

The addition of heat energy and its use in the process of thawing is a balance between three phenomena:

1. Equilibrium of heat loss and gain between the chicken and its environment ~ this includes variation with depth from the surface of the chicken and exchanges between different tissue types, as well as the temperature difference between carcase and environment.

2. The energy movements related to the changes in physical state of water and tissues and the specific heat capacities of the carcase tissues and the heating environment.

CHICKEN TEMPERATURE CONTINUES TO RISE
Water in the chicken flesh melts, except where it is in contact with crystalline ice lining bone surfaces or internal cavity.

Flesh temperature begins to rise. Variable with fat levels.

3. The time of exposure to the temperature difference - i.e. thawing in a fridge takes longer.

● In any case, at an **ambient temperature** of 20°C our chicken is unlikely to rise above 10°C over a 24 hour period, from a frozen start !

> Do it in the fridge and parts could still be frozen after 24 hours !

Minimum energy input: > 15.12 kJ

CHICKEN ACHIEVES +5°C THROUGHOUT
No ice crystals present.

6 Food contamination & its prevention

Contamination of food is a major hazard and may be considered as the occurrence of any objectionable matter in or on the food. Thus, carcases may be contaminated with faecal material, high-risk food may be contaminated with spoilage or food poisoning bacteria and flour may be contaminated with rodent hairs. To prevent the consumption of unacceptable or unsafe food, contamination must be kept to a minimum.

There are three types of contamination:
- *Contamination by bacteria, moulds or viruses (micro-organisms)*
  Usually occurs in food premises because of ignorance, lack of commitment, insufficient time, inadequate motivation/supervision, poor design or because of food handlers taking short cuts. In the early stages it will not be detectable. Contamination of this sort is the most serious and may result in food spoilage, food poisoning or even death.
- *Physical contamination by foreign bodies including insects*
  Physical contamination may render food unfit or unsafe but often involves pieces of paper, plastic, metal or string and is usually unpleasant or a nuisance.
- *Chemical contamination*
  Examples include pesticides on fruit and detergent residues from cleaning.

Food safety legislation requires that food shall not be exposed to risk of contamination. However, for an offence to have been committed it must usually be proved that there is a risk to the health of people consuming the food and regard must be had to the nature of the food, any further treatment, such as cooking, that the food may receive before sale and the manner in which it is packed. Cross-contamination is not defined in legislation but is usually considered to be the transfer of bacteria from contaminated foods (mainly raw) to ready-to-eat foods. This involves direct contact, drip and indirect contact, for example, via hands and food contact surfaces.

In the USA the Food Service Sanitation Manual of the Food and Drug Administration states that:

"At all times, including storage, preparation, display, service and distribution, food shall be protected from potential contamination, including dust, insects, rodents, dirty equipment, unnecessary handling, coughs and sneezes, flooding, drainage and overhead leakage or condensation."

CONTAMINATION BY MICRO-ORGANISMS

Mould spores will be present in the atmosphere, on surfaces, especially damp surfaces, and on mouldy food. Food should always be covered and mouldy food must be segregated. Furthermore, mould must not be allowed to grow on walls, ceilings and window frames. Mould growth often occurs if food is stored at the wrong temperature, at high humidity and in excess of the recommended shelf life. It may also affect cheese stored in vacuum packs that are pierced. Canned foods that are removed from cases opened with unguarded craft knives may become punctured, thus giving rise to mould growth inside the can.

HYGIENE for MANAGEMENT *Food contamination & its prevention*

Mould growth on food due to can punctures.

Viruses are usually brought into food premises by food handlers who are **carriers**, or on raw food such as shellfish that have been grown in sewage-polluted water.

Bacterial contamination is the most significant as it results in large amounts of spoilt food and unacceptable numbers of food poisoning **cases**. Food poisoning bacteria may be brought into food premises by the following **sources:**

- food handlers/visitors;
- raw foods and water;
- insects, rodents, animals and birds; and
- from the environment, including soil and dust.

Food handlers/visitors

People commonly harbour food poisoning organisms in the nose, mouth, intestine and also on the skin. The hands are never free of bacteria and the soiled hands of food handlers are likely to harbour large numbers of moulds, yeasts and bacteria, some of which may be **pathogenic**, for example, *Staphylococcus aureus*. The presence of boils and septic cuts usually guarantees the presence of staphylococci and food handlers suffering with these conditions should be excluded from working in food premises, unless they present no risk.

Carelessness, ignorance of, or disregard for, hygienic food handling may result in contamination and possibly food poisoning. All food handlers must have high standards of **personal hygiene**, wear suitable **protective clothing** and, as far as possible, avoid handling food directly; tongs and other equipment should be used. The importance of thorough and regular handwashing cannot be overstressed.

Raw food

Foods such as raw meat, vegetables, milk, shellfish and eggs are all likely to be contaminated with large numbers of bacteria, which may include food poisoning bacteria. Even when produced under hygienic conditions raw foods must be considered potentially hazardous.

Raw meat and poultry

The contamination of meat starts on the farm and poultry meat is the most common food implicated in **outbreaks** of food poisoning, especially salmonellosis. Eggs from infected breeding stock may result in the introduction of chickens that are already excreting salmonellae. The younger the chicken the more susceptible it is to becoming infected and becoming an excreter. Other possible sources include man, dogs, cats, rodents, insects and wild birds. Overcrowded and badly constructed sheds, which cannot be cleaned and disinfected, encourage the spread of infection throughout flocks.

Animals that are ill should be segregated and infections treated as soon as possible. Stress, excitement, fatigue and overcrowding increase the likelihood of animals excreting salmonellae. Symptomless excreters remain a major problem. In the UK significant improvements in hygiene and biosecurity, and the vaccination of layers has resulted in a major reduction in poultry carcases contaminated with salmonellae (often below 10%). However, the majority of poultry is now contaminated with campylobacter.

Animals and birds should be transported to slaughterhouses in clean and disinfected vehicles. Slaughtering should be carried out in slaughterhouses that comply with the appropriate regulations. Animals should be kept as clean as possible before and after slaughtering. The flesh of healthy, live animals is virtually free of micro-organisms. However, on slaughtering, the walls of the intestine lose the ability to resist bacterial penetration. Consequently, evisceration must be carried out as quickly as possible.

The hide is a possible source of food poisoning organisms, although most contamination of the carcase is likely when the stomach and intestines are removed. All dressing must take place clear of the floor. Areas and procedures that may result in contamination should be eliminated. Poor hygiene amongst slaughterhouse operatives, dirty equipment and unsatisfactory practices will undoubtedly contribute to the amount of contaminated raw meat or poultry leaving the slaughterhouse.

The skin of poultry is usually heavily contaminated with bacteria and S. Enteritidis has been isolated in the muscle. Bruises also harbour many bacteria, a large number of which may be staphylococci. Many birds in the slaughterhouse become contaminated by processes such as scalding, mechanical defeathering and evisceration. Scald tanks operating at 52°C are a much greater risk than those operated at 60°C, as fewer organisms will be destroyed at the lower temperature. Thorough spraying of poultry, with clean water, after evisceration is essential to reduce bacterial loads.

Physical inspection of animals, meat and offal will not detect food poisoning organisms.

Raw meat must be considered potentially hazardous.

However, as one of the best ways of reducing levels of food poisoning is to minimize the number of pathogens on animals and raw food, we must ensure farms and slaughterhouses adopt the highest standards of hygiene. A study undertaken by the Food Safety Research Group of the University of Wales Institute, Cardiff in 2001 confirmed that the packaging of raw poultry was likely to be contaminated with salmonella and/or campylobacter. It was found that 3% of external and 34% of whole packaging was contaminated with campylobacter and 11% of whole packaging (non external) was contaminated with salmonella. The way that packaging is handled in the home means there is a significant risk of **cross-contamination** especially as many people do not wash their hands after handling packaging.

Fish and shellfish

Apart from *Vibrio parahaemolyticus* and *Clostridium botulinum,* which rarely cause problems in this country, most pathogenic bacteria are introduced into fish by poor handling. As fish spoil rapidly they should be gutted and stored under hygienic conditions on ice, in refrigerators or freezers, as soon as possible after removal from the water.

Shellfish are often contaminated with food poisoning organisms, especially bivalves such as cockles, oysters and mussels grown in sewage-polluted water. The amount of heat used to remove the meat from the shells is usually inadequate to destroy the pathogens and recontamination is always likely. A significant number of cases of viral food poisoning and hepatitis are attributed to the consumption of shellfish. Even shellfish that have been relaid in clean water and subjected to UV treatment may still contain viruses. Between 1981 and 1990, 42 outbreaks of illness, in the UK, were associated with eating crustacean species, mainly prawns, and 151 outbreaks associated with molluscs. Of these, confirmatory microbiological or **epidemiological** evidence was obtained in seven and 107 outbreaks respectively.

Eggs

Although the shell of eggs, especially if dirty, may be contaminated with salmonellae, it is considered that infected reproductive tissues are the main **route** of internal contamination, usually of the albumen. Other than in cracked eggs, numbers of salmonellae are likely to be low and the **bacteriostatic** properties of the albumen should prevent significant growth for at least 21 days if temperatures remain below 20°C.

Surveys of eggs carried out by the Food Standards Agency revealed considerable differences in levels of salmonella contamination. In 2003, one in 1740 UK eggs were contaminated compared to one out of 650 UK eggs in 1991. In 2004, surveys found one in 48 imported Spanish eggs and one in 1020 French eggs, were contaminated with salmonella. The Health Protection Agency warned that Spanish eggs were a major source of UK food poisoning.

Most of the food poisoning outbreaks associated with hens' eggs have in fact involved dishes containing raw egg, for example, home-made mayonnaise or lightly cooked egg white, for example, meringue, and are more likely to have been caused by incorrect food handling and failures in temperature control, notwithstanding the source being contaminated eggs. The use of **pasteurized** egg is recommended for dishes that will not be cooked sufficiently to destroy salmonellae. Food handlers must always wash their hands after handling eggs and never take risks with egg products. Cross-contamination must be avoided.

The Advisory Committee on the Microbiological Safety of Foods published their second report on salmonella in eggs (2001) (Contact www.thestationeryoffice.com).

Milk

Milk is an ideal medium for the growth of micro-organisms, which may be introduced from:
- the animal itself, especially if ill or very dirty;
- the person carrying out the milking; and
- unhygienic equipment or storage tanks.

In an effort to reduce the problems from milk, government schemes have been carried out to ensure that all herds are tuberculin tested (accredited) and free from brucellosis. Milk from cows or goats suffering from **mastitis** must not be used for human consumption as it may be infected with staphylococci and streptococci.

Raw or underprocessed milk, including goat's milk, has been implicated in several large food poisoning outbreaks involving *Salmonella spp.* and *Staphylococcus aureus*, in addition to several outbreaks of campylobacter enteritis, *Listeria monocytogenes* and *E. coli* O157. Since 1997 problems of spoilage have resulted because spores of *Bacillus sporothermodurans* have survived UHT processing.

Water supplies and drainage

Although not an ideal medium for the multiplication of bacteria, water can nevertheless transport certain pathogenic organisms considerable distances, for example, salmonellae, *Giardia intestinalis* and *Cryptosporidium spp.* Furthermore, bacterial spores may survive for several months in water. Pollution of water is usually caused by human sewage or faecal contamination from animals. The Aberdeen typhoid outbreak (1964) was associated with cans of Argentinian corned beef that had been cooled in sewage-polluted water.

Cold water supplies used for washing food or equipment require suitable treatment and must be of **potable** quality. Pipework must be properly installed and maintained. Potable and non-potable supplies must never be connected or confused. Regular bacteriological and chemical checks of water used in food production are essential.

Satisfactory drains must be installed to carry away waste water. Flooding, leakage, back-siphonage and blockages must, as far as practicable, be prevented. Food, including canned or packaged food, that has been in contact with sewage-polluted water must be regarded as unfit.

Vegetables and fruit

Vegetable crops, cereals and fruit must be grown on land that is free from toxic materials and must only be sprayed with **insecticide** and other such chemicals in accordance with manufacturer's instructions. Irrigation must not be carried out with sewage-polluted water, which may contain pathogens, tapeworm eggs or other **parasites**.

After harvesting, vegetables must be stored under conditions that do not expose them to risk of contamination, for example, from poisonous chemicals or rodents. Animals and birds must be kept away from storage areas.

Sorting and packing of fruit, and raw vegetables used in salads, may introduce pathogens from food handlers, especially via the hands. All such foods must be thoroughly washed before use. Particular care must be exercised to avoid contaminating vegetables to be eaten raw. An outbreak of salmonellosis in the USA involving coleslaw and onions was traced back to contamination of the vegetables prepared on the same work surface used for cutting up raw chickens. In 1997, raspberries imported into the USA resulted in a major outbreak due to the parasite cyclospora. Outbreaks of salmonella have involved cut melon, bean sprouts and tomatoes. Outbreaks of *E. coli* O157 have involved apple juice (unpasteurized) and lettuce. NLV outbreaks have involved frozen raspberries.

Animals and birds

Both domestic and wild animals are known to carry pathogens on their bodies and in their intestines. Large numbers of *Staphylococcus aureus* are commonly found on the skin and noses of cats and dogs. Salmonellae are often present in the intestines. Consequently, pets must always be kept out of food rooms. Persons handling animals should always change their clothing and wash their hands before touching food. Pet food may be contaminated and should be kept out of food rooms. Terrapins are occasionally implicated in food poisoning cases through contact with infected water. Wild birds have been responsible for an outbreak of salmonellosis in a hospital kitchen.

Salmonella has also been isolated from songbirds in the USA, and there have been outbreaks recorded with the handling of sick birds. Campylobacter carried by pets and *E. coli* O157 carried by farm animals are major contributors to cases of "foodborne disease" from these organisms.

Insects

Insects are capable of contaminating food by defecation, feeding, walking on or dying in the food. Insects of particular concern, with regard to bacterial contamination, are flies, cockroaches and pharaoh's ants.

An ant found in a prawn.

A mouse found in a milk bottle.

Rodents

Rodents, including both rats and mice, commonly excrete organisms such as salmonellae. Contamination of food may occur from gnawing, defecating, urinating or walking over food or food-contact surfaces.

Dust

There are always large numbers of bacteria and spores in dust and moisture droplets floating about in the air. **Open food** should always be removed or covered during cleaning. The ventilation of premises should always be from high-risk areas to low-risk areas unless each area is independently ventilated. All air drawn into food premises should be filtered and

- After checking, remove deliveries immediately to appropriate storage, refrigerator or cold store.
- Keep any unfit food, chemicals and refuse away from stored food. Use only food containers for storing food.
- Keep high-risk foods and ready-to-eat foods apart from raw foods at all times, in separate areas with separate utensils and equipment. Colour coding is useful.
- Maintain scrupulous personal hygiene at all times and handle food as little as possible. Exclude potential carriers.
- Keep food covered or otherwise protected unless it is actually being processed or prepared, in which case bring food out only when needed. Do not leave food lying around.
- Keep premises, equipment and utensils clean and in good condition and repair. Report or remedy defects with the minimum of delay. Disinfect food-contact surfaces and hand-contact surfaces as often as necessary.
- Ensure that all empty containers are clean and disinfected prior to filling with food, for example, empty cans or milk bottles.
- Control cleaning materials, particularly wiping cloths. Keep cleaning materials away from food. Remove food and food containers before cleaning. Care must be taken to ensure that all cleaning residues, including water, are drained from food equipment and pipes.
- Remove waste food and refuse from food areas as soon as practicable. Store in appropriate conditions, away from food.
- Implement **integrated pest management**.
- Control visitors and maintenance workers in high-risk areas. Ensure hygiene disciplines apply to all personnel, including management.
- Inspect food areas and processes frequently, act on any defects or unhygienic practices. Train staff and monitor competence. Food handlers and other operatives must be aware of the bacteriological and physical contamination they may introduce.
- Ensure adequate thawing of foods, separate from other foods.
- Make suitable provisions for cooling food prior to refrigeration.

Food should not be exposed to risk of contamination.

PHYSICAL CONTAMINATION

Foreign bodies found in food may be brought into food premises with the raw materials or introduced during storage, preparation, service or display. It is essential that managers are aware of the types of foreign bodies commonly found in their particular sector of the food industry and that they take all reasonable precautions and exercise all **due diligence** to secure their removal or prevent their introduction. A record should be kept of all customer complaints and steps should be taken to identify the source of the contaminant.

Reported complaints involving foreign bodies (undesirable contaminants that are usually solid matter) are at an unacceptable high level and may pose a threat to consumer health and leave manufacturers and retailers vulnerable to prosecution and lost sales.

Examination of consumer complaints reveals a high incidence of problems associated with processed foods and in particular:

- milk and milk products;
- cakes and confectionery;
- fruit and vegetables (canned, bottled, dried and jams); and
- meat, poultry and fish products.

Contamination of food by **extraneous** matter will cause customer dissatisfaction and may result in bad publicity. If press and media coverage results, the impact on the business can be disastrous leading, in the worst possible case, to loss of product confidence and even company viability. It is therefore in the interests of the manufacturer to minimize the risk of foreign body contamination.

Foreign bodies, such as bone in chicken meat or stalks in vegetables, are **intrinsic** and should be minimized by care in harvesting and processing, although foreign body detection and removal systems, such as inspection belts, will also be necessary. The presence of **extrinsic** foreign bodies in food, such as glass or rodent droppings, is usually of greater concern as this indicates a breakdown in hygiene and will not be tolerated by the consumer. Contamination may occur at any step from storage through to service but is most likely during processing.

Extrinsic foreign bodies may be considered as those that:

- originate from the structure of the building, installations or equipment;
- originate from food handlers;
- originate from the activities of maintenance operatives;
- originate from packaging;
- originate from **pests** or unsatisfactory pest control;
- originate from cleaning activities; or
- are a result of post-process contamination including sabotage by disgruntled employees, blackmail attempts or from the kitchens or equipment of consumers.

Although in the minority, some foreign bodies may be considered as a serious health hazard, such as glass, stones, wire or rodent droppings, which may result in cut mouths, dental damage, choking or illness.

However, all foreign bodies are, at the very least, a nuisance and manufacturers in particular must implement appropriate systems to prevent or remove such contamination. The hazard analysis and critical control point system (**HACCP**) provides the most effective preventive approach, especially for manufacturers, and will be extremely useful if the company wishes to avail itself of the due diligence defence in the event of a prosecution.

EXAMPLES OF FOREIGN BODIES FOUND IN FOOD PRODUCTS

A can lid found in a meat fritter.

Half a mouse found in sausage meat.

The use of HACCP to control foreign body contamination in food processing

Conduct an analysis of all possible hazards associated with product.
- Prepare a flow diagram of the steps in the process, from harvesting through to filling of a closed container or to the purchase of the food by the customer.
- Identify and list the hazards and specify the control measures.
- Set critical limits, target levels and tolerances and establish a monitoring system to ensure effective control.
- Establish appropriate corrective action and satisfactory record keeping.

Having defined the starting and finishing point for HACCP, a team should be selected that, if possible, includes those with experience in production, Quality Assurance, engineering, purchasing of packaging and raw ingredients, hygiene and transport. Extraneous matter may result from poor control in any of these areas and a broad base of experience is vital if HACCP is to be effective. The first task of the team is to clearly define the products under consideration and then construct and verify a flow diagram. The team should consider which foreign body hazards exist and the risk and significance of contamination and then use this risk assessment to identify critical control points. The consequences of a particular hazard (serious health risk or nuisance value) will need to be considered when determining controls and critical control points. Account must also be taken of good manufacturing practices and any industry guides to good hygiene practice.

IDENTIFYING HAZARDS AND SPECIFYING CONTROL MEASURES

Raw ingredients

The variable nature of raw material quality is one of the main sources of problems in food processing. Raw materials have traditionally been a major source of extraneous matter and food manufacturers use a range of cleaning, sorting and grading operations to separate out the offending material. In the manufacture of frozen peas, for example, stones, metal screws, cigarette ends, stalks, sticks, caterpillars and dirt often accompany the vined peas as they arrive at the factory.

Control measures should include specifications to detail maximum permissible levels of contaminants in incoming materials. By agreeing specifications with suppliers and monitoring and evaluating the supplier performance in meeting the specifications, the company has an effective tool in minimizing the risk posed by extraneous matter. Standards of cleaning and sorting in the initial handling of raw materials by the supplier is an important control measure.

Before using raw materials, cleaning or washing and inspection may need to be carried out. For example, vegetables used in food factories must have stones and less-dense material removed. The vegetables should then be conveyed on an illuminated inspection belt of an appropriate colour. The speed of the belt and number of operatives involved in the detection and removal of foreign bodies will depend on the product and the original amount of contamination. Belts should be illuminated to at least 540 lux. Conveyor belts should be designed to eliminate crevices and to allow access to facilitate cleaning and repair. They must be kept clean and inspected frequently for signs of wear or damage. Plastic coated and woven belts should be repaired or replaced if frayed. Wooden sides and rubber strips should not be used and tracking control is recommended to prevent wear. Where practicable, conveyor systems after inspection (and metal detection if present) should be completely enclosed. Food conveyors should be above waist height, as should the tops of boxes, cans and bottles.

Liquids used in food production should be filtered and powders sieved. Filters, screens and sieves should be as fine as possible and must be cleaned and checked regularly. Worn equipment must be replaced. Wooden-framed sieves are unacceptable.

Packaging materials

Packaging may be a source of extraneous matter in the form of warehouse and transport dirt/dust, wood from the pallets, paper and polythene strips from over-wraps and a variety of insects and even rodents. Containers (cans, jars, bottles and plastic pots) may be used directly for filling with minimal cleaning, and any rogue material in the container (metal splinters, glass, dirt, insects, etc.) may end up in the final product.

Staples, cardboard, string, fibres, cloth, rubber, plastic and polythene

Food may be delivered in various containers including paper sacks, cardboard boxes and polythene bags. Particular care is necessary when emptying containers to avoid contamination of food. As far as practicable, all unpacking and packing should be carried out in areas separate from food preparation, if open food is exposed to risk of contamination.

String removed from hessian sacks and ties removed from bags should immediately be placed in suitable containers provided specifically for the purpose. As an extra precaution, coloured string may be specified to aid detection should it end up in the product. Paper sacks should be cut open, although care must be exercised to ensure pieces of paper do not finish up in the food. It is preferable for raw materials to be emptied into suitable lidded containers and not dispensed direct from paper sacks.

Particular care is needed to ensure that staples, which tend to fly considerable distances when boxes are prised open, do not contaminate food. Suppliers should be requested to use adhesive tape to fasten boxes, instead of staples. Many products are delivered in black polythene bags and small pieces of polythene often end up in the product.

Effective measures in terms of good manufacturing practice should be adopted in conjunction with HACCP to minimize the risk of contamination. An example would be the thorough examination of product and the removal of secondary packaging, before delivery to a high-risk a area.

The building, installations and equipment
Wood splinters

As far as possible the use of wood should be eliminated from food production areas. Wooden containers used for transporting raw materials should be phased out. Pallets should not be double stacked over open food.

Bolts, nuts and other pieces of metal

As far as practicable nuts should be self-locking. Bolts, nuts and screws should be non-corroding and positioned to ensure that, should they fall off equipment, they do not drop into the food. In December 2001 250,000 pounds of ham in the USA was recalled by IBP Inc. due to contamination by nails as a result of sabotage by a disgruntled employee. Also in 2001 the Food Standards Agency issued a food hazard warning after stainless steel wire was found in chocolate bars.

Metal washer baked in breadloaf.

Flaking paint or rust

Ceiling structure, pipes or equipment, should be non-flaking and rust-free. This is especially important when such fixtures are positioned directly above open products. In some older factories this problem is very difficult to overcome and consequently additional protection is necessary, for example, enclosed systems for conveying food and empty containers such as cans. New factories should be designed so that fixtures, ducts and pipes are not suspended over working areas or food if the product is exposed to risk of contamination.

Grease and oil

Wherever necessary, food-grade grease and lubricants should be used. It is important that engineers use the minimum amount necessary to lubricate moving parts and that grease is not left on the machine. Careful control will ensure the absence of complaints relating to grease in food. It is preferable for motors not to be positioned above open food. When this occurs, suitable non-corroding, cleansable drip-trays should be fixed underneath to catch oil spillages.

Glass

The use of ordinary glass, porcelain and enamelware in food factories should be avoided. Perspex or wired glass windows should be used. Diffusers, or protective non-breakable sleeves, should be fitted to all fluorescent tubes. All beakers, funnels, etc. used by the quality control staff in production areas should be unbreakable. Glass containers, other than those used for the final product, drinking cups, glass mirrors and gauge covers must all be eliminated from food production areas. Scales coated in vitreous enamel should be replaced, preferably by stainless steel. Particular care is necessary with glass containers used for product as any line breakages may result in glass contaminating other products. Brushes used for sweeping up broken glass may be colour coded and should be discarded after use.

In the event of any glass breakage it is important to ensure that:
- the supervisor is notified;
- production and preparation ceases where contamination is likely;
- food containers should be checked for broken glass and cleaned;
- the glass is cleaned up using a vacuum cleaner and/or brush (discard the brush);
- all potentially contaminated product should be discarded; and
- the area should be fully cleaned and inspected before food preparation starts again.

In factories, all filling hoppers, elevators and belts conveying open food should be protected against overhead contamination. Guards must be easy to remove to facilitate cleaning. Hoppers, silos and tanks containing food must be checked and cleaned before use and protected from contamination with close-fitting lids or covers, which must always be used correctly.

Regular checks should be made of conveyor belts, rubber seals, gaskets, brushes, etc. as deterioration can result in small pieces of rubber or bristles ending up in the food.

In 2001 The Food Standards Agency issued a food hazard warning after glass particles were found in bottles of an orange cocktail drink. In 2002 a major sweet company recalled a batch of mint creams after finding particles of glass in them.

Notices

Notices used for warnings, advice or instructions should be properly fixed and permanent. Sheets of paper sellotaped to equipment or close to open food are unacceptable. Recipe instructions should be enclosed in sealed polythene bags. Drawing pins should never be used.

Notice boards should be kept out of areas where open food is handled and should be covered in perspex or similar sheeting. Velcrose notice boards may be acceptable.

Processing or reworking of product

The manufacturing process is a major source of extraneous material that may find its way into the product during processing as a result of the following:
- cross-contamination resulting from poor layout, unsatisfactory cleaning or inadequate removal of waste material from the processing area;
- inadequate consideration of hygienic design principles when purchasing and installing plant and equipment and subsequently poor maintenance;
- premises not suitable for hygienic food production or poor maintenance of the fabric of the building;
- problems with conveying packaging materials, filling and sealing; and
- delays/breakdowns in processing.

Maintenance operatives

Engineers must be trained/briefed to take extra care when working with food equipment to ensure that they do not leave loose nuts, swarf and pieces of wire in food rooms on completion of maintenance. Written instructions may be useful. Temporary repairs with string should be avoided. It is good practice for managers to check areas where engineers, builders or contractors have been working before food handling commences.

During production, areas that are being decorated, or where repair or maintenance work is being carried out, must be suitably segregated by screens, such as heavy duty polythene, to avoid exposing product to risk of contamination. Maintenance workers should not wear grossly soiled overalls and should not stand on or climb over machinery or open food if there is the slightest risk of introducing contamination. If necessary, all food and food containers should be removed or protected with clean polythene sheeting. The use of ladders over open food or hoppers can result in dirt falling off shoes or rungs and ending up in the final product. After the work has been completed all tools, screws, swarf, grease, etc. must be removed and the area cleaned and, if necessary, disinfected before use. Whenever possible, equipment should be removed from food areas for repair.

Cleaning activities

Care must be taken during cleaning and all staff involved should be trained to ensure they do not expose product to risk of contamination by using worn equipment, especially brushes that are likely to lose their bristles, or by using inappropriate methods such as high pressure spraying in the presence of open food. Particular care must be exercised when using paper towels or cloths to ensure small pieces of paper or cloth do not end up in the product.

Food handlers

Contaminants originating from personnel include earrings, hair, fingernails, buttons, combs and pen tops. Protective clothing, including head covering, must be of a suitable type and worn correctly. The personal hygiene of food handlers must be beyond reproach, and earrings and jewellery, other than plain wedding rings and sleeper earrings, should not be worn. Pencils, pens and pieces of chalk must not be used in situations that expose food to risk of contamination, for example, near filling hoppers and mixing vessels.

Sweet papers, cigarette ends and matches are common contaminants and staff should not eat sweets, chew gum or smoke in food rooms. Continuous coaching and reinforcement, such as posters, should be used together with strict supervision and enforcement of company rules.

Post-process contamination

Contamination of the product may occur during warehousing or distribution. Furthermore, malicious tampering of products in supermarkets continues to pose a threat to manufacturers and retailers. Finally, contamination of the product may occur in the consumer's home and this should always be considered when investigating a complaint.

Control measures for tampering

Tamper-evident packaging is now widely applied in the food industry as a result of sabotage and blackmail by unscrupulous individuals. Most notably, the case of Rodney Witchelo, who was convicted in December 1990 for 17 years after being found guilty on charges of a £3.75 million blackmail plot after contaminating food products and threatening to kill customers. The food industry acknowledges that tamper-proof packaging is difficult to

achieve but tamper-evident packaging is now widely used and examples include:
- roll-on aluminium pilfer-proof enclosures;
- heat shrunk plastic sleeves;
- adhesive labels or strips on the side of the cap and jar;
- vacuum sealed containers with audible click feature; or
- a combination of the above.

Pests and pest control

Rodents, rodent hairs and droppings may be brought into food premises with the raw materials or introduced during the preparation or storage of food in **infested** premises. Food showing evidence of rodent contamination is unfit and should be rejected.

Insects, **larvae** and eggs may also be present in raw materials, although some may find their way into food rooms via openings. Several insects multiply rapidly and infestations can soon spread throughout food premises. Infested food should be discarded and appropriate control measures introduced.

A reputable pest control contractor, experienced at working with food businesses, should be employed to lay rodent bait or traps and control pest infestations should they arise. Bad pest control is likely to result in food contamination. For example, UV fly killers positioned above open food, work surfaces or containers will probably result in dead insects in the food, as will the use of **insecticides** to destroy flying insects in the presence of open food.

Cleaners and other staff must be instructed not to touch bait boxes, unless authorized to do so, and never to put bait trays on shelves above open products whilst cleaning is being undertaken.

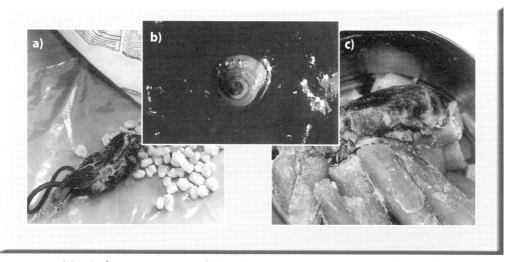

a) Rat in frozen sweetcorn. **b)** Snail in jam. **c)** Mouse in canned green beans.

All reasonable precautions and all due diligence

A food company facing a prosecution as a result of selling a contaminated product will need to demonstrate that they have installed and used an effective, documented detection and rejection system, which is checked regularly, if they are to successfully use the **due-diligence** defence. It will be up to the courts to decide what is "reasonable" having regard to good trade practice, industry hygiene guides and the risk and consequences in relation to cost.

All complaints should be fully investigated. In order to identify the contaminant, and its origin, all relevant details must be recorded. If the complainant will not release the object a photograph may be beneficial.

The presence of the foreign object indicates a breakdown in procedures and control and monitoring systems will need auditing, preferably using the HACCP approach, and modifications made to prevent a recurrence. If the defence of due diligence is to succeed, accurate and complete records will be needed including:

- number and type of specific complaints and action taken to prevent recurrence;
- documentation relating to the HACCP system including good hygiene practices, identification of hazards and controls, monitoring, corrective action, recall procedures and reviews;
- details of staff training/competency;
- details of pest control; and
- cleaning schedules.

If practicable, records should be retained for up to three years depending on product shelf life. If a reasonable precaution has not been taken the defence is unlikely to be successful.

Control systems

When identifying control systems it will be necessary to consider:

- the training of staff. Training programmes should include a training needs analysis, induction training, competence training relating to their activity, reinforcement and refresher training. Records of training should be kept on personal files;
- the design, structure, layout and maintenance of food rooms and the implementation of a clearly defined policy relating to use of glass;
- the design, construction, lighting and maintenance of equipment and especially deterioration with age;
- personal hygiene and practices, including protective clothing (staff and visitors);
- cleaning and disinfection procedures;
- instruction and training of maintenance and pest control operatives; and
- instructions relating to deboxing and packaging.

Foreign body detection and removal

No system can guarantee to remove every contaminant and the effectiveness of a particular machine or system will depend on the type of foreign body, the initial level of contamination and the maintenance of the equipment. The performance of most machines will deteriorate with age and use, and constant testing is essential. There are many contaminant detection and removal systems available including:

- metal detection systems;
- X-ray systems;

- sieves and filtration;
- vision systems including monochromatic, colour and lazer sorters;
- magnets;
- air or liquid separation systems; and
- the use of operatives, for example, as spotters, on bottle lines or illuminated inspection belts.

Metal detection

Metal detection systems should be provided on production lines after visual inspection of the food just before packaging or preferably just after packaging. Metal detection is also useful to screen raw ingredients to avoid damaging processing equipment. Systems must be installed in accordance with manufacturers' instructions and routine maintenance checks are important. Regular checks on performance must be carried out, as must checks on the efficiency of metal detectors and rejection systems. Appropriately sized rods or balls made from different metals should be used, for example, stainless steel, copper and aluminium (1mm to 3mm). Sensitivity is affected by the ratio of the aperture size to product size, the orientation of the metal contaminant, by salt, moisture, vibration, moving metal, the proximity of other detectors, the cleaning methods and by changes in temperature and humidity. The use of a metal detector will only assist a defence of due diligence if there is a controlled system for its operation. If a metal detector is found to be defective during routine testing, all product passing through since the earlier satisfactory check must be retested. All reject product must be examined to identify the reason for the rejection and to trace sources of contamination within the factory.

The latest detectors provide automated self-checking and adjustment, automated compensation for product effect, confirmation that reject product has been deposited in the reject bin and warn of low air pressure, which may prevent effective rejection.

Although products in aluminium foil can be checked for ferrous contamination, non-ferrous metals and non-magnetic stainless steels, commonly used for food equipment, will not be detected.

In 1995, on appeal from Crown Court, a company had its fine reduced to £7,000 with costs of £7,834 for selling chocolate confectionery containing a Stanley knife blade. Although many steps had been taken to avoid contamination, metal detection was not provided at the end of the processing line. The Lord Chief Justice stated that even though this was a first offence the seriousness warranted a large fine. Furthermore, the costs should stand because the company had elected for jury trial.

X-ray inspection systems

Several companies are now using X-ray inspection as a central feature of their contaminant management system. (A documented system that includes comprehensive detection and rejection equipment, that is regularly tested, with records being kept.)

Detection levels of under 1mm can be achieved depending on product speed and presentation. Pipeline systems are claimed to inspect up to 1.5 tonnes per minute and still achieve a 2mm resolution. Product presented in a uniform way facilitates the most accurate detection of contaminants.

Current applications include the detection of bone in meat products, stone and glass in fruit and vegetables, and steel, plastics and natural rubber from processing machinery. Usually

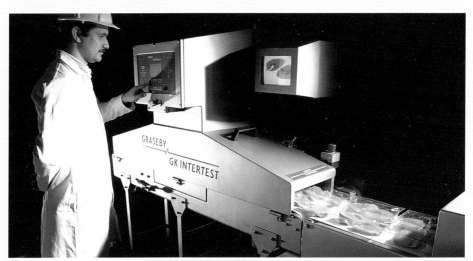

Contaminant management system using X-ray inspection.
(Courtesy of Graseby GK Intertest Ltd.)

the greater the density of the contaminant the easier it is to detect. Contaminants can be identified in plastic, card or glass packaging and even in sealed cans. Because foreign bodies are precisely located and rejected there is a reduction in wastage of non-contaminated product, which may be rejected by metal detection.

X-ray systems can simultaneously be used to inspect the size and shape of product, which enables packages with missing product, or underfills, to be rejected.

Vision systems

Vision systems are primarily used for rejecting unsatisfactory raw materials containing stones, insects, stalks and leaves etc. Cameras are used to measure light reflected in monochromatic and colour sorters. A pre-set threshold on a black and white or colour range determines which products are rejected. When combined with UV or infra-red, monochromatic and colour sorters are used to detect and reject by density.

Lazer sorters are used for product within a narrow colour range. Photo-multipliers, which can detect minute differences in reflected light, are used for rejecting unsatisfactory products. Rejecting systems rely on compressed air guns which require regular maintenance to ensure effectiveness.

REMOVING DIRTY CANS AND BOTTLES

Single-use containers such as cans should be inverted and cleaned immediately before use. Canwashers are preferable but compressed air may also be used. If practicable, some form of inspection of empty containers, for example, using highly polished metal mirrors, should be employed.

Returnable bottles, which are washed and disinfected before refilling, pose much greater problems. Unacceptable dirty or cracked bottles should be removed prior to washing. Filling lines should be fitted with sniffer devices to reject bottles contaminated with pesticides or household chemicals. Removal of pathogens from bottle surfaces depends on optimizing the use of detergents, disinfectants, temperature, pH and additives. The bottlewasher should be well-maintained and always operated in accordance with the manufacturer's instructions.

Spotters should be employed to remove dirty bottles, which pass through the bottlewasher. The number of spotters required will depend on the speed of the line and the number of unsatisfactory bottles leaving the bottlewasher. Lines operating at speeds greater than 300 per minute may require two or more spotters. Black and white backgrounds for empty bottles, and suitable mirrors positioned behind filled bottles assist in the detection of those that are unsatisfactory. Spotters should be changed at regular and frequent intervals, for example, every 30 minutes.

Dirty bottle bases are almost impossible to detect by personnel, and side and base-scanners are essential. These devices consist of several beams of light, which travel through the glass and on to a photo-electric cell. If one of the beams of light is broken, for example, by a piece of dirt, the bottle is rejected automatically. Quality control staff should carry out regular checks to establish the number of dirty bottles escaping detection and how the effectiveness of any of the above measures can be improved.

Cans awaiting inspection.

CHEMICAL CONTAMINATION

Unwanted chemicals can enter foodstuffs during:
- growth, for example, veterinary drugs, fertilizers, pesticides and environmental contaminants such as lead or dioxins;
- processing, for example, oil, cleaning chemicals or insecticides;
- transport, as a result of spillage or leakage; and
- sale, for example, cleaning chemicals, insecticides and leaching of such things as plasticizers from packaging. Chemicals may cause acute poisoning or cause long-term illnesses such as cancer (see chapter 3).

In 1999 a major recall of Belgian products, including poultry, pork, beef, eggs, milk, dairy products and other products made using these ingredients, was instigated because of high levels of the carcinogen, dioxin. It is thought that contaminated animal fat used to make animal feedstuff was the source of the problem.

Cleaning chemicals

To avoid taint, some cleaning chemicals, such as phenols and perfumed soap, must not be used in food premises, especially by those handling dairy/fatty foods. All cleaning materials must be kept in properly labelled containers and stored in a manner that obviates any risk of contamination.

7 Nutritional safety and food quality *by The British Nutrition Foundation*

Introduction to nutrition

All over the world people eat very different types of **foods** depending on many factors, such as where they live, their culture, religion and economic status. Despite these differences the majority of people grow well and stay healthy, provided they get enough of a variety of foods to eat. This is because eating a variety of foods in adequate amounts will supply the body with sufficient quantities of the nutrients it needs. Some nutrients are essential for life.

All foods provide energy and nutrients and it is achieving the correct intake of nutrients that is important for health. Nutrients can be split into two groups:

Energy-providing nutrients (sometimes called macronutrients)

Includes carbohydrate, protein and fat. As they provide energy (calories) they are needed in much larger amounts. Alcohol also provides energy but it is not essential in the diet.
- *Carbohydrates* provide the body with energy.
- *Fats* provide the body with energy in a more concentrated form than carbohydrates.
- *Proteins* provide materials (amino acids) for growth and repair. They can also provide energy.

Vitamins, minerals and trace elements (sometimes called micronutrients)

They are needed in tiny amounts and have specific functions within the body.
- *Vitamins and minerals* are essential for the functioning of different body processes as well as for growth and repair.

Hardly any foods provide only one nutrient. Most are very complex mixtures, consisting mainly of carbohydrates, fats and proteins together with water and a selection of vitamins and minerals. For example, 100g of raw potato provides about 18g of carbohydrate, 2g of protein, 80g of water and less than 50mg of vitamins and minerals. If fried, they will also provide some fat.

Different foods provide different vitamins and minerals, therefore a healthy diet should include a variety of foods. For example, dairy products such as milk and yogurts, are great sources of calcium, but they contain very little vitamin C. Citrus fruits are good sources of vitamin C, but they do not provide any iron.

Energy

Energy is needed for the body to function and be active. In the body, energy is released gradually by a series of carefully controlled steps. Energy is used to perform muscular work and to maintain body temperature and processes such as breathing, but quite a lot is lost as heat. The energy provided by carbohydrate, fat and protein can be measured and used to calculate the energy value of any food. Energy from food is measured in calories (kcal) or kilojoules (kJ).

THE ENERGY PROVIDED BY CARBOHYDRATE, PROTEIN, ALCOHOL AND FAT IN FOOD AND DRINKS

1g carbohydrate provides	3.75kcal	(16kJ)
1g protein provides	4kcal	(17kJ)
1g alcohol provides	7kcal	(29kJ)
1g of fat provides	9kcal	(37kJ)

The micronutrients (vitamin and minerals), fibre and water do not provide energy.

Factors affecting energy requirement include body size and composition, age, physical activity and life stages such as growth, and pregnancy and lactation (breastfeeding) in women. The amount of energy needed by a person who is at complete rest or who is asleep can be measured and is known as the basal **metabolic** rate (BMR). The BMR is higher in relation to body size in infants and growing children than in adults. After adolescence, the needs are proportional to the amount of lean tissue in the body; thus women tend to have a lower BMR than men because they are smaller and have less lean tissue. The BMR is also lower in older people because of the reduction of lean tissue that occurs with age.

Average values for BMR of men and women in different age groups can be estimated from body weight.

Whenever people move, they use extra energy. The heavier they are the more it takes, and strenuous activities require more energy than lighter ones.

THE AVERAGE ENERGY USED WITH DIFFERENT ACTIVITIES

ACTIVITY	ENERGY (KCAL /KJ) USED UP IN 20 MINUTES
Sleeping	16
Reading	20
Driving	30
Walking	80
Swimming	110
Jogging	140
Running fast	200

Someone with a manual occupation and who is active during their leisure time will have a higher total energy expenditure than someone who is in a sedentary occupation and is not active during their leisure time.

The dietary energy required by an individual who is neither gaining nor losing weight equals the energy expended on maintenance (in the body) and physical activity, i.e. a person who uses up about 2,500kcal per day needs to eat about 2,500kcal per day to keep their weight stable. In practice, this balance is achieved over a period of a few days. An excessive intake of only 10kcal (4.18kJ) each day would be equivalent to a weight gain of 0.5kg (1lb) per year.

Obesity

When a person has an excessive amount of body fat they are said to be obese. Currently, one in five of the UK population can be classified as obese. Such people have a greater **risk** of developing various diseases, such as diabetes, cardiovascular disease, high blood pressure and some cancers.

If individuals eat or drink foods that provide more energy than they use up in their daily activities, the fat in the diet is readily converted to body fat. Any kind of food can be "fattening" if eaten in sufficient quantities. However, some foods are more concentrated sources of energy than others. These tend to be foods containing little water and a high proportion of fat, such as butter, margarine, fatty meats, fried foods, cakes and biscuits. Highly palatable foods, rich in fat and/or sugar, can also encourage excessive amounts to be eaten. Obesity may occur even if energy intake is slightly but consistently greater than expenditure over a long period of time. Weight may be lost by decreasing energy intake or increasing physical activity or both. Eating healthily and being regularly active is the safest way to lose excess body fat.

Other constituents of food

Water comprises about two-thirds of the body's weight, and is necessary for all body processes to take place. The need for water by the body is second only to its need for air; adults can survive for many weeks without food but only a few days without water. Water comes from solid food as well as drinks, and it is lost through breathing and sweating as well as in urine. The balance of water retained in the body is normally regulated by the kidneys. Most people need 1 to 2 litres of fluid per day from food and drinks.

Fibre is found in foods such as wholegrain cereals, pulses (beans and lentils), fruit and vegetables. It comprises various plant components that generally cannot be absorbed into the body. Some fibre constituents add bulk to the faeces, which is important for health and in preventing constipation. Other fibre constituents, found in fruit, vegetables, pulses and oats can help to reduce the amount of cholesterol in the blood. However, too much fibre can decrease the absorption of some minerals.

HIGH FIBRE FOODS

Carbohydrates

There are three major groups of carbohydrates in food: sugars, starches and fibre. Sugars and starch are a major source of food energy for people throughout the world. Fibre is important for good gastrointestinal health.

Starch forms the major energy reserve of most plants and is stored in the stems, tubers, roots and seeds. The principle sources in the UK diet are wheat (such as bread and pasta), rice, cereals (such as maize and oats), potatoes and yams. The principle sources of sugars in the UK diet are milk (lactose), fruit and honey (mainly fructose), table sugar (sucrose) and glucose syrups used in cakes, biscuits and confectionery. Approximately 29% of carbohydrate intake in the UK consists of sucrose and glucose, 7% as lactose, and the remainder as starch. On average we eat about 12g of fibre per day, but it has been recommended that we should aim to eat at least 18g per day. Most the fibre consumed comes from vegetables, wholemeal bread and breakfast cereals.

Although all sugars and starches absorbed by the body provide the same amount of energy, they have different physiological effects. Consuming a lot of certain types of sugary foods and drinks at frequent intervals, especially between meals, is associated with increased risk of tooth decay (dental caries). The type of sugar present and its location within a food affect its ability to cause dental caries. Therefore sugars have been classified as follows:

- *Intrinsic sugars* are those contained within the cell walls of food, for example, in whole fruit and vegetables.
- *Extrinsic sugars* are those not contained within the cell structure of a food and include:
    - milk sugars, which occur naturally in milk and milk products (lactose); and
    - non-milk extrinsic sugars, such as table sugar and sugar added to food.

Intrinsic and milk sugars are not considered to have adverse effects on teeth. Non-milk extrinsic sugars, however, can play a significant part in tooth decay particularly when consumed frequently throughout the day. But the effect can be lessened by regular brushing of teeth (twice daily) and the use of fluoride toothpaste.

When it is desirable to reduce the fat content of the diet, a good way of replacing the lost energy is by increasing the intake of fibre-rich starchy foods. There are targets for the amounts of non-milk extrinsic, intrinsic and milk sugars, starch and fibre to be eaten on average by the UK population. Non-milk extrinsic sugars should contribute no more that 10% of dietary energy.

Many individuals, particularly from African, Asian and Indian cultures, have a limited ability to digest lactose. In later life some of these people may develop lactose intolerance, which causes digestive disturbances when a certain quantity of milk or some milk products are consumed. It is rarely found in healthy infants who depend on milk.

Fats

Fats include "visible fats" such as butter and margarine, cooking fats and oils and the fat on meat, and also the "invisible fats" which occur in foods such as cheese, biscuits and cakes, nuts and other animal and vegetable foods.

The components of fat are known as fatty acids, which differ in structure but are classed as saturated fatty acids, monounsaturated fatty acids or polyunsaturated fatty acids.

- *Saturated fatty acids* (saturates) have the most stable structure and are solid at room temperature.

- *Monounsaturated fatty acids* (monounsaturates) are less stable and are liquid at room temperature.
- *Polyunsaturated fatty acids* (polyunsaturates) are also liquid at room temperature and are the most prone to reacting with oxygen in the air and becoming rancid.

Vegetable sources of fat in the diet include vegetable seeds (for example, sunflower and rape used to make oils), peanuts, tree nuts, coconuts, palm kernels, olives, soya beans and avocados. Vegetable sources of fat consist mainly of either monounsaturates or polyunsaturates. The fat content of cereals and most fruit and vegetables is low. Animal sources include meat, fish, eggs and milk and dairy products, and consist mainly of saturates.

OILS AND FATS

Only two fatty acids (both polyunsaturated) are essential in the diet. This is because they cannot be manufactured in the body from other fatty acids. Linoleic acid, found in vegetable seed oils, such as sunflower, maize and soya bean and present in small amounts in animal fat, is one. The other is alpha linolenic acid, found in much smaller amounts in vegetable oils.

A number of other polyunsaturated fatty acids are thought to be particularly beneficial to health (although they are not essential). These are known as long chain omega-3 polyunsaturates and can be obtained from oily fish and sunflower seeds and margarines.

Throughout the world the amount of fat in the diet tends to be higher in affluent than in poor countries. Less than 10% of the energy value of the diet is derived from fat in many of the world's poorest countries. But in developed countries, including the UK, fat provides about 40% of energy.

Diets in poor countries are often low in energy and the World Health Organization has recommended an increase in fat intakes. However, in countries such as the UK and USA there is a need to decrease the fat content of the diet as a means of reducing the risk of various diseases, notably heart disease.

Heart disease

In the UK, in both men and women, heart disease is a serious health problem. The risk of heart disease is increased by various factors such as smoking, high blood pressure and raised heart disease levels of cholesterol in the blood. All, except smoking are influenced by the diet. Blood pressure can be increased by excessive intakes of alcohol and by obesity and in certain people by high intakes of sodium (a major constituent of salt). Obesity and high intakes of saturates can lead to increased blood cholesterol levels, particularly in susceptible individuals.

Cholesterol is mostly made in the body by the liver and is carried in the blood mainly by two carrier proteins – low density lipoprotein (LDL) and high density lipoprotein (HDL). LDL cholesterol is considered to be undesirable because if it increases to a high level in the blood, it can be deposited within the walls of the blood vessels forming plaques that may eventually lead to narrowing of the arteries, which supply the heart with blood. If the arteries become blocked completely with further plaque or by a large blood clot, the blood supply to the heart is interrupted leading to a heart attack and, in severe cases, death. HDL cholesterol is desirable as the HDL carries cholesterol from parts of the body where there is too much to the liver where it can be disposed of. Replacing some of the saturates with monounsaturates and polyunsaturates in the diet can lower LDL cholesterol levels in the blood. The cholesterol in food has an effect on blood cholesterol levels, but this is much smaller than the effect of saturates. The polyunsaturates in fish oils (sometimes known as omega-3s) do not have a major effect on blood cholesterol levels, but may help to prevent heart disease by decreasing the tendency of the blood to clot and keeping the heart cell membranes stable.

Current recommendations for fatty acid intake are that, on average, total fat and saturates should provide not more than 35% and 11%, respectively, of dietary energy intake. Monounsaturates should provide 13% of dietary energy. Polyunsaturates should provide 6.5% of energy and a daily intake of 0.2g (1.5g per week) omega-3 polyunsaturates is also recommended. The essential fatty acids, linoleic acid and alpha linolenic acid should provide at least 1% and 0.2%, respectively, of dietary energy.

Proteins

Proteins are essential constituents of all cells, where they regulate body processes or provide structure. Protein must be provided in the diet for growth and repair of the body, any excess is used to provide energy. Proteins consist of chains of amino acid units. Only 20 different amino acids are used by the body, but the number of ways in which they can be arranged is almost infinite. It is the specific and unique sequence of these units that gives each protein its characteristic properties.

Amino acids can be divided up into two types: indispensable and dispensable. Indispensable amino acids cannot be made in the body, at least in amounts sufficient for health, and must therefore be present in the diet. Dispensable amino acids are equally necessary as components of proteins in the body, but they can be made within the body.

EIGHT INDISPENSABLE AMINO ACIDS FOR ADULTS

- Isoleucine
- Leucine
- Lysine
- Methionine
- Phenylalanine
- Threonine
- Tryptophan
- Valine

Histidine is indispensable for infants because of their rapid growth rate.

THE REMAINING AMINO ACIDS

- Alanine
- Arginine
- Aspartic acid
- Asparagine
- Cysteine
- Glutamic acid
- Glutamine
- Glycine
- Proline
- Serine
- Tyrosine

The overall proportions of amino acids in any single vegetable food (cereals, nuts and seeds, potatoes and legumes, such as peas and beans) differ from that needed by humans. For example, wheat and rice are comparatively low in lysine, and legumes, such as lentils, are low in tryptophan and methionine. These proteins are therefore said to have a low biological value, because the quality of a protein depends on its ability to supply all the indispensable amino acids in the amounts needed. Mixtures of such foods, however, complement each other and result in greatly enhanced values so that, even among those who eat little or no animal protein, deficiency is rarely a problem provided they have enough food to eat.

A supply of the indispensable and dispensable amino acids can be achieved by eating a mixed diet at each meal. Mixtures of vegetable protein foods such as beans on toast, or of animal and vegetable foods such as fish and chips, bread and cheese, and breakfast cereal with milk complement each other. On average, about one-third of the protein in the UK diet comes from plant sources and two-thirds from animal sources.

Vitamins

Vitamins are complex substances needed in very tiny amounts for many different body processes. They have numerous functions in the body but as they cannot be made in the body they must be provided by diet.

Minerals

Practically all of the inorganic elements or minerals can be detected in the human body, but only some of them are known to be essential and must be derived from food.

The balanced diet

No single food contains all the essential nutrients the body needs to be healthy and function efficiently. The nutritional value of a person's diet depends on the overall mixture or balance of foods that are eaten over a period of time as well as on the needs of the individual eating them. That is why a balanced diet is one that is likely to include a large number or variety of foods, so adequate intakes of all the nutrients are achieved.

We need energy to live, but the balance between carbohydrate, fat and protein must be right for us to remain healthy. Too little protein can interfere with growth and other body functions, too much fat can lead to obesity and heart disease. Adequate intakes of vitamins, minerals and dietary fibre are important for health, and there is growing evidence that a number of substances found in fruit and vegetables, termed phytochemicals, are also important in promoting good health. This information is presented in the **"Balance of Good Health"**.

The table below gives the main functions and sources of each vitamin.

VITAMIN	MAIN FUNCTIONS	SOURCES
Fat soluble vitamins (these vitamins are provided by fat-containing foods)		
A Exists as the substances retinol and beta carotene	Maintains and repairs tissues needed for growth and development. Essential for immune function, normal and night vision.	Retinol: milk, cheese, eggs, liver, oily fish. Beta carotene: vegetables and fruit, especially carrots, tomatoes, mangoes, apricots and green leafy vegetables.
D	Promotes calcium and phosphate absorption from food. Essential for bones and teeth.	Sunshine, fortified margarines and breakfast cereals, oily fish and eggs.
E	Acts as an antioxidant, protects cell membranes from damage by oxygen.	Vegetable oils, margarines, wholegrain cereals, nuts and green leafy vegetables.
K	Essential for blood clotting.	Dark green leafy vegetables, fruit, vegetable oils, cereals and meat.
Water soluble vitamins		
C	Needed for the production of collagen, which is used in the structure of connective tissue and bones. Helps wound healing and iron absorption. Acts as an antioxidant.	Fruits, especially citrus fruits, fruit juices, green vegetables and potatoes.
B_1	Involved in the release of energy from carbohydrate. Important for brain and nerves.	Cereals, nuts, pulses, green vegetables, pork, fruits and fortified breakfast cereals.
B_2	Involved in energy release, especially from fat and protein.	Liver, milk, cheese, yogurt, eggs, green vegetables, yeast extract and fortified breakfast cereals.
Niacin	Involved in the release of energy.	Liver, beef, pork, lamb, fish, fortified breakfast cereals and other cereal products.
B_{12}	Necessary for the proper formation of blood cells and nerve fibres.	Offal, meat, eggs, fish, milk, fortified breakfast cereals. No plant foods contain bioavailable B_{12} naturally.
Folate	Involved in the formation of blood cells. Reduces risk of neural tube defects in early pregnancy.	Liver, orange juice, dark green vegetables, nuts, wholemeal bread and fortified breakfast cereals.
B_6	Involved in the metabolism of protein.	Widely distributed in foods; potatoes, beef, fish, chicken and cereals.

The table below gives the main functions and sources of each mineral.

MINERAL	MAIN FUNCTIONS	SOURCES
Calcium	Has a structural role in bones and teeth. Also essential for cellular structure, inter- and intra-cellular metabolic function and signal transmission.	Milk and milk products, bread, pulses, green vegetables, dried fruits, nuts, seeds, soft bones found in canned fish.
Magnesium	Involved in skeletal development, nerve and muscle function. It is also necessary for the functioning of some enzymes involved in energy utilization.	Cereals, particularly wholegrain and wholemeal products, nuts, seeds, green vegetables, milk, meat and potatoes.
Phosphorus	Has a structural role in bones and teeth. Also a constituent of all the major classes of biochemical substances in the body.	Milk, milk products, bread, meat and poultry.
Sodium	Involved in maintaining the water balance of the body and is also essential for muscle and nerve activity. However, a high sodium intake has been linked to increased blood pressure.	Processed foods: bread, cereal products, breakfast cereals, meat products, pickles, canned vegetables, sauces and soups, packet snack foods, spreading fats, cheese and salt added to food.
Potassium	Complements the action of sodium.	Vegetables, potatoes, fruit, especially bananas, juices, bread, fish, nuts and seeds.
Iron	Important for the formation of red blood cells.	Meat and meat products are a rich source of well-absorbed iron. Plant sources are cereals, bread, breakfast cereals, green leafy vegetables, beans, lentils and dried fruit. To help absorption from plant sources, a source of vitamin C should be consumed at the same meal as the iron-containing food.
Zinc	Involved in the metabolism of protein, carbohydrates and fats, and formation of cells in the immune system.	Meat, meat products, milk, milk products, bread, cereal products, especially wholemeal, eggs, beans, lentils, nuts, sweetcorn and rice.
Copper	A component of a number of enzymes.	Shellfish, liver, meat, bread, cereal products and vegetables.

MINERAL	MAIN FUNCTIONS	SOURCES
Selenium	Acts as an antioxidant by being an integral part of one of the enzymes that protects against oxidative damage.	Nuts, especially Brazil nuts, cereals, meat and fish.
Iodine	Forms part of the thyroid hormones that help control metabolic rate, cellular metabolism and integrity of connective tissue.	Fish, seaweed, milk, milk products, beer and meat products.
Fluoride	Protects against tooth decay and has a role in bone mineralization.	Fish and water.

The Balance of Good Health shows the proportion and types of foods needed to make up a balanced diet. The guide is divided into five food groups: fruit and vegetables; bread, other cereals and potatoes (starchy foods); meat, fish and alternatives; milk and dairy foods; and foods containing fat and sugar. The guide is shaped like a dinner plate, which makes it simple to understand and interpret.

It is not necessary to achieve this balance at each meal but the model can apply to the food eaten over a day or even a week. The amounts that should be consumed will vary with energy needs (based on age, sex and physical activity levels) and appetite. Dishes containing more than one food can also fit into the model. A pizza, for example, has a dough base with toppings. The dough base counts as a starchy food so having a thick base is a good idea. If the pizza is home-made the topping could be made with a reduced fat cheese or less cheese and more tomato. Including a side salad with the pizza would increase the amount of vegetables eaten and fruit could be eaten to complete the meal.

BALANCE OF GOOD HEALTH

What about salt?

A high salt intake may contribute to the development of hypertension (high blood pressure). Salt consists of the two minerals, sodium and chlorine, known together as sodium chloride. A high sodium intake is thought to be harmful to health. Approximately 70% of the sodium we eat is derived from processed foods. Looking for lower salt options in the supermarket, avoiding adding salt during cooking, and using alternative seasonings at the table can help to reduce sodium intake.

What about drinks?

Everyone should drink plenty of fluid to make sure they are properly hydrated. Being just 2% dehydrated can affect mental concentration and 5% dehydration can affect physical performance. All drinks count in providing fluid (even tea and coffee, contrary to popular belief). However, large intakes of alcohol have a dehydrating effect.

What does this mean in terms of food?

The Balance of Good Health is based on the UK government's 'Eight Guidelines for a Healthy Diet', which are:
- enjoy your food;
- eat a variety of different foods;
- eat the right amount to be a healthy weight;
- eat plenty of foods rich in starch and fibre;
- don't eat too many foods that contain a lot of fat;
- don't have sugary foods and drinks too often;
- eat plenty of fruit and vegetables; and
- if you drink alcohol, drink within sensible limits.

The guide applies to all people including those who are overweight, vegetarians and people of all ethnic origins. It does not apply to children under two years of age.

Enjoy your food

It is not necessary to eliminate foods from the diet in order to eat healthily. Eating should always be enjoyable as well as nutritious. Foods should not be classified as either "good" or "bad"; it is the balance that is important.

Eat a variety of different foods

By including a variety of different foods in the diet, a wide variety of nutrients will be obtained and it will be unlikely that a deficiency of any particular nutrient will occur.

Eat the right amount to be a healthy weight

The body mass index (BMI) is a simple calculation that has been developed as a guide to achieving a healthy weight for your height. There is no perfect weight, rather a weight range, which allows for differences in people's build and frame size. There are limitations to its use, for instance, people who have a large amount of muscle in their body, such as body builders, have a heavy weight, but they are not obese. For the majority of people it can be used as an effective guide.

CALCULATING BODY MASS INDEX

$$BMI = \frac{\text{WEIGHT IN KILOGRAMS}}{(\text{HEIGHT IN METRES})^2}$$

RANGES
- below 20 .... underweight
- 20/25 .... healthy weight
- 26/30 .... overweight
- above 30 .... obese

(There are 2.2lbs in 1kg and 14lbs in 1stone)
(There are 2.5cm in 1inch and 100cm in 1metre)

Eat plenty of foods rich in starch and fibre
Most of our dietary energy should be provided by starchy foods such as bread, rice, breakfast cereals, pasta and potatoes. Wholegrain varieties, which are higher in fibre, are the best choices.

Eat plenty of fruit and vegetables
It is important to include lots of different types of fruits and vegetables in the diet to provide vitamins, minerals, fibre and antioxidant phytochemicals. A minimum of five portions of fruit and vegetables should be included in the diet every day.

Don't eat too many foods that contain a lot of fat
Current fat intakes are higher than the recommended levels. Avoid eating too many fried foods, choose lean meat and lower fat dairy foods, and eat cakes, biscuits and chocolates only in moderation. This will help to ensure fat intakes are within healthy eating guidelines.

Don't have sugary foods too often
Frequent consumption of sugary foods can increase risk of tooth decay. Sugar-containing foods and drinks can be part of a healthy balanced diet. However, it is advisable to consume these foods as part of a meal rather than constantly throughout the day.

Putting the advice into practice
Eat plenty of foods rich in starchy carbohydrates and fibre
Eating more starchy foods such as bread, potatoes, rice and pasta will help reduce the amount of fat and increase the amount of fibre in the diet. Changing the balance of carbohydrate to fat can be achieved at each meal, for example, by having more rice or pasta with less sauce.

Don't eat too many foods that contain a lot of fat
Choosing leaner cuts of meat and lower fat versions of foods such as milk and cheese will also help reduce fat intake. All visible fat and skin should be trimmed from meat and poultry. Cooking methods should be employed that do not add fat, for example, grilling instead of frying. Some fat in the diet is necessary; olive oil and rapeseed oil are good sources of monounsaturates, nuts, seeds and vegetable oils provide polyunsaturates, as does oily fish.

Eat more fruits and vegetables

All fruits and vegetables count except for potatoes (which are classed as a starchy food). The fruits and vegetables do not need to be fresh or raw; canned, dried, frozen and juiced are just as good. It is recommended that at least five portions are eaten per day and that fruit juice counts only as one portion, however much is consumed in a day. The same rule applies to beans and pulses. Fruits and vegetables are low in fat and high in fibre so help to achieve the right balance between the nutrients in the diet. They provide, vitamins, minerals, phytochemicals, which, may help protect from diseases such as cancer and heart disease

Breakfast

Because of the length of time since the previous meal, and the consequent low blood sugar level in the morning, it is desirable to eat breakfast before going to school or starting work. It is particularly important for children and people in demanding jobs to have a good breakfast as this helps to keep them alert during the morning. Young children may not be able to satisfy their needs during the rest of the day if breakfast is missed; others may be tempted to fill up on nutrient-poor foods during the morning. A person who feels unable to eat breakfast immediately on getting up should be encouraged to take a nutritious snack as soon as possible during the morning.

The link between diet and health

Scientists have agreed for many years that some common health problems, including heart disease and strokes, can be diet-related. Too many saturates, for example, is recognized as a contributor to heart disease, and so in this sense a "balanced" diet also means one with a limited content of fat.

National targets have been set for reductions in the number of people suffering heart disease and strokes and, to help achieve these aims, related targets to reduce fat and saturates in the nation's diet have been set. Scientific evidence for the relationship between diet and disease is regularly reviewed by expert committees. A person's genes influence his or her risk of developing diet-related diseases, and other lifestyle factors are also regarded as being very important, particularly exercise and smoking. Diet can play a part in both the development and prevention of stroke, since blood pressure is affected by obesity and excessive alcohol intake, and there is evidence to suggest that in certain people risk can also be affected by sodium intake. Obese people also have an increased risk of heart disease because they are more likely to have raised levels of blood pressure and blood cholesterol. However, sound nutritional practice can help in preventing these problems, by developing eating habits conducive as far as possible to maintaining good health throughout life. For example, when someone is obese through persistent overeating and/or low levels of physical activity, only a steady reduction in energy intake or increase in energy expenditure, and not sporadic bouts of starvation or exercise, will lead to weight loss that can be maintained. A vitamin deficiency will not result from a diet that is low or lacking in that vitamin for a few days. Nevertheless, the diet is much more likely to contain enough vitamins, especially vitamin C, if fruit, fruit juice or vegetables are eaten every day than if they are eaten only at infrequent intervals. Although heart disease will not result from eating the occasional fat-rich meal, it is wise not to consistently include large amounts of fat, especially saturates, in the diet. The maintenance of a sensible and regular eating pattern is important for everyone, so that over the long term people should ensure that their overall diet is balanced, particularly those whose needs are high or whose appetites may be small, such as a young child, a pregnant woman or an elderly person. Food is more likely to be enjoyed, rather than treated simply as something to fill up on in a hurry, if people ensure that their diet is made up of a wide range of different foods so that it is varied and interesting.

The nutritional needs of different groups of people

People differ from each other in the amounts of energy and nutrients they need. Factors that affect energy and nutrient need include:
- age (stage of growth);
- sex;
- body size;
- health;
- pregnancy and lactation; and
- activity.

In addition to this individuals have different preferences and beliefs regarding food and diet. The needs of some different groups of people are discussed below.

Infants and young children

Infants are unique in that they must rely on a single food, milk, to satisfy all their nutritional needs. Breast milk is ideal for several reasons:
- all the nutrients are present in the right amount for human infants, and in a readily absorbed form. Those nutrients that are low, such as iron and copper, are those that are already stored in large amounts in the infant's liver;
- it contains several natural agents that protect against disease; and
- it is clean, cannot be prepared incorrectly, and does not trigger allergies.

A mother should therefore try to breastfeed her baby for at least two weeks and ideally for four to six months. Few mothers are unable to breastfeed. Those who cannot do so for medical reasons or who prefer not to can use formulas, usually based on cow's milk, which has been modified so that it is more like human milk. Since the immature kidneys of young infants are unable to adapt to high concentrations of protein and some minerals it is very important to make up these feeds exactly according to the instructions so that they are not too concentrated. In hot weather, extra drinks of water may be needed but sugary drinks and juices can harm the developing teeth.

Solid foods should not be introduced before four months of age. There is no advantage to the baby to do so and there may be some risks of allergies developing in genetically prone infants or of the child putting on too much weight. Drops containing vitamins A, C and D are provided free for infants and fluoride supplements may be important for dental health in areas where the drinking water is low in fluoride.

Schoolchildren

Schoolchildren are growing fast and are also very active, therefore energy requirements are high in relation to their body size compared with those of adults. The big appetites of some older children usually reflect a real nutritional need rather than greed. Because of their smaller size compared with adults, and correspondingly smaller stomachs, it is important that young children should eat meals that are not too bulky. Bread, milk, cheese, meat, fish, liver, eggs, fruit, green vegetables and potatoes are all excellent sources of a number of nutrients. Milk, whether whole, semi-skimmed or skimmed, is one of the best sources of calcium, riboflavin and protein. Children should be taught sensible eating habits from an early age: biscuits, sweets, soft drinks, chips and crisps should not displace other foods too often, either at home or at school. Sweet and sticky foods and snacks eaten frequently between meals are one cause of dental decay. Children should be encouraged to clean their teeth twice every day with a fluoride toothpaste.

Adolescents

The nutrient needs of adolescents are higher in many respects than those of any other group. Healthy adolescents have large appetites and it is important that they should satisfy them with food of high nutritional value in the form of well-balanced meals rather than by too many snacks rich in fat, sugar or salt. Obesity among schoolchildren is common and this may continue into adult life. It is more sensible to prevent obesity than try to correct it by periodically eating little or no food; excessive dieting can be dangerous. There is also evidence that adolescent obesity is partly due to a general decrease in physical activity and hence in energy expenditure rather than just to an excessive energy intake. A knowledge of nutrition and the incentive to apply this knowledge in practice is likely to benefit the health of young people for the rest of their lives.

Adults

Many adults in Britain are more likely to be at risk of overnutrition than of undernutrition. In general, healthy, well-balanced diets are high in starchy foods and fruit and vegetables, contain moderate amounts of meat (or alternatives) and milk and dairy foods and only small quantities of foods containing fat and sugar. Sodium intake can be reduced by adding little or no salt at the table and in meal preparation. It is also important to keep alcohol intake within sensible limits. The Department of Health suggest intakes are kept to 27 units of alcohol per week for men, and up to 21 units for women spread throughout the week, with one or two drink-free days. A unit of alcohol is half a pint of beer, a small glass of sherry, a glass of wine or a single measure of spirits.

Pregnancy and lactation

A woman's nutritional needs increase during pregnancy and lactation. This is not only because her diet must provide for the growth and development of her child, but also because other physiological changes occur to ensure that sufficient nutrients are available for the child (such as the laying down of new tissues in the woman's own body) and that the mother has enough energy to carry the extra weight. Much of the weight gain during the early part of pregnancy is due to the accumulation of fat, which provides an energy store to meet the additional demands of the growing foetus and the breastfed infant.

It is most important that the mother's diet contains sufficient energy, protein, iron, calcium, folate and vitamins C and D (and liquid during lactation) for building the baby's muscular tissues, bones and teeth, and for the formation of **haemoglobin**; if it does not, her own stores of nutrients may be reduced. In practice most of these extra nutrients will be obtained simply by satisfying the appetite with a good mixed diet including plenty of bread, fruit and vegetables, dairy products and meat or its alternatives. Extra folic acid/folate before and during pregnancy is needed by some women to decrease the risk of occurrence of neural tube defects in their babies. As it is not known who is at risk, in the case of first-time pregnancies or those with normal children, all women planning a pregnancy are advised to take a daily supplement of 0.4mg of folic acid before becoming pregnant as well as eating folate-rich foods. Those who have already had an affected child should take 4 or 5mg folic acid each day. However, pregnant women are advised not to take supplements containing vitamin A or eat foods (particularly liver) that may be extremely rich in vitamin A, except on the advice of their doctor, due to the possible risk of birth defects. A number of other foods should be avoided during pregnancy to reduce the risk of **food poisoning** and excessive alcohol intake should also be avoided.

Older people

There is very little difference between the nutritional requirements of most older people and the younger adult. However, as age progresses and body weight and energy expenditure decrease, people tend to eat less and hence may find it difficult to satisfy all the nutrient requirements. It is important, therefore, that older people are encouraged to maintain a good energy intake unless they are obese. They should also have foods that are concentrated sources of protein, vitamins and minerals. A healthy weight can be more readily maintained and recovery from illness or injury will be more rapid if older people are also encouraged to take regular gentle exercise. Older people would also benefit from eating plenty of fruit and vegetables (these may have to be stewed, juiced or pureed if the person experiences a problem with eating). Fruit and vegetables are often lacking in the diets of older people, but they are important in order to prevent vitamin C deficiency. Foods rich in fibre can also help to prevent constipation.

For those who do not have regular exposure to sunlight because they are completely housebound good dietary sources of vitamin D such as margarine, eggs or fatty fish (for example, sardines and mackerel) are important in the diet.

Slimmers

Planning a slimming diet is a matter of individual preference. Essentially, the energy intake needs to be cut down by 500 to 1,000kcal (2,092–4,182kJ) each day to achieve a weekly weight loss of 0.5 to 1.0kg (1 to 2lb). Other nutrients should still reach recommended levels. It is often convenient to cut out fatty and sugary foods such as cakes, sweets, preserves, biscuits and some puddings as well as alcohol, as these tend to be sources of energy rather than nutrients. Low- or reduced-fat and sugar products, now readily available, can be substituted for traditionally high-fat and sugar foods. Fat can be trimmed from meat, and foods can be boiled or grilled rather than fried. Foods high in water or fibre can induce feelings of fullness and so help to reduce the desire for more food. Weight loss is improved by being physically active.

Vegetarians

Vegetarians do not eat meat and most do not eat fish, but the majority consume some animal products – the most important of which are milk, cheese and eggs. Such diets may be rather bulky and lower in energy than a mixed diet because most vegetables have a high water content but, in general, their nutritional values are very similar to those of mixed diets.

A much smaller group, known as vegans, eat no foods of animal origin at all. Human nutrient requirements, with the exception of vitamin B_{12}, can be met by a diet composed entirely of plant foods, but to do so it must be carefully planned using a wide selection of foods. A mixture of plant proteins derived from cereals, peas, beans and nuts will provide enough protein of good quality, but special care is needed to ensure that sufficient energy, calcium, iron, riboflavin, vitamin B_{12} and vitamin D are also available. **Yeast** extract is a good source of some of the B-vitamins, including vitamin B_{12}, which are otherwise found mainly in foods of animal origin. In extreme cases, such as Zen macrobiotic diets where little but wholegrain cereals are eaten, intakes of calcium, iron, vitamin B_{12} and vitamin C are likely to be too low for health.

Ethnic groups

In general the traditional diets of ethnic communities provide adequate nourishment to those who consume them. Recent immigrants to the UK who have difficulty adapting their traditional diets and customs to local circumstances may have special dietary problems. In particular, Asian vegetarian groups may have very low intakes of vitamin D. As exposure to sunlight (especially by women and children) may also be low due to customs of dress and because they tend to remain indoors, rickets (children) and osteomalacia (softening and weakening of the bones in adults) sometimes develop. Good sources of vitamin D should therefore be included in sufficient quantity in the diet and vitamin supplements may be necessary. Iron deficiency anaemia sometimes occurs among women and children, particularly in the Asian community, since the iron content of certain traditional diets may also be low.

DIETARY RESTRICTIONS PRACTISED BY RELIGIOUS AND ETHNIC GROUPS

Hindus	No beef. Mostly vegetarian; fish rarely eaten; no alcohol.	Period of fasting common.
Muslims	No pork. Meat must be Halal*; no shellfish eaten; no alcohol.	Regular fasting, including Ramadan for one month.
Sihks	No beef. Meat must be killed by "one blow to the head"; no alcohol.	Generally less rigid eating restrictions than Hindus and Muslims.
Jews	No pork. Meat must be kosher**; meat and dairy foods must not be consumed together; only fish with scales and fins must be eaten.	
Rastafarians	No animal products except milk may be consumed. Foods must be "I-tal" or be alive, so no canned or processed foods eaten; no salt added; no coffee or alcohol. Food should be organic.	

*Halal meat is dedicated to God by a Muslim present at the killing.
**Kosher meat must be slaughtered by a Rabbinical-licensed person and then soaked and salted.

People with diabetes

Diabetes is a metabolic disorder that reduces the ability of the body to control the amount of glucose in the blood. It is important for diabetics to avoid the large rises in blood glucose by controlling blood glucose levels through the use of medication (in the form of injections of insulin or tablets) and/or diet. It is particularly important for diabetics to control their weight since obesity reduces the body's ability to metabolize glucose and can therefore worsen diabetic control. Otherwise they should eat diets similar to those recommended for other adults.

Allergic reaction

The principles set out in this chapter hold in general for all healthy individuals. There are, however, a few people who possess personal idiosyncrasies causing reactions to certain foods, for example, eggs, shellfish or strawberries. In some cases, such as **allergy** to nuts, inadvertent consumption can result in the rapid onset of anaphylactic shock, which, unless treated immediately, can be fatal. In conditions such as coeliac disease or lactose intolerance, a special diet should be followed. These allergies and illnesses are a medical rather than a nutritional problem.

FOOD ADDITIVES

Many modern day foods contain **additives**. They are needed to produce the types of food we have now come to expect. For example, margarine could not be produced without emulsifiers and stabilizers to keep the water and fat together. If there were no additives present many foods would have much reduced **shelf lives** and some foods may become hazardous. Additives are also used to provide food with characteristic flavours, colours and textures. The term additive is a broad one and covers such products as **preservatives**, antioxidants, colourings, emulsifiers and stabilizers and artificial sweeteners. Perhaps the group of food additives we come across most are the preservatives. These compounds can help reduce food wastage, give a wider choice of foods and help to keep food safe. Most responsible food companies have a policy as regards the use of food additives to only make use of them when they are technically necessary, and where possible to use natural additives, for example, vitamin E as an antioxidant, in preference to man-made additives.

All of the compounds allowed in food are tabulated on permitted lists, and often the maximum amount of the additive allowed is also specified. They have all had a good deal of research carried out on their safety and many continue to be studied to make sure that they are safe. A few people are allergic to some food additives, for example, some children are allergic to the colour tartrazine, but in the main the amount of additive used in a product causes no problem. Perhaps the potential problems lie in the effects of combined additives. If a person consumes a variety of foods, they may take in a whole range of different additives and the long-term effects of this are less well understood. This is one of the reasons that many manufacturers have decided to limit the use of additives.

Approved food additives usually have a special number. Any number that also has an E designation means that the additive has been approved by the European Community. The purity of the additive is also described in EC legislation. If an additive is present in food it must be declared on the label of the packaged product. The additive may be properly named, for example, ascorbic acid (vitamin C) or may be listed by its code number E300.

Conclusion

The food we eat has a strong influence on our health and risk of developing diseases including heart disease and some types of cancer. This chapter provides a basic understanding of how to choose a balanced diet that will help a person stay healthy. However, it is important to recognize that food is more than just energy and nutrients. We choose the foods we eat for various reasons including personal beliefs, their cultural meaning and social significance and our environmental surroundings.

The body needs different amounts and types of nutrients to keep healthy. The amount of each nutrient needed varies; protein, fat and carbohydrate are needed in larger quantities than vitamins and minerals. We all need the same nutrients but in different amounts depending on age, sex, level of activity, body size and state of health.

To make informed choices about our diets we need to be aware of the various ways in which food may have been treated and how such processing impacts on its nutritional content.

DIGESTION

Most of the foods we consume must be broken down by the body into a simpler form, which can be used or stored. The process for achieving usable products is known as digestion. On entering the mouth, solid food is cut and torn by the teeth and at the same time moistened with saliva to aid swallowing. In addition, saliva contains an **enzyme**, salivary amylase, which converts starch to sugars. Food is passed to the stomach by contraction and dilation of the muscular walls of the oesophagus; a process known as peristalsis.

The stomach is a muscular bag, which churns the food and also secretes gastric juices from glands in its mucous membrane lining. These juices contain hydrochloric acid and also pepsin, an enzyme which breaks down proteins into peptides. The acid destroys some **micro-organisms** and stops others from multiplying. The food normally remains in the stomach for three to five hours when it is gradually released via the pyloric valve into the first part of the small intestine, the duodenum. The second and third parts of the small intestine are known as the jejunum and ileum, respectively.

Food in the small intestine is further broken down by digestive juices produced by the mucous membrane of the small intestine, bile and pancreatic juice. Trypsin, an enzyme produced by the pancreas, and erepsin, produced in the small intestine, complete the breakdown of proteins and peptones into amino acids. Bile salts from the gall bladder **emulsify** fats, which are acted on by another pancreatic enzyme, lipase, to form fatty acids and glycerol. The carbohydrates are finally converted into single sugars, such as glucose, by the action of the enzymes, amylase and sucrase.

These basic food components are absorbed through the small intestine and into the capillaries. They eventually end up in the portal vein and are transported to the liver.

The remains of the food, consisting of fibre, cellulose, dead bacteria and cells, pass into the large intestine, which is divided into three: the caecum, the colon and the rectum. Water is absorbed from the contents of the large intestine along with sodium chloride. The remaining mixture is expelled as faeces.

Should food containing large numbers of food poisoning organisms or **toxins** be consumed, the normal body defences will be unable to cope. The body will react by attempting to expel these poisons by vomiting or diarrhoea. Diarrhoea results when the food has been hurried through the intestines because of the food poisoning irritants and, consequently, there is insufficient time to extract the water, which passes out with the faeces. Excessive water loss may cause dehydration.

THE HUMAN BODY

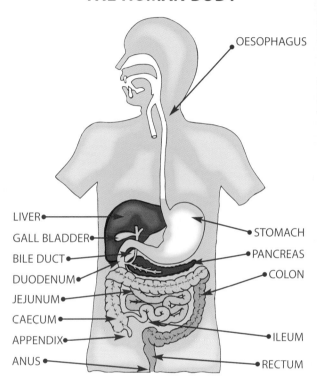

8 The storage & temperature control of food

The storage of foods is important to ensure adequate provision throughout the year and to overcome fluctuations in supply. Furthermore, bulk purchasing is normally cheaper and more convenient. However, it is preferable for fresh produce, particularly perishables, to be supplied daily. All suppliers should be approved and deliveries monitored as part of the HACCP system.

Correct storage of food is fundamental to the hygienic and profitable operation of any food business. Failure to ensure satisfactory conditions of temperature, humidity, stock rotation and the integrity of packaging can result in problems of unfit or spoiled food but will, at the very least, result in a considerable reduction in shelf life. Inadequate temperature control of high-risk food is the most common cause of food poisoning. The time and temperature of storage, thawing, preparation, processing, cooking and cooling is critical to food safety. Critical limits, target levels and tolerances should be established, controlled and monitored. Control systems will need to take into account:

- the nature of the food including a_w, pH, preservatives, and the initial level and types of micro-organisms present;
- the proposed shelf life unless intended for immediate consumption;
- the methods of packaging and processing;
- any further treatment the product is likely to receive, for example, cooking; and
- the intended consumers and the potential for abuse by the consumer.

Raw materials

Most raw food of animal origin, shellfish, vegetables and some fruits may be contaminated with pathogenic micro-organisms and should be considered as potentially hazardous. Inadequate cooking/processing or cleaning, and cross-contamination of high-risk foods are common causes of food poisoning.

Food premises must have adequate and suitable storage facilities to keep all raw materials satisfactorily. The quality and bacteriological condition of raw materials has a significant influence on the finished product. Furthermore, unsatisfactory deliveries may introduce problems such as insects, rodents or mould into the storage area.

To demonstrate due diligence and to ensure that deliveries meet the agreed specification an effective, documented checking system must be implemented. The inspection required will depend on the type of product and the condition of other deliveries from the same source. Checks may include quantity, temperature, date code and quality. Damaged or discoloured packaging must be regarded with suspicion and the consignment thoroughly examined by trained staff, preferably before unloading from the delivery vehicle. Leaking cartons, rusty cans and old stock should all be rejected. A documented system detailing reason for rejection should be maintained.

Perishable food must be checked quickly and removed to cold stores, freezers or refrigerators without delay. Unloading of vehicles should take place, as far as practicable, in covered bays screened from adverse weather. Raw food and high-risk, or ready-to-eat food should be completely segregated to avoid **risk** of contamination. Non-food items and strong-smelling foods, which may cause taint problems, should normally be delivered separately. Once unloaded, batches of food should be clearly marked to ensure they are used in strict rotation and to identify the date of delivery in the event of subsequent complaints, for example, mould growth.

Raw meat and poultry

Raw meat joints should be stored between −1°C and +1°C, with a relative humidity of approximately 90%. Humidities above 95% encourage unacceptable microbial growth and humidities below 85% result in excessive evaporation and subsequent weight loss. Air speeds of around 0.3 metres per second minimize drying of meat but are only intended to remove heat coming through the insulated wall, or introduced when opening the door. Higher air velocity is required for cooling meat. Only cooled meat should be placed in the refrigerator/cold store. Meat should not touch the wall surface.

Given the correct temperature and humidity the shelf life of meat depends on the initial bacterial loading. For example, meat with a **total viable count** of less than 100 organisms per gram will have a shelf life of at least double that of meat with a count of approximately 1,000,000 organisms per gram. This fact clearly illustrates the need for the hygienic handling of meat in the slaughterhouse. The avoidance of contamination and multiplication during distribution is also important to maintain low counts of micro-organisms. Joints of beef packed in a temperature-controlled atmosphere and stored at −1°C to +1°C should keep for up to a week. Beef usually has a longer shelf life than pork, lamb, poultry and fish. Processed raw meats, offals and such products as sausages have the shortest life.

The unsatisfactory storage of meat.

Eggs

As soon as possible after lay, eggs should be stored in a **clean**, dry place, at a constant temperature below 20°C (ideally between 10°C and 15°C). Temperature fluctuations result in condensation and must be avoided. Eggs should not be stored near heat sources such as fridge motors or in direct sunlight. After delivery to consumers and caterers, eggs should be stored below 8°C under refrigeration and should be used within three weeks of lay. Stock rotation is essential and cracked or dirty eggs should not be sold. Large numbers of salmonellae have been isolated from contaminated eggs stored for 21 days above 20°C. Hands must always be washed after handling raw eggs.

Meat pies, pasties and sausage rolls

These products must be obtained from a reliable source and should preferably be stored under refrigeration. Temperatures of around 7°C, with good air movement, are recommended so that the pastry remains crisp. As these foods are cooked to temperatures as high as 90°C very few **bacteria** survive and, consequently, they remain bacteriologically safe at this higher than normal temperature of refrigeration. However, stock rotation is important and such products should be sold on the day of production or the following day. Pies or pasties to which something has been added after baking, such as jelly, should be stored at or below 5ºC. Fluctuations in temperature result in condensation and mould growth.

As these products may be consumed without further cooking they must never be stored with raw meat or vegetables. Strong smelling foods such as cheese may introduce taint or mould problems. Price tickets must not be stuck into pies or any other food. If pies are to be sold hot from retail outlets they must be cooked thoroughly and if stored, maintained above 63ºC. Alternatively, pies may be microwaved as required. They must never be rewarmed.

Fruit and vegetables

The quality and shelf life of fresh fruit and vegetables depend on the way in which they are harvested, handled, transported and stored. Deterioration may be physical, physiological, chemical or pathological. Storage conditions should reduce the rates of respiration and transpiration (loss of water by evaporation) to ensure the maximum shelf life. Although each fruit and vegetable has its own optimal storage conditions, a general guide is to keep them refrigerated (0ºC to 2ºC) at a relative humidity of 90%. However, tropical fruits such as pineapples and bananas should be stored around 10ºC to 13ºC to avoid "chill injury". Rapid turnover and practical difficulties such as lack of separate refrigerators usually results in catering premises storing fruit and vegetables in cool rooms without undue problems.

Care must be taken to avoid warm moist conditions and condensation, which will encourage bacterial spoilage and mould growth. Low humidities and excessive ventilation result in dehydration and must also be avoided. Fruit should be examined regularly and mouldy items removed to avoid rapid mould spread. Transit wrappings may need to be removed to avoid condensation.

Ice cream

During production and at ambient temperatures, ice cream is an ideal food for bacterial growth and must be handled with care. Ice cream should be kept in a clean freezer, preferably by itself, and open ice cream must never be stored with raw products. It should always be kept frozen and if defrosted must be discarded. One tub of a particular variety should be used completely before opening a fresh one and nothing should be placed on top of an open tub. Lids should normally be kept on. The texture of ice cream is more temperature sensitive than other frozen foods and it may be held at –12ºC for one week before use.

Clean scoops should be used for serving ice cream and these should be rinsed in clean, running water after each use and wiped dry using disposable paper. After service the scoop must be cleaned, disinfected, allowed to dry and stored in a protected place.

Milk and cream

Milk used to be a major source of disease and even now raw or inadequately heat-treated milk continues to cause problems. Milk and cream should be stored under refrigeration (below 5ºC) and should be placed in the refrigerator or cold store as soon as received. Imitation cream should also be stored under refrigeration. Raw milk will be labelled untreated and should not be used by any food business. Crates of milk should not be stored below raw meat.

Between 1992 and 1996 there were 20 general outbreaks (600 cases) of food poisoning associated with dairy products. Two outbreaks (32 cases) implicated home-made ice cream and two outbreaks (eight cases) implicated cheese.

Flour and cereals

Flour and cereals should preferably be stored in mobile stainless steel containers. Lids must be tight-fitting. Large stocks of flour kept in original sacks must be stored clear of the ground and free from damp. Condensation may result in mould growth on wet flour and must be prevented. Regular cleaning of containers used for flour storage must not be overlooked. Frequent **inspections**, at least weekly, should be carried out for rodents or insects. Permanent bait points, laid by trained operatives, are recommended.

Canned foods

The risk from canned foods is very small compared with the number produced and the excellent safety record of cans filled in this country will be maintained if:

- blown cans are not used;
- badly-dented, seam-damaged, holed or rusty cans are not used; and
- stock rotation of cans is carried out. Stocks should be marked with the date of delivery to ensure older cans are used first.

Canned goods should be examined regularly and blown or leaking cans must be rejected. Cans from the same batch should not be used until clearance has been received from the manufacturer or the local environmental health department.

The shelf life of canned foods

High-**acid** cans of fruit, such as prunes, rhubarb, tomatoes and strawberries, may blow if kept longer than recommended by the manufacturer. The **acid** nature of these foods enables them to attack the tin and iron of the can, especially at the seams, and hydrogen gas is released causing the can to blow. An unacceptable amount of tin and iron may be present in the food and even if the can is not blown, high levels of these metals, with or without a metallic taint, will render the food unfit for human consumption. Fortunately, new canning techniques and lacquers for these products has reduced this problem and increased their shelf life. However, some imported canned fruit could still be affected.

Some large cans of meat, especially ham, may only have been **pasteurized** and therefore need to be stored under refrigeration. If this is the case, the can should be clearly labelled.

Once the can is opened the contents should be placed in a suitable plastic or stainless steel container, if the food is not for immediate use. If the food is discoloured or has an unusual smell or texture, it must be rejected. The inside of the can should be inspected and if there is any rust or discolouration, the food must not be used.

Wrapping and packaging

Packaging of food is important not only to assist marketing but also to:

- prevent chemical, physical and bacteriological contamination;
- retain quality and nutritional values;
- prolong shelf life; and
- provide a water vapour and gas barrier.

Packaging materials must be economic, non-toxic and easy to remove. Chemicals from packaging materials must not leach into the food or expose the food to risk of contamination when opened or removed. Materials must be capable of withstanding the storage conditions to which they could be exposed, for example, freezing. The integrity of seams is particularly important and seams must withstand the pressures, handling and distribution systems that they are likely to experience.

The use of laminates and plastics is increasing whilst the more traditional materials such as wood are declining in popularity. Laminates may consist of several materials including duplex board, polyethylene and aluminium. Plastics include polyethylenes, polypropylene, polystyrene and polyvinyl chloride (PVC). Greaseproof paper or aluminium foil can also be used.

Packaging and wrapping materials must be stored in clean areas where they are not exposed to risk of contamination. Wrapping of food must be carried out under hygienic conditions by staff who observe the same high standards as employees involved in food preparation. Suitable handwashing facilities should be available close to wrapping and packaging areas.

Cling film

Cling film is useful for preventing the dehydration of food and protecting it against contamination (special breathing films are available for raw meats). Under certain conditions, however, it can speed up spoilage and mould growth by trapping moisture against the surface. It is important therefore that:

- raw meat or wet food is unwrapped when removed from the refrigerator; and
- food wrapped in cling film is not left in bright light or sunlight.

In 1995 the Department of the Environment, Food and Rural Affairs advised that because of the risk of chemical migration, cling films should not be used where they could melt into food, such as in conventional or microwave ovens, or for wrapping foods with a high fat content, such as cheese, meats with a layer of fat, fried meats, pastry products and cakes with butter icing or chocolate coatings, unless the manufacturer's advice indicated their suitability for this purpose.

Vacuum packing

Vacuum packing is widely used for an increasing number of foods, particularly fresh meat, bacon, poultry, cooked meat and cheese. The use of a vacuum pack for fresh meat was pioneered by the Cryovac Division of W. R. Grace Limited in the USA in the 1950s. (Cryovac is a registered trade mark of W. R. Grace & Co.)

Materials such as nylon/polyethylene laminate, which are used in vacuum packing, prevent the entry of oxygen and moisture, thus reducing spoilage of food by bacteria, moulds and rancidity. Furthermore, the food is protected against contamination, excessive drip and weight loss due to dehydration. The storage life of food is prolonged considerably, often being doubled or more. Immediately after opening the vacuum pack the contents should be removed. The slightly darker colour of meat and the acid odour will disappear shortly after being removed.

Vacuum packing will not improve the quality of a poor product and strict observance of hygiene and temperature control throughout the operation is essential. After the air is removed closure may be affected by heat sealing or the application of a metal clip.

Vacuum packs may allow the growth of anaerobic food poisoning organisms and if not stored under refrigeration there is a possibility of consuming unspoiled but unsafe products. Cooked meat that is contaminated with food poisoning organisms just prior to vacuum packing is particularly hazardous. The absence of competition from other organisms will, if temperatures permit, enable the rapid growth of these pathogens.

Vacuum packing of fish, especially smoked salmon and trout, is a worrying development as conditions are ideal for the multiplication of *Cl. botulinum* which is commonly isolated from the intestines of these fish. Storage of these products should always be below 3ºC.

Care must be taken to avoid puncturing packs, for example, with sharp bones or rough handling. Defective seams commonly result in the loss of pack integrity. However, air-tight vacuum packaging may blow because of the presence of fermentative lactobacilli. Semi-permeable films favour the growth of **aerobic** bacteria, which produce slime and off-flavours. It is advisable to purchase branded vacuum packs from reputable suppliers to avoid receiving low-grade meat of dubious origin. Unmarked packs without "**use-by**" dates should always be regarded with suspicion.

Modified atmosphere packing (MAP)

Modified atmosphere packing is being used increasingly for products such as raw meat and fish, to reduce spoilage rates and maintain a more attractive colour. Gases used include carbon dioxide, nitrogen and oxygen, which must each be at the correct concentration depending on the product. A mixture of gases is injected at sealing. Storage under refrigeration remains essential to retain product quality, safety and shelf life.

Damaged stock

All damaged stock should be segregated and thoroughly examined before use. If there is any doubt regarding the fitness of food it should be discarded, or the local environmental health department may be asked for advice. The practice of some supermarkets of selling damaged packets of food or dented cans is not recommended. In the event of a justified complaint of unfit food the company is unlikely to be able to avail itself of the "due-diligence" defence.

The greenhouse effect

Food, especially high-risk food, should not be stored in windows or in glass display cabinets that are exposed to direct sunlight. The heat within the cabinet will build up in the same way that it does in a greenhouse, with the result that ideal temperatures for bacterial multiplication are provided. The greenhouse effect may even occur in chill cabinets without proper chilled air circulation, and fluorescent tubes may exaggerate the problem.

Storage systems and equipment

Safe food storage and stock rotation are facilitated by the use of systems and associated equipment.

- *Mobile racks* – wheel-mounted stainless steel shelving on adjustable supports allow free standing storage, shelf spacing to be set for specific stores and easy inspection.
- *Slotted trays* – available in high-density coloured plastics, frequently used by bakeries for bread deliveries. They enable stacks to be set up to store a variety of different items. They can be dismantled and stacked in space-saving nests when not in use.
- *Bins* – bulk dry-foods may be stored in mobile plastic or stainless steel bins and wheeled from the store to the point of use.
- *Mobile silos* – as above but larger, with a gravity fed hopper discharge allowing substances such as flour to be stored in safety and transported to the site of use.

♦ *Pallets* – large quantities of bulk goods can be stored on pallets and moved around with fork-lift trucks or hand-operated pallet trucks.

Stock rotation

Satisfactory rotation of stock, to ensure that older food is used first, is essential to avoid spoilage. Stock rotation applies to all types of food. Daily checks should be made on short-life perishable food stored in refrigerators whereas weekly examination of other foods may suffice. Stock that is undisturbed for long periods will encourage pest infestations. Good stock rotation has the added advantage of assisting in the maintenance of the correct levels of stock. Remember the rule: **"First in, first out."**

Codes

Stock rotation has been much easier since the advent of open-date coding but some products do not require a "use-by" date and, in these cases, retailers should adopt their own code to identify the date of delivery. A colour code system is one of the easiest to use; a blue line for Monday, a red line for Tuesday, etc. The practice of selling old stock cheaply is not recommended and it is an offence to sell foods bearing an expired "use-by" date or to change this date (see the food labelling regulations, chapter 17).

Dry-food stores

Rooms used for the storage of fruit and vegetables, dried, canned and bottled foods should be dry (relative humidity of 60 to 65%), cool, well-lit, well-ventilated and large enough to facilitate the tidy packing and rotation of stock. They must be kept clean and spillages should be cleared away promptly. Internal walls should be free of cavities. External walls should be north-facing. Wall surfaces should be smooth and impervious. Joints and cracks harbour dust, prevent efficient cleaning and encourage pest infestations. Floors should be impervious, jointless and capable of being kept clean. Sufficient strength is required to support loads and any vehicles used in the store. The method of cleaning must be considered when selecting finishes. Where practicable, wall/floor joints should be coved. Ceilings should be easily cleaned and not encourage condensation, if this is likely to be a problem. Ledges collect dust and should be avoided. Wooden pallets should be replaced, for example, by reinforced plastic pallets, as they allow a build-up of dirt and dust, which provides ideal conditions for mites and insects.

Stores should be rodent- and bird-proof; doors should be provided with metal kick-plates, and airbricks, if present, fitted with wire gauze. Holes around pipes passing through walls should be effectively sealed. As far as possible food should be kept in rodent-proof containers. All goods should be stored clear of the wall and floor to allow cleaning and pest control. Stores should be close to the areas in which it is intended to use the food and also be readily accessible to delivery men, who should not have to pass through food preparation areas.

Natural ventilation will assist in keeping the temperature of the room low, but in the summer, mechanical ventilation and circulation will probably be necessary. The absence of windows will prevent solar heat gain. Temperatures of around 10°C to 15°C should be maintained for the best storage conditions and stores should not be situated next to a source of heat, for example, adjacent to the boiler room.

Shelves should have a non-absorbent, cleansable finish. Tubular mobile racking made of non-corroding metal is recommended. Cupboards should be avoided. Non-food items, including cleaning equipment and chemicals, and strong smelling foods, should not be stored in dry-food stores. Packs of food should be handled carefully to avoid damage. Part-used packs should be adequately resealed to prevent contamination. Alternatively food may be transferred to clean containers with pest-proof lids.

The storage of perishable foods

High-risk perishable foods may be contaminated by pathogenic organisms, which can multiply to dangerous levels without altering the appearance or taste of the food concerned. For this reason all premises should have adequate cooling and cold storage facilities to keep perishable food that is not intended for immediate consumption.

REFRIGERATORS

Refrigeration must be considered as a means of delaying and not preventing food spoilage by bacteria and moulds. Furthermore, although the common food poisoning organisms are incapable of multiplication or **toxin** production below 5°C, there is concern regarding certain **psychrotrophic** pathogens capable of growth below 5°C, including *Yersinia enterocolitica* (1°C), *Listeria monocytogenes* (–1.5°C), *Aeromonas hydrophila* (1°C) and *Clostridium botulinum* type E (3°C). Fortunately, growth and toxin production at temperatures just above 0°C takes several days or weeks and should not usually cause problems during short-term refrigerated storage. The consequences for prolonged refrigerated storage of food, even in vacuum packs, are of more significance. The nearer the refrigerator operates to 1°C the safer it will be.

Siting

Refrigerators should be readily accessible and should not be positioned near any heat source. Ideally they should be in well-ventilated areas away from the direct rays of the sun. Multi-deck cabinets may be affected by draughts from doorways, high room temperatures and ventilation and heating grills fitted in ceilings, which may result in a significant increase in temperatures. High humidities will eventually result in problems of condensation and sweating. The siting of the unit should allow the cleaning of the surrounding area as well as the cooling coils.

Construction and design

Refrigerators should be constructed to facilitate easy cleaning. Large motors are best positioned outside as they generate heat and collect dust. (This also ensures a good circulation of air.) The inner skin and shelves should be impervious, non-corrosive and easy to clean. The number, position and size of shelving are critical in ensuring the effective distribution of cold air to all parts of multi-deck units. Lighting inside cabinets can raise temperatures and, if used, should be of the cold cathode type. Floors should preferably be jointless, laid to a suitable fall and coved for ease of cleaning. Door seals should be checked regularly as they can become perished and difficult to clean. Large chill stores should be fitted with self-closing doors and protected by air-locks, air-curtains or plastic screens. Storage capacity must be adequate to cope with peak demands. Damage to the external or internal cladding of units, which enables moisture to penetrate the insulation, will reduce the effectiveness of the refrigerator by creating a "warm bridge". Domestic refrigerators are not designed or suitable for most commercial operations.

Defrosting and cleaning

Each unit must have an adequate number of defrost cycles to prevent ice build-up on the **evaporator** coil. This would reduce air flow across, and the surface area of, the coil and result in decreased efficiency and increased temperatures and costs. Assisted defrost systems prevent ice build-up and product warming.

Both exterior, especially door handles, and interiors should be cleaned regularly and internal surfaces of fans disinfected. Base plates and baffle plates should be removed to check for debris such as price tickets and packaging, which can obstruct air flow and block drains in multi-deck units. Drip trays, if present, should be emptied and cleaned. Dust on evaporator coils and air ducts must be removed to maintain performance.

Cooked and raw food stored on same shelf. Unsatisfactory shelving.

After cleaning and disinfection, units should be completely dried and allowed to attain the correct operating temperature, usually 1°C to 4°C, before being reloaded. Stock should be kept in back-up chill stores during cleaning operations.

Food that experiences a significant increase in temperature, above 10°C, due to breakdown or power cut should be destroyed. Regular cleaning and maintenance will improve efficiency and reduce breakdowns.

Operating temperature

To prevent the growth of most pathogens and reduce spoilage, the majority of high-risk foods, including cans of pasteurized ham, should be stored below 5°C. The optimum temperature for multi-use refrigerators is therefore between 1°C and 4°C. Higher temperatures allow greater bacterial activity and also prolong the time taken to reduce the temperature of food placed in the refrigerator, i.e. prolonging the time the food remains in the **danger zone**.

Door seals on refrigerators must be kept clean.

From a food safety perspective the operating temperature of the refrigerator over a 24 hour cycle is much more important than a spot temperature taken once or twice a day. Constant use of the refrigerator, opening the door and stocking with warm food is likely to result in temperatures in excess of 8°C for a short time. However, provided the refrigerator operates throughout the night/early morning below 5°C, pathogenic growth is very unlikely to be a problem. In fact as spoilage organisms grow faster at lower temperatures, the food is likely to spoil before it becomes unsafe.

The storage of some foods just below freezing results in the formation of large ice crystals and consequential loss of food quality and texture. Fan assisted cooling is recommended as the circulated cold air maintains uniform temperatures, and quickly establishes correct temperatures after the door has been opened.

Cook-chill foods with a maximum shelf life of five days must be kept between 0°C and 3°C to ensure safety and reduce the risk from *Listeria monocytogenes*. It is recommended that raw meat, poultry, fish and raw meat products such as sausages should be stored at −1°C to 1°C to maximize shelf life. Furthermore, regulations originating from EC legislation specify that carcases and cut meat in slaughterhouses and cutting premises should not exceed 7°C, offal 3°C and poultry meat 4°C.

Air velocity

Many refrigerators have a low air velocity, around 0.3 metres per second, to avoid drying the food. However, if reasonable cooling rates are to be achieved a minimum air velocity of around 1.7 metres per second is recommended. This level of air velocity is only likely to be found in blast chillers.

Hot food

Hot food should be cooled rapidly and not placed directly into a refrigerator as this will raise the temperature of food already stored, as well as increasing the ice build-up on the cooling unit. Condensation may also occur on some foods, resulting in an unacceptable drip on to food stored below. However, depending on the amount of food and the capacity of the refrigerator, it may be preferable to place a small amount of food, which has cooled to around 60°C, into a large capacity unit instead of leaving it in a warm place overnight. During the day a fan can be used to cool food prior to storage in the refrigerator.

Contamination

It is essential that precautions are taken to avoid contamination. Raw food must always be kept separate from ready-to-eat food. Ideally two refrigerators should be used, one for raw meat and the storage of food prior to cooking and the other refrigerator to store ready-to-eat food. Each refrigerator should be clearly labelled to indicate its intended use. If only one refrigerator is available, high-risk food must be stored on shelves above raw food. Shelves

Unsatisfactory storage in a walk-in refrigerator. Raw meat and cooked meat on the same shelf. Shelving and support corroded and filthy.

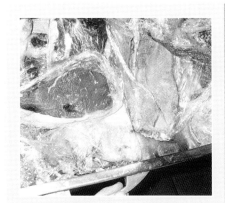

Unfit steak and cooked steak stored with raw meat.

Packing and rotation of food

Refrigerators should not be overloaded and they need packing in a manner that allows good air circulation. Food should not be placed directly in front of the cooling unit as this reduces efficiency. Non-perishables should not be stored in a refrigerator as this takes up valuable space. Good stock rotation is essential to avoid the spoilage of foods and daily checks should be made for out-of-date stock.

Staff responsibilities

Staff should be given clear instructions on how to use the refrigerator and, in addition, should ensure that the door is opened as little, and for as short a time, as possible. Acid foods should not be stored in opened cans within the refrigerator because of the acid attack on the internal surface of the can. The temperature of the refrigerator should be checked regularly throughout the day and recorded at least twice a day. Spillages should be cleared up immediately.

The shelf life of perishable food under refrigeration

The storage life varies considerably depending on the operating temperature of the refrigerator and the specific perishable food. For example, at 1°C most vegetables will keep for at least two weeks, butter for two months and fresh beef up to a week. A reduction in the initial numbers of bacteria on food can, provided it is stored correctly, increase its shelf life significantly.

Chilled display cabinets

Display cabinets should comply with BS 3053. Cabinets should be loaded correctly so that cold air inlets are not obstructed and load-lines are not exceeded. Only chilled products should be loaded.

Radiant heat can have a significant effect on food stored in chilled cabinets, particularly on the upper shelves. Air temperatures may vary from 1.5°C at the back to 6.5°C at the front, and the temperature of such product as cartons of yogurt stored at the front of such units may vary from between 7°C and 11°C. Cabinets should be sited away from heating units, high-intensity lights and out of direct sunlight.

Monitoring temperatures

Cabinet thermometers rarely reflect product temperature accurately. Refrigerator temperatures should be checked using an accurate digital tip sensitive thermistor or an electronic thermocouple. Bare wire thermocouples provide a quick response and can also be used for testing between packs. **Probe** diameters should be 15mm or less and have a grounded junction at the probe point. Indicating or recording thermometers are essential and premises with a large number of units will need automatic monitoring equipment and alarms to warn of unacceptable temperatures. Bimetallic coil thermometers are inaccurate and often difficult to use.

Thermometers will need calibrating, as often as recommended by the manufacturer and always when involved with court proceedings and/or due-diligence defences. Calibration can be undertaken in slush ice and boiling water, or by electronic calibration devices. From a food safety perspective accuracy for measuring cooking temperatures is more important than refrigeration temperatures. For example a 1.5°C error can double the time necessary to destroy salmonella and make food safe. The salmonella kill takes 52 seconds at 65.5°C and 104 seconds at 64°C. (Dr P. Snyder Jr.)

The Food Law Code of Practice suggests a three stage approach to determining compliance with statutory food temperatures:

- an examination of daily records of air temperature monitoring at the premises;
- if records are unsatisfactory then non-destructive testing, between packs, should be carried out; and
- if concern remains then probing of the food should be undertaken.

The correlation between air and product temperature can be obtained by pre-chilling a probe thermometer (leave it in a refrigerator for five minutes), taking the product temperature, after at least 24 hours storage, then checking the unit's air temperature.

Each section in a long, refrigerated cabinet should be checked for defective fans and it should be noted that multi-deck cabinets vary by up to 2°C with the cold air being at the top. Variations of 2°C can also occur due to the **compressor** cutting in and out. Temperatures may be raised slightly during defrost and this can be checked as any highly polished surfaces are usually misted up. **"Air on" temperatures** are the most reliable indicator of performance as this is the warmer air at the return air vent. (**"Air off"** refers to the cold air leaving the evaporator coil where it enters the cabinet.)

Temperature monitoring should be recorded at least twice a day, preferably first thing in the morning, and at the end of the day. Accurate records of in-house checks should be retained and available for enforcement officers. They will be very useful to establish a **due-diligence** defence in the event of complaint.

Infrared thermometers

Infrared thermometers, which work by measuring the amount of radiant energy, are assuming increasing importance for monitoring food temperatures. Products can be scanned very rapidly and any identified problems can be checked immediately with a digital thermometer. Additional advantages include the fact that it is non-destructive testing and there is no risk of cross-contamination. It is particularly useful for scanning retail cabinets, frozen food, deliveries and despatch as large consignments can be checked for hot spots.

Although simple to use, staff require training to ensure accuracy of measurement. The object to be measured needs to fill the field of view, or the reading will be an average of the object and its surroundings. Infrared is unsuitable for measuring temperatures of foods packaged in reflective foil unless a matt black spot is painted on the surface. Furthermore, care must be taken because the emissivity of surfaces, for example, packaging and food varies. It is inappropriate to use for checking centre temperatures during cooking, thawing or cooling as only the surface temperature is measured. Infrared thermometers may also be sensitive to the surrounding air temperature, for example, they can be inaccurate up to 6°C for around 15 minutes in a refrigerator. After this time they should come to equilibrium with the case.

HYGIENE for MANAGEMENT *The storage & temperature control of food*

Transportation of fresh lettuce that has to be kept within a set temperature band. The loggers are being used to ensure that this happens and have had their alarms set in order to highlight any discrepancies outside these limits. In the first picture you are also able to see the information being downloaded in situ using a palm unit. (Courtesy of Gemini Data Loggers (UK) Ltd.)

Temperature/data loggers

Data loggers and printers will provide a variety of useful information about refrigerated storage including current temperature, maximum, minimum and trends of temperature over a specified period. Should temperatures rise above a predetermined level they can also trigger an alarm. Loggers may be preferable to routine manual recording of temperatures.

The selection of effective refrigeration units

When purchasing refrigeration units for commercial use, it is important to consider a whole range of factors apart from the obvious one of capacity. Choice must not be based solely on price, and units built for domestic use are unlikely to have the life or qualities that could be expected from a refrigerator purpose-built for commercial organizations. When selecting a unit, regard must be had to the following:

- *Storage temperatures* – units must be capable of maintaining satisfactory storage temperatures in the high ambient temperatures of commercial kitchens. Some busy kitchens, with inadequate ventilation, have been recorded as high as 43°C. Some units may need to operate at 10°C, for example, if used for wine storage, or others at temperatures of −1°C to 1°C, for example, for the storage of fresh meat or fish. Fan assisted cooling is essential to circulate cold air throughout the storage compartment and maintain uniform temperatures. After opening and closing the door the correct temperature will be quickly re-established.
- *Construction* – door(s) and floor should be stainless steel to provide the greatest resistance to impact damage. Heat-reflecting aluminium sides and interior surfaces are often used to save energy. White painted finishes are prone to chipping, cracking and rusting. Plastic liners are easily damaged.
- *Automatic defrost* – ensures that the coil remains free of insulating ice and keeps down running cost. Staff do not have to remember to defrost the unit on a regular basis. Air temperatures of units should not rise above 10°C when defrosting.
- *Thermometers* – all units should be fitted with an indicating or recording thermometer positioned to show the temperature of the warmest part of the unit.

Thermometers should be located externally and should be easily readable with the door(s) closed. It is also useful if there is a light that indicates when the unit is on automatic defrost.
- *Fittings* – a comprehensive choice of fittings is available including: adjustable, nylon-coated, wire shelves; gastronome pans; locks; meat rails and fish drawers.
- *Labour saving options* – an increasing number of modifications are being provided to save staff time and at the same time increase safety, for example:
  - pass-through cabinets, which reduce staff fatigue and prevent congestion between the kitchen and the servery;
  - roll-in cabinets, which enable fully-loaded trolleys to be wheeled directly into storage so reducing handling;
  - heavy-duty drawers, which are used to minimize stooping and reaching;
  - door opening foot pedals, which leave hands free for carrying;
  - glass doors, which enable stock checks without opening doors; and
  - counter models, which can be sited under workbenches so reducing walking and minimizing the time high-risk food is kept at ambient temperatures.

Cooking

Cooking is a form of preservation but is essentially used to make food more palatable and safe for immediate consumption. Temperatures achieved during cooking are usually sufficient to ensure an effective reduction, or the elimination, of vegetative pathogens, although some preformed toxins and spores may be unaffected. In some cases, cooking activates spores and a significant multiplication of vegetative bacteria may occur during subsequent cooling. Internal temperatures of 75°C should usually be achieved to ensure bacteriological safety, although heating food to a lower temperature for longer periods of time may be equally as effective. Faecal streptococci may survive these temperatures, especially in cooked ham.

The core temperature of cooked meat should be checked regularly with an accurate thermometer, which is always disinfected before use. The external surface of, for example, a joint may give the appearance of being thoroughly cooked but closer examination often reveals unacceptable centre temperatures. Rare beef will always have received a terminal centre cooking temperature of less than 62°C (at 62°C/63°C the myoglobin changes from red to brown/grey due to the denaturing of the albumen, which loses its transparency) and this may cause problems if pathogenic contamination is introduced to the centre of rolled joints during boning.

Particular care must be exercised when using the probe thermometer to ensure you are measuring the

An Evolution logging food thermometer and a flexible type T thermocouple probe being used to check the temperature of cooked food as it emerges from a travelling oven. (Courtesy of Comark Ltd.)

lowest temperature of food achieved through cooking. Food cooked in microwave ovens is notorious for cold spots. Food in microwave ovens should be fully enclosed in a suitable container and periodically stirred or rotated to ensure even heat distribution. After cooking food should be left for two minutes to achieve even temperature distribution.

Dr Brad Berry, USA advised that variations of up to 19°C were measured in cross sections of beefburgers cooked to 71°C. Burgers cooked from frozen had the greatest variations and were pink after cooking whereas thawed patties were brown. Slow cooking at low grill temperatures is usually safer.

Some products such as cakes and pork pies may be cooked satisfactorily but are exposed to risk because of the subsequent addition of contaminated cream or gelatin. Storage of these products must always be under refrigeration.

Safe cooking temperatures in the USA are based on a 71% reduction of salmonella, for example 1.6 seconds at 75°C, 12.5 seconds at 70°C or 95 seconds at 65°C. The United States Department of Agriculture and the US Food and Drug Administration recommend the following minimum temperatures:

Minced beef, lamb and pork	**71°C**
Minced chicken and turkey	**74°C**
Beef and lamb roasts and steaks (rare)	**63°C**
Beef and lamb roasts, pork and steaks (medium)	**71°C**
Whole poultry	**82°C**

60°C for 13 minutes also gives a 71% reduction of salmonella. However, salmonella can triple in resistance to kill if it is held for some time around 43°C. This would mean that a temperature of 60°C for 39 minutes would be required to give the same salmonella kill. Potential increase in resistance of organisms after cooking may also be a reason for ensuring reheating temperatures of cooked food should be higher.

The colour of meat should not be relied on as a definite indicator that safe cooking temperatures have been achieved. Beefburgers can turn brown after cooking to 57°C or may still be pink after cooking to 74°C (Dr P. Snyder Jr.).

Generally speaking the safest cooking methods include roasting for long periods, casseroles or cooking thin slices under a hot grill or in a frying pan. If stir frying it is important to cut any meat into small cubes.

Slow cooking

Cooking of foods at lower temperatures for prolonged periods can be dangerous if the food is between 20°C and 50°C for between ten and 15 hours. This would allow production of heat resistant toxins by *S. aureus* or *B. cereus*, which would not be destroyed by the final temperatures of cooking of around 65°C to 75°C.

Monitoring food temperature in a blast chiller.
(Foster Refrigerator UK.)

Cooling of hot food

Rapid cooling of cooked foods to be chilled or frozen is extremely important. Bacteria may survive cooking as spores or even in the vegetative state and there is always a risk of contamination after cooking. By whatever route bacteria enter the food it is essential that their multiplication during cooling is minimized. It is recommended that food is cooled below 10ºC in less than 1.5 hours and blast chillers are available that can achieve this specification, but only for relatively thin portions of food. Furthermore, cooling units must be cleaned regularly to avoid problems with mould growth.

Cooling methods vary depending on the type of food being cooled. They include:
- use of shallow containers, 2.5cm maximum recommended depth of food;
- using smaller or thinner portions;
- blast chillers;
- ice baths, increase cooling cooling rate by three to five times. Little difference in rates between metal and plastic;
- stirring food;
- using fans, food should be loosely covered and supported, for example, on racks, as about 75% of heat is extracted from the bottom of the pan;
- adding ice as an ingredient;
- cold water sprayers, for example, manufactured joints;
- cold running water, for example, rice, potatoes and pasta; and
- large walk-in refrigerators when there is no risk of a temperature rise of any other food in the unit. This should only be used as a last resort but is preferable to, for example, leaving food out in a warm room overnight.

During cooling, food should be loosely covered or protected from contamination to facilitate heat transfer from the food surface. If the joint is sliced or hot liquids drained off into shallow containers to assist cooling, extreme caution is necessary to avoid the risk of introducing pathogenic bacteria such as salmonella. Recently cooked food is an ideal medium for the growth of pathogens because there are few competing organisms.

Unfortunately, it is very difficult to cool large quantities of thick sauces and even relatively small joints of meat (or poultry) within the recommended times. A study undertaken by the AFRC Institute of Food Research, Langford, Bristol found that:

Rolled Joints	wt (kg)	Approximate Cooling time (min) 70ºC to 10ºC (50ºC to 10ºC)		
		Convection	Immersion	Pressure vacuum
Beef silverside	2.7	291(237)	186(153)	39(36)
Beef forequarter	2.7	304(261)	216(185)	24(21)
Boneless turkey	5.9–6.5	396(325)	335(275)	16(13)

Convection cooling was in air at 0ºC and 1.2ms^1.
Immersion cooling was in water at 0ºC.

A study by Dr Peter Snyder Jr. demonstrated that it took 6.5 hours for Alfredo cream sauce 2.5cm deep in a 30cm x 50cm stainless steel dish to cool from 60ºC to just above 5ºC in a refrigerator operating at 5ºC with an air flow of around 0.25 metres per second. Sauce that was 5cm deep would take up to 14 hours. The same amount of sauce in a 10cm diameter jar would have taken ten hours to cool to 5ºC.

As can be seen from these studies most catering/retail premises will not be able to achieve anywhere near the recommended cooling rate of below 10°C in 1.5 hours unless they have a very expensive blast chiller and use small joints.

The danger of slow cooling is the **germination** of spores activated during cooking and the multiplication and/or toxin production of the vegetative bacteria. *Clostridium perfringens* is the pathogen of most concern. However, a study by Juneja *et al.* demonstrated that only 1 log multiplication of *Clostridium perfringens* occurred when a cooked hamburger was cooled continually from 55°C to 7°C over 15 hours.

Forced cooling does not need to start immediately food is removed from the oven. The highest temperature from the multiplication of known pathogenic bacteria is 52°C (*Clostridium perfringens*). Food can be cooled for around 1.5 hours using, for example, a fan and then placed in a large capacity refrigerator operating at around 5°C.

Blast chillers

To avoid the formation of large ice crystals and consequential loss of quality, it is important that chilled foods do not freeze during the chilling process. Usually, chilled air at 2°C to −7°C is circulated around the product but some blast chillers use the vapour from liquid nitrogen and solid carbon dioxide.

Chillers are intended for the rapid removal of heat and not for the storage of food that is already chilled. Consequently, they operate at air velocities well in excess of those used in refrigerators, for example, 5ms[1].

Distribution of high-risk food

Vehicles used for the distribution of high-risk food must always be insulated and preferably refrigerated, even for short journeys. Insulation of the roof and floor is just as important as the insulation of the walls. Properly located thermometers should be fitted to all vehicles. The maximum temperature will be recorded by a sensor fitted in the returned air system and the minimum temperature can be measured near the outlet of the evaporators.

Deliveries of high-risk foods should be in refrigerated vehicles.

Refrigerated vehicles are designed to maintain temperatures of chilled or frozen food and must not be loaded with warm food. Vehicles should be precooled prior to loading. The refrigeration unit should be switched off whilst the doors are open to avoid icing-up of the cooling units. During loading or unloading breaks, the doors must be closed and the unit switched on. Loading must be carried out quickly to avoid unacceptable temperature rises.

Vehicles must be maintained in a clean and tidy condition. Raw food should never be transported with high-risk food unless they are completely segregated to avoid any risk of contamination. Stacking of vehicles should facilitate sufficient air circulation around the food.

Deliveries of high-risk food

The temperature of food should be checked on arrival. Food requiring refrigerated storage should be rejected if above 8ºC and frozen food rejected if above –12ºC. Deliveries must be placed in refrigerated storage as quickly as possible to avoid temperatures exceeding the rejection temperatures. The Quick-frozen Foodstuffs Regulations 1990 require quick frozen food to be delivered below –15ºC.

THE STORAGE OF FROZEN FOOD

At temperatures of around –40ºC, most frozen food should keep for several years without noticeable deterioration. However, most domestic and retail freezers operate at –18ºC and at this temperature a gradual loss of flavour and a toughening of texture occurs. Above –10ºC spoilage organisms, especially osmophilic yeasts, moulds and halophilic bacteria, commence growth and together with biochemical reactions cause spoilage problems including souring, putrefaction and rancidity. As the lowest recorded temperature for the growth of a pathogen, *L. monocytogenes,* is –1.5ºC, food stored below this temperature should not become a health problem.

Spores and significant numbers of pathogenic bacteria are usually able to withstand freezing and prolonged frozen storage. After a suitable lag phase, survivors will commence multiplication on thawing. Strict precautions must therefore be taken in the manufacture of frozen foods to ensure the absence of pathogens. The storage life of frozen food depends on the initial number of micro-organisms, the final storage temperature and the temperatures, and times, of distribution and manufacture. Short periods at a temperature of approximately –15ºC will have little effect on quality or shelf life.

The manufacturing temperature of frozen food is usually –23ºC or below. Commercial cold stores are usually operated at an air temperature of between –25ºC and –30ºC.

The quality of frozen food will be affected if:
- food is frozen too slowly or not to a low enough temperature;
- there is a breakdown in the cold store, which results in a significant temperature rise. Cold stores should always be fitted with temperature recording charts so that fluctuations in temperature, and times, are accurately recorded;
- distribution vehicles are incapable of maintaining temperatures of –18ºC. Problems may occur if there are many small delivery points necessitating the opening of doors a large number of times;

Freezers must be defrosted regularly to avoid ice building up.

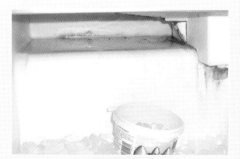

Condensation and mould in an ice-making machine.

HYGIENE for MANAGEMENT *The storage & temperature control of food*

Ice build-up in a machine room freezer.

Poor stacking and stock control.

A well organized cold store.

- there are excessive delays in loading or unloading that result in frozen food being left at ambient temperatures for considerable periods of time; and
- storage facilities at retail or catering outlets are inadequate.

All frozen food should be coded to enable the recall of suspect batches that have escaped manufacturing controls.

The storage of frozen food at retail premises

All cabinets used for the storage and display of frozen foods should comply with BS 3053: 1983. To ensure that customers receive the highest quality of frozen food, managers must:

- only use reputable suppliers;
- reject frozen food deliveries above −12°C, or −15°C for quick frozen foods, or which show signs of thawing or having been refrozen, for example, packs of peas that have welded solid;
- not allow frozen food to remain at ambient temperatures for longer than 15 minutes. Food will, of necessity, be at ambient temperatures during unloading of deliveries and stocking display units from back-up stores;
- not use display freezers for freezing fresh food, as they are only capable of maintaining the temperature of food that is already frozen;
- ensure that display units are not filled above the load-line;
- carry out regular inspections of freezers and check temperatures at least daily but preferably more frequently. Electronic probe thermometers should be used to ensure the accuracy of the indicating thermometers, which should be fitted to all units in an easily readable position;
- ensure that back-up stores are fitted with strip-curtains or air blowers and the doors are opened as little as possible to avoid unacceptable fluctuations of temperature. Ice build-up on the walls or floor of units must not be allowed;
- implement effective systems of stock control and stock rotation. It is advisable to code food on delivery to assist rotation; and
- ensure that food is not mishandled. Damage to packaging may result in loss of product, contamination and freezer burn. For similar reasons packaging, designed to protect frozen food, must not be removed. Stable air temperatures are important as

warm air will result in a fall in vapour pressure and the sublimation of water vapour from the food to restore the balance. Dehydration of the product (freezer burn) results in a pale discoloured food surface, rather like balsa wood, that cannot be restored.

Storage times

All food should be used within the time recommended by the manufacturer. However, a general guide for food kept at −18°C is:

	Months
Vegetables, fruit, most meat	up to 12
Pork, sausages, offal, fatty fish, butter and soft cheeses	up to 6

Salad vegetables, non-homogenized milk, single cream, eggs and bananas should not be frozen. Cream can be whipped and stabilized to overcome the separation in desserts.

The star marking system is used to indicate the temperature and storage times of food in a frozen storage compartment.

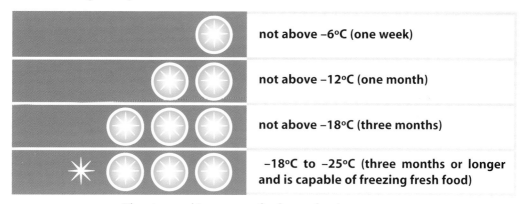

The star marking system for frozen food compartments.

Freezing and refreezing

Freezing of food will not improve its quality. The slow freezing of food in domestic freezers results in the formation of large ice crystals, which rupture cells, leading to a slight deterioration in quality, due to changes in the composition of proteins in the presence of enzymes. This deterioration is much more noticeable if food is thawed and refrozen, apart from the obvious dangers if the thawed food is maintained above 8°C for a considerable time. Large ice crystals are particularly noticeable in ice cream that has been refrozen. However, food that has been frozen, thawed and thoroughly cooked may be refrozen quite safely, although flavour and texture will be altered and nutritional value lowered.

Effect of fluctuating temperatures

Clear plastic packaging of food may act as a greenhouse and the radiant heat from fluorescent tubes and air conditioning may increase the temperature of frozen food significantly. Fluctuations from −3°C to −18°C have been observed with the lower temperatures only being achieved during the night when heat sources have been switched off.

The presence of ice in packets of frozen food usually indicates fluctuating temperatures have allowed water to evaporate from the product and subsequently freeze. Individually quick frozen products, such as peas, weld together in a solid mass if exposed to fluctuating temperatures.

Freezer breakdown

If the freezer breaks down or food becomes thawed, for example, due to a power failure, the food may occasionally be treated as fresh. In certain circumstances the food may be cooked and refrozen. If the food has a solid core of ice it may be safe to refreeze without cooking. If in doubt advice should be obtained from the local environmental health department.

In the event of breakdown, the lid of the freezer should be left closed and the unit covered in newspapers and blankets until repaired. Food may remain frozen for at least two days in a well-stocked, well-insulated freezer.

Multichannel data logger monitoring product temperatures inside a freezer. The logger has been mounted on the freezer door, with the probe inside, attached to a food item. The logger can be removed at required intervals for downloading on to a computer. (Courtesy of Comark Ltd.)

Thawing of frozen food

Most food taken from the freezer can be cooked immediately but poultry, joints of meat and other large items must be completely thawed before cooking. The manufacturer's instructions should always be followed. If food is not completely thawed, ice is likely to be present at the centre and the heat from subsequent cooking will be used to melt the ice and not to raise the internal temperature above that required to destroy pathogens.

However, thawing of food poses several problems and inadequate thawing of poultry frequently results in undercooking and subsequent food poisoning. Thawing at room temperatures (21°C to 25°C) for around 12 hours results in a small multiplication of spoilage bacteria on the surface of food. Thawing in a small refrigerator can be extremely hazardous. A 20lb turkey will take several days to thaw and the surfaces of the refrigerator, and ready-to-eat food, may become contaminated with thawed liquid containing pathogenic bacteria. Furthermore, the temperature at which the refrigerator is operating has a significant effect on thawing times. Work carried out by Mr T.C. Grey at the Bristol AFRC Institute of Food Research showed that a 1.1kg broiler took around eight hours at 20°C, 11 hours at 15°C, 13 hours at 10°C, 40 hours at 5°C and 70 hours at 1°C for the temperature to go above 0°C in the deep thigh. Birds left in polythene bags took longer especially at the colder temperatures. It is therefore essential for advice regarding thawing times under refrigeration to state actual temperatures. Thawing of frozen poultry is best carried out at 10°C to 15°C in an area entirely separate from other foods, in a thawing cabinet, or at room temperature preferably at less than 21°C.

In 1968 a study by Klose, *et al*. concluded that it is satisfactory to thaw turkeys at room temperatures at or below 21°C. It takes a 10kg turkey 15 hours or less to get to 0°C while the surface temperature rises to 12°C or less. During this time spoilage organisms may multiply by four times but no detectable multiplication of salmonella occurs. In 1999, Jiménez, et al., confirmed that chickens weighing over 3kg thawed to an internal temperature of 5°C within the breast in ten hours at room temperature. There was no increase in inoculated *Salmonella Hadar* or spoilage organisms during this time.

THAWING AND COOKING TIMES OF FROZEN POULTRY

Oven ready weight kg (lbs)	Approx. thawing time at room temp (21°C) (hours)	Cooking time at 180C/350F Gas 4* (in foil) (hours)
2.25 (5)	15	2.5
4.5 (10)	18	3.5
6.75 (15)	24	4.75
9.0 (20)	30	5.75

*These are minimum times. The bird is cooked when the juices run clear, and should be checked with a disinfected probe thermometer.

Food thaws slowly because of the low thermal diffusivity of food at 0°C. As soon as the surface thaws it acts as an insulating layer and this reduces the rate of heat flow into the product. Thawing in refrigerators is even slower because of the small temperature difference between the refrigerator operating at 1°C to 4°C and frozen food. Consequently, it takes a considerable time to provide the latent heat of thawing, i.e. the heat necessary to change ice to water.

Several techniques have been suggested for thawing food and some of these have been used commercially. Electrical methods of thawing, namely dielectric, microwave and resistive, are not affected by the low thermal diffusivity of food but unfortunately may give rise to an effect known as runaway heating: thawing does not occur evenly throughout the frozen food and areas that thaw absorb energy preferentially to areas that remain frozen. This results in some parts of the food cooking before the whole mass of food is thawed.

Rules for thawing and cooking frozen poultry
- Segregate from high-risk food.
- Thaw completely in a cool room at less than 15°C, in a thawing cabinet, or at room temperature (less than 25°C). Cold running water and refrigerators may be used but cross-contamination is likely. Furthermore, thawing may not be complete in the refrigerator, which is the major hazard. Poultry will be ready for cooking when the body is pliable, the legs are flexible and the body cavity is free from ice crystals.
- Remove giblets (care because of risk of cross-contamination).
- Do not wash the birds as you will contaminate the sink, taps, dishcloth and adjacent work surfaces, possibly with campylobacter, salmonella *E. coli* O157 or *Clostridium perfringens*.
- Once thawed cook immediately, or keep in the refrigerator and cook within 24 hours.
- Cook thoroughly and cook the stuffing separately.
- All utensils and surfaces used for the preparation of raw meat and poultry should be thoroughly cleaned and disinfected before being used for high-risk food. It is preferable to use separate work surfaces.

- Eat immediately after cooking or, if the bird is carved cold, cool it quickly and store in the refrigerator. As with all meats **refrigerated storage is essential within 1.5 hours.**
- Avoid handling the cooked bird unnecessarily.

In the USA it is suggested that frozen poultry can be safely cooked without thawing, cooking times are increased by around 50%.

COOK-CHILL

Cook-chill is the name given to a catering system in which food is thoroughly cooked and then chilled rapidly in a blast chiller to a temperature of 3ºC or below within 1.5 hours. The food is stored between 0ºC and 3ºC until required for reheating. In mass catering, some form of cooking and chilling is often essential. There is no other way of bringing together all the components of a meal simultaneously without a disproportionately large labour force and an excessive cooking equipment provision. As such the chilling of cooked food is not a completely new concept. The rudiments are practised by virtually every large catering establishment. Sometimes, the regime is as basic as "chilling" the food to ambient in a cool part of the kitchen but in most operations there tends to be an initial prechilling at ambient followed by retention in a cold store or refrigerated cabinet. Such uncontrolled practices have significantly contributed to a number of food poisoning cases in the UK.

The formalization of cook-chill as a specific form of catering has, in this context, done nothing more than introduce specialist equipment for chilling and reheating, enabling that which was already practised to be undertaken safely and more efficiently. Large-scale schemes are, however, more than just cook-chill. To meet the costs of re-equipping with expensive equipment, a number of operations have centralized production in a single kitchen, known as a central production unit (CPU). The most cost-effective results are achieved using large-scale equipment such as bratt pans, tilting kettles, large steamers, computer controlled fast fryers and convection ovens. A major distribution system is then organized to get the food

The storage of cook-chill meals.

to the former producing kitchens, now known as end kitchens, where it is reheated.

There are usually eight stages in a cook-chill system:
- bulk storage;
- preparation;
- cooking;
- portioning, packaging and labelling;
- blast chilling;
- storage at or below 3°C;
- distribution at or below 3°C and;
- reheating (regeneration) and serving.

With the introduction of cook-chill schemes there have been coincidentally, and necessarily in respect of centralization, developments in production aimed at systemizing and standardizing the catering process. The fundamental nature of these operations is that they have more in common with food processing than catering. However, there are significant differences. Batch size tends to be smaller than would be handled in food processing and the menu is considerably more varied. An absolutely key difference, however, is that the CPU can rarely, if ever, cook all the food used in a scheme. Anything up to 60% of total consumption may simply pass through after being portioned and packed. The CPU, therefore, as well as being a manufacturing plant is also a distribution centre, often handling a high proportion of precooked foods. These high-risk foods must be integrated into the mainstream production line. Composite foods such as salads and sandwiches may also require integration. In effect, the CPU is two separate operations under one roof, combining to form a single operation at

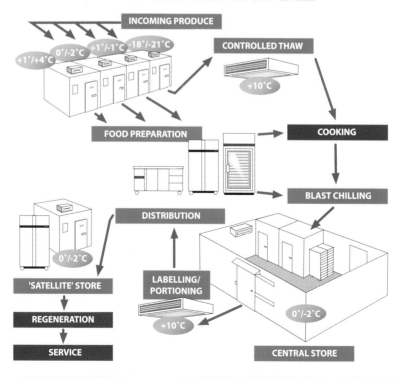

the end. This mixing of functions has a far-reaching effect on the layout. The simple "dirty in one end – clean out the other" flow, favoured in operations as diverse as food factories and sterile supply units, cannot usually be achieved. A far more sophisticated flow and layout is required.

The responsibility of the designer is even greater than for conventional kitchens. The operations heavily influence the design requirements, but the design controls the operation to an even greater degree. The designer should seek to impose a specific regime and set methods of working to ensure a systemized operation. The aims are to ensure that the operation works in the most efficient and safe way and to prevent it working in any other way.

The following benefits are claimed for the cook-chill system:

- cost-effectiveness – savings are possible because:
    - of centralization and better utilization of equipment;
    - centralized purchasing and accurate portion control results in less wastage;
    - careful design and planning can reduce energy consumption;
    - less floor space is required;
    - fewer staff are required in both the CPU and satellite kitchens; and
    - a reduction in overtime. Careful planning results in a steady workload instead of troughs and peaks throughout the day;
- better staff conditions and less work in unsociable hours. Staff turnover is usually reduced;
- flexibility – orders for meals can be accepted at much shorter notice; and
- improved quality and palatability compared with meals kept hot, above 63°C, for long periods. Complaints of dried-up and overcooked food should not occur. Meals are also more consistent.

The safety of cook-chill

Because of the emphasis on strict temperature control throughout the system and the provision of recording thermometers on all units, a properly controlled cook-chill system is safer than most conventional systems of mass catering. Furthermore, as staff are not working under immense pressure for short periods of time, there is less rushing about and mistakes are less likely. However, the reduction in risk tends to be offset, at least on a statistical basis, by the larger scale of production. Where more meals are produced, a unit has to be safer than a smaller operation, just to maintain the same statistical risk of food poisoning. There is also the essential matter of extended shelf life: foods to be consumed some days after cooking have obviously to be at a higher standard than food produced for immediate consumption. The general approximation that cooking kills germs, which governs conventional catering, does not apply. The cooking process is seen as a reducing factor only. Where initial bacterial loading of the raw product is too high, the cook-chill process cannot produce food at an acceptable standard, no matter how well it operates or how hygienic it is. Raw produce standard assumes a much greater importance.

Spores of *B. cereus* and *C. botulinum*, will not be destroyed during cooking. As some strains of *B. cereus* can grow below 8°C and some strains of *C. botulinum* can produce toxin under refrigeration, strict attention to cooling, storage temperatures and shelf life is essential. The potential of the low temperature pathogens such as *Yersinia enterocolitica* and *Listeria monocytogenes* to cause problems must not be overlooked, notwithstanding that it appears that multiplication and toxin production over a five day period is insignificant at temperatures below 3°C. Furthermore, most low temperature pathogens should be destroyed during the reheating process.

In order to ensure the safety of cook-chill the following rules should be observed:

- all raw materials should be of good microbiological quality. Purchasing contracts should include product specifications and suppliers should be monitored to ensure compliance. Raw products should be stored at appropriate temperature and humidities, which are monitored;
- cross-contamination must be avoided; the preparation of raw materials must be in areas physically separated from cooking and post-cooking areas. Personnel handling raw foods should be confined to the raw material area. Separately identifiable equipment should be used for raw products;
- controlled thawing equipment is required if frozen raw materials are used;
- the highest standards of hygiene, especially **personal hygiene**, must be observed throughout all stages of preparation, cooking, storage and reheating. Staff will require additional training;
- immediately prior to cooking, excess prepared food must be held below 10°C;
- to ensure the destruction of *Listeria monocytogenes,* food should be cooked to a temperature above 70°C for not less than two minutes, at the slowest heating point (checks should be made with a probe thermometer). Recording thermometers should also be used;
- food should be portioned and chilled to below 3°C within two hours of cooking. Chillers must be capable of reducing the temperature of a 50mm layer of food from 70°C to 3°C in not more than 90 minutes when fully loaded. Automatic controls are required including an accurate (±0.5°C) indicating thermometer and temperature recorder. Product depth during chilling should not exceed 50mm and may need to be reduced to achieve chilling specification. Joints of meat should not exceed 2kg and 100mm in thickness. Uniform shapes of joint should be used. Handling of food should be minimized and the use of disposable gloves does not remove the need for frequent handwashing;
- reusable lidded containers, capable of being cleaned and disinfected, or disposable containers may be used for cooked food portions. (Use of lids during chilling reduces dehydration but slows down chilling.) Containers should be date marked to ensure strict stock rotation. The label can also indicate type of food and reheating time (with or without the lid);
- the refrigerated store, fitted with indicating and recording thermometers and alarms, should maintain food between 0°C and 3°C and should only be used for cook-chill products. The maximum life of cooked products is five days, including the days of cooking and consumption;
- should the temperature of the food exceed 10°C during storage or distribution and before reheating, it should be destroyed. If the temperature exceeds 5°C the food should be eaten within 12 hours;
- insulated containers, chilled before use, may be used for short distribution runs but refrigerated vehicles are preferred and are essential for long journeys, especially during warmer months;
- food should be reheated as soon as possible after removing from chill and never longer than 30 minutes. The centre temperature of food should reach at least 70°C

and be maintained for at least two minutes. Service of food should commence as soon as possible and within 15 minutes of reheating. Temperatures must not be allowed to fall below 63°C. Unconsumed, reheated food must be destroyed;
- reheating is preferable using infrared units, forced air and steamer convection ovens (traditional ovens may cause dehydration);
- foods intended to be eaten cold should be consumed as soon as possible and within 30 minutes of removal from chill; and
- to avoid health hazards documented HACCP should be implemented. Action plans should be formulated to deal with deviations at critical control points.

Management

Cook-chill demands considerable management and supervisory skills. Menu planning must take account of the keeping quality of different types of food. Foods with a high unsaturated fat content may oxidize more quickly and may not be suitable for the full five day storage. Work scheduling so that chilling times are rigidly observed, taking into account the differing rates of chilling of food for different types and densities, is also challenging for managers.

Equipment failure and abnormal temperature readings can lead to difficult decisions and high losses. Increased complexity of equipment demands increased expertise from both managers and maintenance staff.

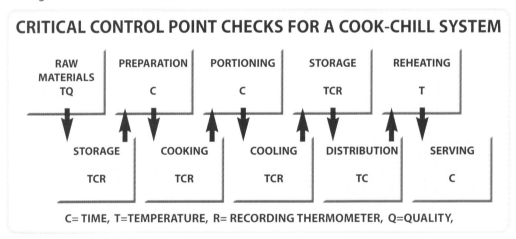

Microbiological guidelines to be used for setting up the system (Chilled and Frozen Guidelines on Cook-Chill and Cook-Freeze Catering Systems: DH 1989)
Total aerobic colony count (48 hours at 37°C) < 100,000 per gram.
Salmonella and *Listeria monocytogenes* not detected in 25g.
Staphylococcus aureus (coagulase+ve) and *Clostridium perfringens* < 100 per gram.
Escherichia coli < 10 per gram.
(100g samples to be taken immediately prior to reheating.)

SOUS VIDE COOK-CHILL

Sous vide cook-chill, also known as cuisine sous vide, is an interrupted catering system in which raw or par-cooked food is sealed in a vacuumized plastic pouch or container, heat

treated by controlled cooking, rapidly chilled and then reheated for service after a period of chilled storage. The advantages include:
- extended shelf life;
- enhanced sensory quality; and
- improved nutritional quality.

Disadvantages include:
- significant capital and operating costs;
- limited number of products are suitable for this operation; and
- poorly designed systems and badly controlled systems may result in serious consequences, such as the growth of *Clostridium botulinum* or *B. cereus* in the product.

The sous-vide process

Ingredient purchasing, delivery and storage

All raw materials should be of the highest quality and stored at correct temperatures and humidities for the minimum time.

Preparing and portioning

High standards of hygiene must be maintained at all times. Prepared food must be held, or if pre-heated, chilled to below 10ºC prior to vacuum packaging. Single portion pouches are preferred but if multi-portion pouches are used product thickness should not exceed 50mm, to ensure acceptable heat transfer. Adequately sized pouches are essential. It should be possible to pinch 12.5mm of plastic together on all four sides of the product and the opening of the pouch should be folded back to prevent soiling during filling.

Vacuum packaging

After the removal of air and heat sealing of the pouch, it should mould tightly to the contours of the food. All pouches should be checked and any with defects such as air or product movement should be repackaged.

Prime cooking/pasteurization

Controlled heating may be undertaken using atmospheric steamers, combination ovens (set on "steaming" mode) or water baths. Food should be heated as soon as possible after preparation and always within two hours. Pouches of differing thickness should be processed separately and pouches must not be stacked on top of each other.

One of the following time/temperature treatments should be used:

Core temperature (ºC)	Minimum holding time (minutes)
80	26
85	11
90	4.5
95	2

Chilling

Chilling may be achieved by blast, immersion or cryogenic chillers and should commence immediately after pasteurization and always within 30 minutes. Food must be chilled to at least 3ºC within two hours of cooking.

Chilled storage and distribution

Products must be labelled with their name, date of production (or batch number), **use-by date** and reheat instructions. In the UK the maximum shelf life is usually considered to be eight days including days of production and consumption. Storage and distribution must be between 0ºC and 3ºC. If the food temperature rises to between 5ºC and 10ºC the food must be consumed within 12 hours, if it exceeds 10ºC the food should be destroyed.

Reheating and service

Prior to reheating (regeneration), any defective pouches should be destroyed. Reheating should take place immediately after removing from chilled storage and always within 30 minutes and never more than 15 minutes prior to service. A minimum centre temperature of 75ºC should be achieved and the temperature of the food must not fall below 63ºC before being served. Unconsumed reheated products must be destroyed.

THE SAFETY OF CHILLED FOODS

In 1992 The Advisory Committee on the Microbiological Safety of Food recommended that in addition to maintaining the temperature of prepared, chilled foods with a shelf life of more than ten days below 10ºC, at least one of the following controlling factors should be used, to prevent the growth and toxin production of psychrotrophic *Cl. botulinum*:

- a heat treatment of 90ºC for ten minutes or equivalent lethality;
- a pH of 5 or less throughout the food;
- a minimum salt level of 3.5% in the aqueous phase throughout the food;
- an a_w of 0.97 or less throughout the food; or
- a combination of heat and preservative factors, which can be shown consistently to prevent growth and toxin production of psychrotrophic *Cl. botulinum*.

A well organized large chiller operating below 3ºC.
(Courtesy of Marks and Spencer.)

The Advisory Committee also recommended that a code of practice for the manufacture of vacuum and modified atmosphere packaged chilled foods should be drawn up containing detailed guidance on:

- raw material specification;
- awareness and use of HACCP;
- process establishment and validation (including thermal process);
- packaging requirements;
- temperature control through production, distribution and retail;
- factory auditing and quality management systems;
- the requirements to establish a safe shelf life;
- the control factors necessary to prevent the growth of bacteria and toxin production;
- the application of challenge testing;
- equipment specifications, particularly with regard to heating and refrigeration; and
- training.

COOK-FREEZE

Traditional catering involves considerable fluctuations in workload with a peak of around two hours, based on the time of meals. Cook-freeze is a system of catering devised to ensure a planned daily workload that utilizes division of labour techniques and reduces wastage. The system takes advantage of economies of scale and operates on the same principle as a factory production line. Requirements can be forecast months in advance but fluctuations in demand can be dealt with as they occur. Furthermore, skilled staff are fully utilized throughout the day and a greater proportion of unskilled staff can be employed, for example, in serving kitchens.

The first four stages of cook-freeze are the same as cook-chill, namely: bulk storage; preparation; cooking and portioning; packaging and labelling. The fifth stage is blast-freezing. Precooked, lidded packs are loaded on to trolleys, which are wheeled into tunnel-type blast-freezers that reduce the temperature to −20ºC in less than 90 minutes. Rapid freezing is essential to avoid the formation of large ice crystals, which result in poor texture and loss of nutritional value on regeneration. The frozen containers are kept at −20ºC and may be stored for up to 12 months.

The exact number of meals required can be removed from storage on demand and regenerated to a temperature of at least 75ºC in serving kitchens using, for example, forced-air convection ovens.

Cook-freeze systems may be installed as purpose-built, self-contained production units, which ensure a positive flow line, or, alternatively, existing equipment can be used with the addition of portioning, wrapping, blast-freezing and low-temperature storage. Cook-freeze is most suitable for operations requiring at least 1,000 meals per day, although some units have proved viable serving 200 meals per day.

Hot storage of food

Temperatures used for keeping food hot, prior to service, must be high enough to prevent bacterial multiplication, especially pathogenic bacteria. However, appliances such as hot cupboards and bains-marie are designed for storing hot food that has been thoroughly cooked; they must not be used for warming up cold food. Storage temperatures must not be relied on to destroy pathogens.

Very few bacteria are able to multiply above 55ºC, including those that commonly cause

Hot food storage.
(Courtesy of Marks and Spencer.)

Hot food storage.
(Courtesy of Marks and Spencer.)

problems of food poisoning. Consequently, the law requires all food businesses to keep hot food at or above 63°C.

It is important to remember that temperatures of cooking may cause the germination of *Clostridium perfringens* spores and the vegetative bacteria will commence rapid multiplication at approximately 50°C. Consequently, a slight reduction in storage temperatures can have disastrous consequences.

VENDING MACHINES

The use of automatic vending machines continues to increase. The complexity and types of machine and range of food dispensed are considerable. The hygiene of vending machines is controlled by Chapter 111 of the Regulation EC No 852/2004. Vending machines must be sited, designed, constructed and maintained to ensure the safety and wholesomeness of all dispensed food. Machines must be kept in good condition to avoid the risk of contaminating food. They must be durable and capable of being kept clean and, where necessary, disinfected, both internally and externally. Pests must be denied access. Food-contact surfaces should be smooth, impervious, non-toxic, corrosion resistant and be able to withstand the repeated cleaning to which they will be subjected.

Drinks vending machines must be connected to a supply of water of acceptable quality. Storage of water in machines must not result in an unacceptable deterioration of quality and carbon filters should be changed regularly. Plumbing of drinks vending machines must comply with any relevant local water byelaws. Hot drink vendors must be designed to prevent steam from affecting dry ingredients.

Machines must not be sited in direct sunlight or adjacent to sources of heat, for example, boilers, which could affect the temperature of food. Lights and other heat-generating mechanisms within machines must be fitted so that the temperature of food is not significantly affected. Location of machines should facilitate the thorough cleaning of adjacent wall and floor surfaces. External or internal sites that expose food to risk of contamination, including dust, condensation or odour, should not be used for vending machines. The name and address of the machine operator should be clearly visible on the machines. Adequate litter containers, which are emptied as frequently as necessary, should be provided next to machines.

Stock rotation of food is essential. All items should be given a use-by date, or a suitable code, to ensure that unsatisfactory food is removed. It is extremely desirable for vending machines dispensing high-risk foods, which are stored under refrigeration or above 63°C, to be fitted with chart recording thermometers, to ensure that food that has been subjected to unacceptable temperatures, for example, during a power cut, is not sold. At the very least, an indicating thermometer, visible externally, should be provided.

Vending machines must not be used for heating up cold food or for cooling down high-risk food from ambient temperatures. Hot food placed in vending machines must be thoroughly cooked and placed in the unit above 63°C.

A supply of hot and cold potable water must be available for cleaning machines vending open food. Regular and thorough cleaning and, if necessary, disinfection of vending machines is essential to ensure food safety and quality. Machines should display precise cleaning instructions on the inside of the front service door, and operators must always follow these instructions. The date and time of each cleaning should be noted in the machine.

Vending machines are increasing in popularity.
(Courtesy of Marks and Spencer.)

Appropriate facilities must be available to maintain adequate personal hygiene for personnel involved in loading open food or cleaning food-contact parts.

Adequate arrangements and/or facilities must be available for the hygienic storage and disposal of hazardous and/or inedible substances and waste.

9 Food spoilage & preservation

FOOD SPOILAGE

Immediately vegetables and fruit are harvested or animals slaughtered, they usually commence to decompose due to the release of autolytic (self-splitting) enzymes from dying cells. The enzymes break down the largest food molecules and spoilage is then accelerated by bacterial exoenzymes, which degrade the complex molecules into their constituent molecules, which are then absorbed by the bacteria. Spoilage usually starts with aerobic and facultative bacteria and then obligate anaerobes take over as oxygen levels are reduced. Moulds and yeasts are responsible for some spoilage, especially when conditions do not favour bacteria. The rate of decomposition depends on the condition of the food, the pH, a_w, temperature, oxygen tension, the presence of inhibitory substances and the type and number of spoilage organisms present. As off-flavours and odours develop and a breakdown in texture occurs, the food will eventually become unfit for consumption.

Spoilage may also be caused by insects or vermin, parasites, chemical contamination, physical damage, such as freezer burn, and oxidation. Perishable foods such as meat, poultry, fish, dairy products, fruit and vegetables, unless preserved in some way, are the most susceptible to microbial spoilage and must be handled carefully. Some foods such as sugar, flour and dried fruit are often described as stable or non-perishable and are unlikely to be affected by spoilage unless they are handled badly, for example, by storing under damp conditions.

As spoilage bacteria multiply more rapidly than moulds and use up the available surface oxygen, meat at ambient temperatures or stored in refrigerators of high humidity will usually be spoiled by aerobic bacteria, such as *Pseudomonas*, growing on the surface. Off-odours and slime will be produced. However, in low humidities the surface of meat may be too dry for rapid bacterial growth and consequently moulds such as *Cladosporium herbarum* (Blackspot) and *Thamnidium elegans* (Whiskers) will develop. The presence of mould usually results in food having a musty odour and flavour and, although usually considered to be harmless, increasing concern is being expressed because of the possible presence of mycotoxins. The acidity of fruit ensures that most primary spoilage is caused by moulds and yeasts, which are able to multiply at a lower pH than bacteria. Although several species of bacteria do cause spoilage of vegetables, moulds are much more commonly involved, for example, those belonging to the genera *Botrytis*, *Rhizopus* and *Penicillium*.

Vegetables stored in vinegar, such as beetroot, may be attacked by yeasts, which gain access to jars after processing, for example, if the lid on the jar is defective, damaged in transit or removed and replaced by inquisitive customers. The beetroot turns a rusty-brown colour and a pinky-white precipitate may be observed on the base of the jar, which is attributable to fermentation by the yeast. Yeast spoilage of food can often be detected by the alcoholic taste and smell and the presence of bubbles in liquid.

Moulds are responsible for most of the spoilage of baked products, especially bread and pies. As mould spores are destroyed by normal cooking temperatures, post-process contamination from airborne spores and contact with contaminated surfaces must be prevented. Particular care must be taken to remove debris and waste food that would support mould growth. Cleansable surfaces and effective ventilation to stop condensation are also important to prevent growth. Products should be cooled quickly and not wrapped whilst warm as the resultant condensation will encourage rapid mould growth, especially if the food is stored at ambient temperatures. Moulds commonly involved include *Rhizopus* and *Mucor spp.* (white growth with black spots), *Penicillium spp.* (green) and *Monilia sitophilia* (red/pink).

Staleness of bread usually develops with prolonged holding due to physical changes in the carbohydrates. Refrigeration increases the rate of staling, however, staling does not occur during frozen storage at –18°C.

Rope

Rope in bread is caused by *Bacillus subtilis* and spores of this bacterium may be present in flour. They are able to withstand baking temperatures, and slow cooling and warm, humid storage ensure rapid germination and multiplication. Affected bread develops a fruity, sickly smell and a soft, sticky texture. Internally the loaf discolours, becoming yellow or brown. Rope is prevented by the use of proprionates. Other products such as cakes and doughnuts may occasionally be affected by rope.

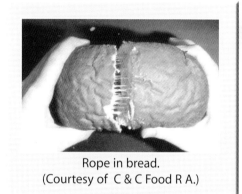

Rope in bread.
(Courtesy of C & C Food R A.)

Rancidity

Fat in dairy products may be broken down into free fatty acids by micro-organisms or by naturally occurring enzymes, lipases, with the production of off-odours and flavours. This process is described as hydrolytic rancidity. Heating may destroy lipase-producing bacteria but not any lipase already formed and rancidity may still occur.

Oxidative rancidity may also occur in dairy products, often due to the presence of copper or iron contamination. Off-odours and flavours are due to the formation and decomposition of peroxides. The prolonged cold storage of fatty fish, bacon and pork results in rancidity unless vacuum packed.

COMMON FOOD SPOILAGE BACTERIA

FOOD	SPOILAGE BACTERIA Commonly isolated	TYPICAL SIGNS OF SPOILAGE
Fresh meat and poultry	*Clostridium, Flavobacterium, Pseudomonas* spp., *Micrococcus, Achromobacter, Acinetobacter* and *Psychrobacter immobilis, Brochothrix thermosphacta*, psychrotrophic Enterbacteriaceae, and occasionally cold-tolerant clostridia	Slime, greenish discolouration, white spots (bacterial colonies), souring, putrefaction, off-odours and flavours
Processed meats	*Achromobacter, Micrococcus, Pseudomonas, Clostridium Lactobacillus, Streptococcus**	Souring, gas production, discolouration, surface slime
Bacon**	*Streptococcus*, Micrococcus, Lactobacillus*	Slime formation, white spots, discolouration, off-odours. Souring of vacuum packs (which may blow)

Fish	*Acinetobacter, Pseudomonas* spp., *Shewanella* spp. (marine fish), *Acinetobacter/Moraxella, Aeromonas* spp. or other members of the Vibrionaceae if fish held above 5ºC, *Photobacterium phosphoreum* when packed in modified atmospheres	Off-odours, discolouration
Vegetables ***	*Pseudomonas, Corynebacteria Erwinia, Leuconostoc Bacillus, Clostridium*	Soft rot, foul odour, discolouration, black spots
Raw milk	*Streptococcus, Micrococcus Lactobacillus, Bacillus, Pseudomonas*	Tainting, off-flavours and odours, souring (above 15ºC), rancidity
Pasteurized milk	*B. cereus, Streptococcus,**** Lactobacillus****	Bitty cream (sweet curdling), off-odours and flavours

Streptococcus faecalis is quite salt tolerant.
**Packs of bacon with faulty seams are vulnerable to mould spoilage.
***Frozen green vegetables, which defrost, become yellowish and khaki, odours and slime develop. The organism commonly responsible is *Leuconostoc mesenteroides*.
****Thermoduric strains, which survive **pasteurization**, e.g. *Streptococcus faecalis*.

Spoilage of poultry

The spoilage rate is dependent on the storage temperature and the number and type of spoilage organisms present. Spoilage is evident at around 10^8 organisms per cm^2.

The smell of spoilt chicken is due to the production of hydrogen sulphide by spoilage bacteria. The hydrogen sulphide diffuses into the muscle tissue and combines with the haem pigments of blood and muscle in the presence of air to form the green pigment sulphaemoglobin just under the skin.

SPOILAGE RATE OF POULTRY*

Temperature	100 organisms/cm^2	10,000 organisms/cm^2
0ºC	11.5 days	7.6 days
5ºC	6.2 days	4.1 days
10ºC	3.9 days	2.6 days
15ºC	1.8 days	1.2 days

*Mead G. C. (personal communication).

Soft drinks and fruit juices

Spoilage of soft drinks and fruit juices is caused by bacteria, Gluconobacter (Acetomonas) and *Alicyclobacillus acidoterrestris*, yeasts, *Zygosaccharomyces bacilli* (often resistant to preservatives) and heat-tolerant **fungi**, *Byssochlamys fulva, B. nivea, Neosartorya fischeri* and Talaromyces species.

Alicyclobacillus acidoterrestris is an aerobic bacterium that grows between 45ºC and 70ºC. Its optimum pH is 2.5 to 4.5. Reported **D values** (decimal reduction time) range from 60.8 to 94.5 minutes at 85ºC, 10 to 20.6 minutes at 90ºC and 2.5 to 8.7 minutes at 95ºC. Spoilage from *Alicyclobacillus acidoterrestris* is easily detected due to a distinct **antiseptic** off-odour attributed to guaiacol, a **metabolic** by-product of the bacterium off-flavour due to, 2, 6-dibromophenol. Spoilage can be prevented by the addition of lysozyme.

FOOD PRESERVATION

Preservation is the treatment of food to prevent or delay spoilage and inhibit growth of pathogenic organisms, which would render the food unfit. Preservation may involve:
- the use of low temperatures or high temperatures;
- the use of **dehydration**, i.e. moisture control;
- the use of chemicals;
- controlled atmospheres and the restriction of oxygen; and
- physical methods including smoking and irradiation.

FOOD PRESERVATION BY THE USE OF LOW TEMPERATURES

This form of preservation is based on the fact that all metabolic reactions of micro-organisms are enzyme catalyzed. The speed of enzyme reaction depends on temperature and the colder it is, the slower the reaction. Temperatures used may be:
- above freezing (refrigerator);
- at freezing (commercially used with chilled beef); or
- below freezing (freezer).

Temperatures above freezing

Refrigerators are used both commercially and domestically and should usually operate at between 1°C and 4°C. They are suitable for the storage of most perishable foods over a relatively short period of time. Most common pathogenic organisms cease multiplication below 5°C, although increasing concern is being expressed about *Yersinia enterocolitica*, *Aeromonas hydrophila* and *Listeria monocytogenes,* all of which are capable of growth under refrigeration. **Psychrophilic** spoilage bacteria that cause problems include those belonging to the genera *Pseudomonas, Acinetobacter, Flavobacterium* and *Alcaligenes.* Mould genera capable of growth at low temperatures include *Penicillium, Mucor, Cladosporium* and *Botrytis.*

Temperatures below freezing

As well as the inhibition of enzyme reactions, freezing also relies on reducing available moisture to ensure effective preservation. The a_w of water at 0°C is 1.0 whereas at –15°C the a_w of ice is approximately 0.85. The freezing of food destroys some bacteria, including pathogens, and a gradual reduction of survivors occurs during storage. Generally, the lower the storage temperature of frozen food the greater is the survival rate of micro-organisms. Spores and toxins are practically unaffected by freezing or frozen storage. The greatest number of bacteria are killed between –2°C and –5°C; unfortunately, many enzymes are extremely active at around –2°C and food stored at this temperature would soon deteriorate.

Some parasites can be destroyed by freezing, for example, trichina cysts in pork stored at –18°C for 21 days, *Cysticercus bovis* in beef carcases stored at –10°C for 14 days and the fish nematode anisakis, which is destroyed at –20°C in 24 hours.

Moulds and yeasts are more likely to grow on frozen food than bacteria as they are better able to withstand lower a_w and temperatures. In practice very few organisms grow below –8°C. Even at temperatures of around 0°C it takes the most rapidly growing **psychrophiles** a day to achieve a tenfold increase in numbers.

On thawing, after a short lag time, a rapid growth of those bacteria that survive freezing will soon compensate for those destroyed, especially if food is allowed to reach temperatures of 20°C to 30°C.

Before vegetables are frozen they must be blanched. **Blanching** is carried out by immersion in hot water for a short period, approximately one minute. Its function is to destroy enzymes, such as peroxidases, which produce off-odours and flavours, reduce bacterial load, fix colour, remove trapped air and induce wilting in some vegetables to aid packing. Overblanching will result in excessive loss of vitamin C by leaching and must be avoided. Care must be taken when operatives handle frozen food to avoid contamination, especially by staphylococci.

As the temperature is reduced, pure water will commence freezing at 0°C. (The freezing point is depressed by the presence of solutes such as salts and sugars.) However, the rate of cooling slows due to the loss of the latent heat of fusion prior to ice formation. As most of the water is converted to ice, the rate of cooling once again increases. Quick freezing involves passing through this zone of maximum crystallization of ice as quickly as possible. The faster the process the smaller the ice crystals, the better the food quality and the smaller the amount of liquid (drip) produced on thawing.

The actual rate of freezing depends on:
- the shape and weight of the product;
- the initial and final temperature;
- the surface heat transfer coefficient and thermal conductivity;
- the amount of heat to be removed (enthalpy);
- the temperature of the refrigerant; and
- the type of packaging.

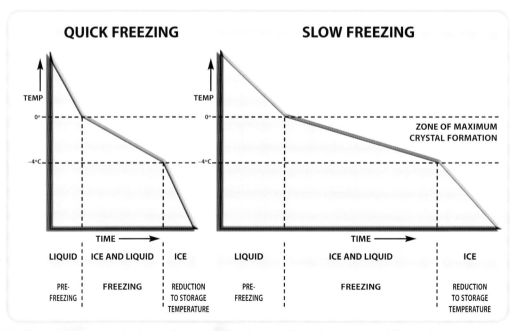

Rates of freezing: the latent heat zone and zone of maximum crystal formation overlap and, if food is frozen slowly, considerable damage and loss of quality occurs.

Most foods will keep for prolonged periods in a freezer, although a recommended **shelf life** is given because of loss of texture, flavour, tenderness, colour and overall nutritional quality. Foods must be properly wrapped to avoid loss of moisture from the surface, i.e. freezer burn. The oxidation of food is slower at −18ºC and this also assists in preservation. However, vacuum packing is essential to extend the shelf life of frozen food susceptible to oxidative rancidity, for example, bacon, and the shelf life of kippers can be extended from three to nine months. Very sophisticated laminate packaging is used for some **ready-to-eat** meals and boil-in-the-bag products. Films of low-density polyethylene, polyesters, aluminium foils and polypropylene are all used to improve packaging characteristics. Some products such as fish fillets are given a thin coat of ice to avoid dehydration during storage.

Freezing systems

Fluidized-bed freezing

Particulate vegetables such as peas are moved along a tunnel on a perforated tray, through a blast of very cold air. They are normally borne along by a cushion of freezing air forced up from below. This ensures that each item is individually quick frozen, so ensuring that the product does not weld together. The process takes about three to eight minutes.

Air-blast freezing

Air-blast freezers are the commonest method of freezing and include: static tunnels where trolleys of boxed product such as beef and cakes are pushed through; solid continuous belt freezers, which are used for fish fillets, patties and pizzas; and spiral belt freezers, which are relatively small and allow the refrigerated air to pass through the open link belt. They are used for pizzas, fish fillets, meals and cakes. Air circulates around the food at temperatures of −30ºC to −40ºC and as with all freezing systems, the freezing time depends on the dimensions of the food. The process usually takes two to three hours.

Plate freezing

This method is used for food packed in flat cartons, for example, fish blocks and meals. The cartons are placed between narrow metal shelves in which a very cold refrigerant circulates, so ensuring freezing. Revolving plate freezers are used, for example, for boil-in-the-bag products. Plate freezing usually takes two to three hours.

Cryogenic freezing

The food is sprayed with, or dipped into, a refrigerant such as liquid nitrogen. It is a very quick technique and, although capital costs are relatively low, high refrigerant costs restrict its use to high-cost, small products such as raspberries and prawns. Studies have shown that after a few months of frozen storage there is no significant difference between liquid nitrogen frozen and air-blast frozen foods.

Pellofreeze system

The "pellofreeze" system is an interesting development in freezing technology, which involves freezing liquids and semi-solids in pellet form. Spinach, cream, orange juice, egg and soup have all been frozen commercially by this method.

FOOD PRESERVATION BY THE USE OF HIGH TEMPERATURES

Heat energy is used to destroy both spoilage and pathogenic organisms and so preserve food. However, heat resistant bacteria, some toxins and spores may survive heat treatment. The number of organisms that survive heating is related to the initial numbers and strain of organism, the time and temperature used, the pH and the presence of protective substances such as proteins and fats. For example, the thermal death point of *E. coli* in cream is 73ºC and in skimmed milk 65ºC. The destruction of micro-organisms depends on their D value, which is the time required at a specific temperature to destroy 90% of the organisms present. Therefore if the D value of an organism is one minute at 75ºC and there are 10^6 organisms present then after one minute there will be 10^5, after two minutes 10^4, three minutes 10^3, etc. The D value of *salmonella spp.*, *S. aureus*, *L. monocytogenes* and *E. coli* is less than ten seconds at 72ºC. Therefore, heating food at 72ºC for one minute will result in reducing one million of any of these bacteria to one. Any equivalent heat energy will produce the same effect. However, the D value of spores of *Clostridium perfringens* and *Bacillus cereus* may be up to 20 minutes at 100ºC and spores will not be destroyed at 72ºC. Examples of equivalent cooking time/temperatures for the destruction of bacteria include: 60ºC for 45 minutes, 65ºC for ten minutes, 70ºC for two minutes, 75ºC for 30 seconds, and 80ºC for six seconds.

This log reduction of bacteria by heating clearly demonstrates the need to minimize the number of pathogenic bacteria in raw food.

Care must always be taken to avoid the recontamination of food after processing, or the multiplication of bacteria not destroyed during heating will occur. This involves the use of suitable packaging, for example, bottles or cans, or storage under conditions that inhibit bacterial growth, usually refrigeration.

Pasteurization

Pasteurization normally involves heating food at a relatively low temperature for a short time. For example, milk may be heated at 72ºC for 15 seconds. The time and temperature combination chosen depends on the particular type of food and must be sufficient to destroy vegetative pathogens and a considerable proportion of spoilage organisms. Toxins and spores generally survive pasteurization and, to avoid the growth of heat-resistant organisms and vegetative bacteria that may be present because of spore germination during heating, refrigerated storage is often essential.

Organisms preferring to multiply at temperatures above 45ºC are known as thermophiles and those capable of withstanding heat treatment, but not necessarily multiplying at the higher temperatures, are known as thermoduric.

The main advantage of pasteurization is that food is rendered safe with the minimum effect on flavour and nutritional value. Foods that are pasteurized include milk, ice-cream, eggs and liquid egg, wines, canned fruit and large cans of ham.

Guidelines to the types of food products stabilized by pasteurization treatments and recommendations for the design of pasteurization processes are provided in "Food Pasteurization Treatments: Technical Manual No. 27. Campden and Chorleywood Research Association, 1992".

Sterilization

Sterilization involves the destruction of all micro-organisms and as this is sometimes difficult to achieve, heat treatment adequate to destroy all viable organisms may be utilized.

In this case food may be considered **commercially sterile**. This means that any organisms remaining after treatment will be of no significance under normal methods of storage. **Low-acid** canned food is given such a treatment. Temperatures required for sterilization normally exceed 100ºC and are usually achieved by means of steam under pressure.

There are several factors affecting the heat-resistance of organisms.

- *Humidity* – the lower the moisture the more resistant the organism, i.e. dry heat is not as lethal as steam at the same temperature.
- *Fat, sugar and proteins* – the presence of these substances usually increases heat resistance.
- *pH* – neutral pH is preferred, acid and **alkaline** conditions increase heat sensitivity. For this reason acid foods such as fruit require a much lower temperature to render them commercially sterile.
- *Chemicals* – the presence of certain chemicals such as nitrites may decrease heat resistance.
- A_w – the heat-resistance of salmonellae and other bacteria increases significantly by reducing the a_w.

The main advantage of sterilization is the prolonged shelf life. The main objections are a lowering of nutritional value, including loss of vitamins, and a marked difference in texture and flavour.

Ultra heat treatment (UHT)

The ultra heat treatment of milk is a result of research carried out to find a product with an extended shelf life without the organoleptic changes caused by sterilization. Milk is heated to a temperature of not <135ºC for one second before filling **aseptically** into sterile containers. This reduces the amount of caramelization and also enhances keeping quality. UHT milk will usually keep for several months without refrigeration.

Ohmic heating

Ohmic heating involves heating food by passing a high-voltage electric current through a liquid and solid type food such as chilli con carne. The **commercially sterile food** is then cooled and packed under aseptic conditions to provide a product that has a prolonged shelf life at ambient temperatures. Products must be capable of being pumped and can have a solids content of up to 60% with particulates having a diameter of about 25mm.

DEHYDRATION AS A MEANS OF PRESERVATION

This method of preservation relies on the fact that all micro-organisms require moisture to facilitate their metabolic reactions. Dehydration reduces the amount of available water, and dried foods normally contain less than 25% moisture with an a_w of less than 0.6. Most bacteria require an a_w of at least 0.95 and very few micro-organisms can exist at levels around 0.6. However, some bacteria form spores that will germinate on the reconstitution of the dried product, for example, *Bacillus cereus*. Gram-negative rods are the most sensitive bacteria to drying and Gram-positive cocci are usually the most resistant, with some able to grow down to an a_w of 0.85. Yeasts and moulds usually grow at lower levels of a_w than do bacteria. Organisms growing under dry conditions are usually referred to as **xerophilic** (dry loving).

The role of dehydration in food preservation became more important in the food industry with the development of packet convenience foods. Mould and enzyme spoilage is also prevented and, provided that food is stored in suitable air-tight packs, it will keep for a considerable period of time.

Sun drying was the earliest method of dehydration and is still practised in hot climates, for example, for drying currants, raisins and figs. Artificial drying is quicker and normally more effective than natural means. Unfortunately, food undergoes irreversible changes to the tissue structure during drying, which affects both texture and flavour.

Methods of drying may be divided into those using hot air, for example, tunnel drying, fluidized-bed drying, roller drying and spray drying, and those using warm air such as accelerated freeze drying. Hot air **denatures** protein; foods are more difficult to reconstitute and it induces greater changes in flavour. However, the choice often depends on the type of foodstuff and the degree of dehydration required. Blanching of vegetables must be carried out before drying to obviate enzyme activity during storage.

In spray drying a solution, paste or slurry is dispersed as small droplets into a stream of hot air. The small droplets result in a rapid loss of moisture and a large proportion of the colour, flavour and nutritive value of the food is maintained. However, because the evaporative cooling effect keeps the temperature of the droplets low, a pasteurization process is usually required prior to spray drying.

In roller drying, a paste of the food is made and fed on to a heated drum from which the moisture is driven off and the dried cake is scraped from the drum. Generally, the quality of the product is not as good as the spray dried product.

For fruits and vegetables a tunnel drier is used. This consists of a tunnel 10 to 15 metres long in which trays of product are passed along. Hot air is blown across the trays and a continuous process leads to a gradual loss of moisture.

Total removal of moisture may not be necessary. Water that does remain forms a strong solution of salts and soluble proteins. This water remaining is not available to micro-organisms and thus the food has a low a_w.

Accelerated freeze drying

Food is frozen quickly and then subjected to a mild heat treatment under vacuum. The ice in the food changes directly to water vapour, which is then removed, a process known as **sublimation**. The cells of the food are substantially less affected than by other methods of drying and the product reconstitutes with very little change.

The advantages and disadvantages of dehydration

The advantages of dehydration over other methods of preservation include:
- the food is lighter and takes up less space; this results in cheaper transport and storage;
- costly insulated or refrigerated distribution vehicles and storage facilities are not required; and
- the relatively long shelf life of the food.

The disadvantages include:
- considerable care is necessary in packaging to eliminate oxygen; if the packaging is damaged or permeable then spoilage will occur and oxidative rancidity may be a problem in foods containing fats;

- non-enzymic browning may occur in foods;
- there may be a significant alteration in flavour and a lowering of nutritional value, especially vitamin C;
- spore-forming bacteria survive and on reconstitution will germinate;
- many foods do not rehydrate very well and may become tough; and
- rehydration damages cell walls, which increases the rate of decomposition.

To minimize chemical changes, particularly browning, in dried foods it is important to:
- keep the moisture content as low as possible;
- minimize the level of reducing sugars;
- use clean blanching water with low levels of leached soluble solids; and
- use sulphur dioxide as a preservative.

All dried food should be stored at low levels of relative humidity and kept dry in order to prevent spoilage, especially mould growth.

CHEMICAL METHODS OF PRESERVATION

A wide range of chemical additives is available for food preservation and may be used to prevent microbial spoilage, chemical deterioration and mould growth. The use of additives is strictly controlled by legislation and maximum permitted levels are usually specified. Ministers are advised by the Food Standards Committee and the Food Additives and Contaminants Committee and are empowered, under the Food Safety Act, 1990, to make regulations and orders to control their use. However, a preference for milder flavours, and in some instances increasing public resistance, has resulted in a reduction in the amount of some preservatives such as nitrite used in food manufacture. Although this is desirable, extreme caution must be exercised if preservatives are being used not only to increase shelf life but also to prevent pathogen multiplication. As the amount of preservative is reduced, satisfactory handling and refrigerated storage are becoming even more important. Initially, therefore, the safety of an altered or new recipe containing reduced preservative must be microbiological challenge tested. (The "new" product is inoculated with appropriate pathogens and then subjected to various temperatures to ascertain the effect on, for example, previous recommendations for cooking and storage.)

Any chemical used in food processing must be used in the best interests of the consumer and not as a device to deceive the consumer or to hide faulty processing and handling techniques. All food additives used must be shown to be safe for the consumer. In the UK a positive list principle is applied, that is, only those compounds specifically listed can be used for food and, in most cases, the maximum amount of material allowed in specific types of food is also specified.

Salt

Salt has been used as a preservative since ancient times. In part, preservation is due to osmosis. When salt is added to food, water passes out of the cells by diffusion to create an equilibrium concentration. This moisture becomes unavailable for micro-organisms. When added to 100g of water, 1.7g of sodium chloride will depress the a_w by 0.01, 3.4g by 0.02, 5.1g by 0.03, etc. The effectiveness depends on the concentration and this is related to many factors including water content, contamination levels, pH, temperature, protein content and the presence of other inhibitory substances. The salt tolerance of micro-organisms is usually decreased by lowering the temperature or the pH. Moulds are less exacting in their water requirements than bacteria and consequently, for effective preservation the available moisture must be less than will permit mould development. Salt may be rubbed into the meat, used as a brine solution, injected into the muscular tissue as a concentrated solution or added directly, for example, in the manufacture of sausages.

Organisms that can grow in high concentrations of salt are known as **halophiles** (salt loving). Those that can withstand high levels but do not grow are termed **haloduric**. Staphylococci will grow in relatively high salt concentrations and are often associated with food poisoning from semi-preserved salted meats.

In preservation the use of salt, with the addition of other chemicals such as sodium nitrate, is termed curing whilst its use for flavour or colouring is termed brining.

The salt used for food preservation must be pure and free from contaminants and extraneous matter, for example, sea salt can often contain a range of mineral substances in addition to sodium chloride.

Nitrates and nitrites

Sodium nitrate ($NaNO_3$) and sodium nitrite ($NaNO_2$) are used in curing meat to stabilize red pigmentation and reduce spoilage. They are also essential for use in such products as pasteurized ham to stop the production of botulinum toxin by preventing the germination of spores. Traditionally, salt and nitrate solutions were injected into the meat, which was then immersed in the brine solution to enable salt tolerant bacteria to convert nitrate to nitrite. However, the current trend is to use nitrites direct as they are much more effective than nitrates. Unfortunately, nitrites react with amines to form nitrosamines, many of which are said to be **carcinogenic**. For this reason the levels allowed are strictly controlled. Both nitrates and nitrites gradually disappear during storage and heating.

The effectiveness of curing salts depends on various factors, including the pH of the meat, the number and types of micro-organisms present and the curing temperature.

Sugar

Sucrose acts in a similar manner to salt but concentrations need to be about six times higher. Moulds and yeasts are less susceptible than bacteria and can withstand up to 60% of sucrose. Organisms that are able to grow in high concentrations of sugar are designated **osmophiles** (loving high osmotic pressures) and those that withstand high levels without multiplying are **osmoduric**. The substitution of sucrose by artificial sweetening agents must be carefully evaluated because of the implications for reduced product stability and increased potential for growth of pathogens.

This method of preservation is commonly used for jam and other preserves, candied fruit and condensed milk. Certain types of cake have increased shelf life due to the effect of sugar.

Benzoic acid/sodium benzoate

This chemical occurs naturally in some foods, for example, cranberries. It is used to inhibit the growth of moulds and yeasts in high-acid foods such as fruit juice, pickles and dressings, but is not generally effective in neutral foods. It acts by inhibiting the cellular uptake of amino acids. Excess benzoate can cause an unpleasant burning taste.

Sulphur dioxide/sulphite

Sulphur dioxide may be used in gaseous or liquid form or as a salt. It is an antioxidant and also inhibits growth of bacteria and moulds. It is utilized in the dehydration of some foods to prevent enzymatic browning. Sulphur dioxide is used in wine, beers, fruit juice and comminuted meat products including sausages, where it is allowed up to 450µg/g. Apart from reducing the growth of spoilage organisms, sulphur dioxide also limits the growth of salmonellae.

Sorbic acid/potassium sorbate

Like benzoate, sorbate is only effective in acid food. It is particularly effective against moulds and yeasts and it is also reported to inhibit the growth of salmonellae, faecal streptococci and staphylococci but not clostridia. It may be used in hard cheese, as lactic acid bacteria are unaffected. It is also used in bread, jam, syrups and cakes.

Fermentation

Fermentation involves the anaerobic oxidation of carbohydrates by micro-organisms and the production of acid which lowers the pH and preserves the food. For example, starter cultures of lactic acid bacteria ferment glucose in milk to produce lactic acid and yogurt. Yeasts ferment carbohydrates to produce ethanol (beers and wines) and moulds are involved in the production of blue cheeses and soy sauce. Fermented foods also include butter, sauerkraut, bread, olives, salami and pepperoni. Sodium chloride may be added to inhibit the growth of competing spoilage organisms. Sugar may be added, especially to meat, to increase the amount of lactic acid produced.

The production of vinegar (acetic acid) involves the fermentation of sugar into alcohol by yeast and the fermentation of the alcohol into acetic acid by vinegar bacteria.

The pH limit for bacterial growth is around 4.5 for *Cl. perfringens, Cl. botulinum* and *Staphylococcus aureus*; 4.2 for most *Bacillus spp., Salmonella spp.* and *E. coli*, and below 4.0 for *Lactobacillus*. Some salmonella can grow as low as 3.8. However, many yeasts and moulds will grow below a pH of 3.0.

Acidification or pickling

This process involves lowering the pH of food by using an organic acid, for example, citric acid or acetic acid, i.e. vinegar. It is essential that the acidification process is controlled so that the pH of each part of the product is below 4.5. Acetic acid is used in the production of mayonnaise and salad dressings. Sufficient time must be allowed for any pathogens, such as salmonella, to die off after the acid has been added.

Sodium and calcium propionate

Propionates are active in low-acid foods and very useful to prohibit mould growth. They are used in bread, cakes, cheese, grain and jellies.

Antibiotics

These chemicals have a preservative role in addition to their normal function. Their use is strictly controlled by regulations to avoid the build-up of resistance by pathogenic organisms. An example of an antibiotic is nisin, sometimes used in cheese and canned foods. Nisin is heat-resistant but is destroyed in the stomach by trypsin and should not cause problems of pathogen drug resistance.

PHYSICAL METHODS OF PRESERVATION

Controlled atmospheres
Modified atmosphere packaging

One of the simplest food processing/preservation methods using a physical method is to change the atmosphere around the food. This is termed **modified atmosphere packaging** (MAP). The air around a product is modified to contain a different proportion of the gases normally present, for example, lower levels of oxygen and higher levels of nitrogen and carbon dioxide, and will slow down the growth of many spoilage organisms thus giving an extended shelf life to the product. MAP should also be combined with correct chilled temperature control in order to further guarantee the control of microbial proliferation. Concentrations of 10% carbon dioxide may be used to keep chilled beef free from spoilage for up to 70 days. Comprehensive guidelines for the good manufacturing and handling of modified atmosphere packed food products have been produced.*

The restriction of oxygen

The development of oxidative rancidity and the growth of strict aerobes such as moulds can be prevented by vacuum packing. However, removal of oxygen allows the growth of anaerobes such as *Clostridium perfringens*, and sufficient oxygen normally remains in so-called vacuum packs of meat to facilitate the growth of some aerobes.

Regular sampling shows that vacuum packs of ham are generally no better bacteriologically, and may even be worse, than ham sliced off the bone. Vacuum packs of cooked meat must be stored under refrigeration to achieve a reasonable shelf life. A vacuum pack of cooked ham with an initial count of 10^2 organisms per gram will have a count of 10^6 after 39 days if stored at 2°C; 20 days at 5°C; and seven days at 15°C.

Smoking

Smoking is used primarily with meat and fish, after brining or pickling, by suspending the food over smouldering hardwoods, such as oak and ash, which should be free from chemical preservatives. The principle purpose nowadays is to enhance the flavour. The smoking process also has some dehydrating effect and, when properly controlled, there may be some preserving action due to the presence of bactericidal chemicals such as phenols, alcohols and aldehydes, which are absorbed by the food. Most non-sporing bacteria will be destroyed but moulds and *Cl. botulinum* type E may survive, especially if there is a low salt concentration. Smoked products should, therefore, always be stored under refrigeration, preferably below 3°C to prevent toxin formation if *Cl. botulinum* is present. Poor control of the process may allow incubation and hence an increase in numbers of spoilage organisms.

Another way of adding a smoke flavour to the food is to use a liquid, produced by trapping smoke in water and spraying this on to the food. The preserving effect of this process is limited and no dehydration of the food takes place.

Food irradiation

There has been a considerable debate regarding the use of irradiation as a method of food preservation. It involves subjecting the food to a dose of ionizing radiation, for example, gamma rays emitted from an isotope such as cobalt 60. (X-rays with energies up to 5MeV (million electron volts) or accelerated electrons with energies up to 10MeV could also be

---

*Source: Technical Manual No. 34, 1992. Campden and Chorleywood Research Association.

used.) Several countries irradiate food, and from the scientific studies carried out it seems to present an effective and safe method of extending the shelf life of food. It destroys parasites, insects including eggs, bacteria, moulds and yeasts thus reducing the risk of food spoilage and foodborne illness. However, microbial spores and toxins remain unaffected at the levels used. Foods most commonly irradiated include chicken, fish, onions, potatoes, spices and strawberries. The irradiation of burgers started in the USA in 2002.

One particular advantage of food irradiation as a food preservation method is that it introduces virtually no temperature rise in the treated product and is often referred to as a "cold pasteurization process". Fish, fruits and vegetables remain "fresh" and unchanged. Irradiation of packaged food is possible and may be particularly important in areas where hygiene is difficult to maintain or control. Frozen food can be treated without thawing but higher doses are required. However, irradiation must never be considered as a substitute for good hygiene practices. Irradiated foods must be handled and stored with the same hygiene considerations as non-irradiated foods.

The dose of irradiation used (which is measured in Grays (Gy) or kiloGray (kGy) is a measure of absorbed energy by the food. One Gy is equivalent to one joule per kilogram. (1Gy=100 rads.) It should be noted that doses of around 5Gy can be lethal to man. Irradiation may be used as part of a continuous process or for controlled batches.

Spores are least sensitive to radiation, then viruses, Gram-positive bacteria and moulds with Gram-negative bacteria being the most sensitive. Type E botulinum spores have survived the irradiation of fish and produced toxin more rapidly than those in non-irradiated fish.

Food irradiation as a method of preservation has the same limitations as many other food preservation technologies. Vitamins may be destroyed and enzymes are not deactivated. Other disadvantages include the encouragement of oxidative rancidity in fatty foods, the possible production of free radicals in food, which stimulate a range of chemical reactions, and the softening of some fruit.

The dose of irradiation received varies throughout the product depending on thickness, orientation and packaging. It is important to ensure that the minimum dose necessary to achieve the end result is delivered to all parts of the food but that the maximum average dose of 10kGy is not exceeded in order to limit the adverse effects.

DOSE RANGES NEEDED FOR EFFECTIVE TREATMENT

	kGy
Inhibition of sprouting in onions and potatoes	0.03-0.1
Sterilization of insects and parasites	0.03-0.2
Killing insects and parasites	0.05-5.0
Reduction by 10^6 of vegetative bacteria	1-10
Complete sterilization of food (would result in unacceptable off-odours and flavours together with a cooked texture)	20-40

The following matters need to be considered when determining the acceptability of the food irradiation process.

Production of toxic agents

Irradiation of food will result in some chemical reactions in the food but it is alleged that no toxigenic compounds have been demonstrated in treated foods. In 2002 the Centre for

Food Safety and the Public Citizen urged the USA Food and Drugs Authority to halt the approval of irradiated food until it could be proved that a new class of chemical, cyclobutanones, formed during irradiation were harmless.

Microbiological interactions

A high dose of irradiation sufficient to achieve commercial sterility should cause no health problems, however, at lower dose levels when some micro-organisms survive, problems may arise. An increase in radiation-resistant organisms will occur but normally only on repeated exposure to irradiation.

A second problem in the use of sub-lethal treatments is that they might result in shifts in microbial flora and the possibility of spoilage being unrecognized through odour before toxins reach dangerous levels. In practice this does not seem to occur and irradiated foods appear to follow the normal spoilage pattern.

Possible carcinogenicity

There is no proof that food that has been irradiated shows any carcinogenic activity.

Induced radioactivity

Only at very high doses could the food become radioactive. At the levels used for food processing there is no danger of the food developing radioactivity.

Nutritional quality

The changes produced in the nutrient content of food will depend on many factors including the type of food, the radiation dose and the processing temperature. Generally, vitamins A and E are the most susceptible to irradiation damage. Although the shelf life of food is prolonged, some bacteria survive and irradiated food must be stored at correct temperatures. Irradiation cannot replace refrigeration.

THE MAIN USES OF IRRADIATION

- To destroy a percentage of spoilage and pathogenic bacteria on food.
- To control mould growth particularly on soft fruits such as strawberries.
- To disinfest fruit and grain (limited to a depth of around 8cm).
- To inhibit the sprouting of potatoes, onions or garlic.
- To inactivate parasites such as *Trichina spiralis*.
- To delay the ripening of fruit such as bananas.

The main disadvantages claimed by opponents of irradiation include:
- inadequate information exists to assess the long-term effects of consuming irradiated food;
- some reduction in the nutritional value of food, particularly vitamin loss;
- the production of undesirable flavour changes in some foods, for example, eggs and some dairy products. Some foods are softened to a mulch;
- food of poor bacteriological quality and shelf life may be irradiated to disguise this fact;
- hygiene standards will be reduced as some manufacturers will rely on irradiation to produce an acceptable final product;

- spoilage bacteria are more easily destroyed than pathogens and reducing competition may allow more prolific growth of pathogens in certain circumstances; and
- more dangerous mutants of pathogens may be formed.

The Food (Control of Irradiation) Regulations, 1990 authorized the irradiation of food in the UK, subject to licensing controls on treatment facilities administered by Central Government. The Food Labelling (Amendment) (Irradiated Food) Regulations, 1990 require food that has been irradiated to be so labelled.

In the UK Isotron PLC has a licence that allows the irradiation of a seasoning blend to a maximum overall average dose of 10kGy. However, surveys undertaken by the Scottish Universities Research and Reactor Centre at East Kilbride and Suffolk Trading Standards have discovered several irradiated foods on sale, for example, seven out of 18 samples of prawns.

Further information on the irradiation of food is available in the WHO Technical Report Series no. 890 (1999) High-dose Irradiation: Wholesomeness of food irradiated with doses above 10kGy. (WHO, Geneva).

The symbol indicating food has been irradiated.

ANGSTRÖM UNITS

One angström unit = 1/10,000,000mm

Part of the electromagnetic spectrum of interest to the food industry.

10 Personal hygiene

High standards of personal hygiene are essential to reduce the risk of foodborne illness and a prerequisite to the implementation of HACCP. The risks associated with food handlers contaminating high-risk foods with low-dose organisms such as *E. coli* O157, norovirus, hepatitis C, *S.* Typhi, *Shigella spp.* and campylobacter should not be underestimated. Personal hygiene of staff is a management responsibility and must be dealt with proactively. It must not be assumed that good personal hygiene is common sense that will be implemented automatically by all staff.

Documented procedures and rules for staff hygiene should be part of the food safety policy and brought to the attention of new employees during induction training. Staff must then be trained, motivated, supervised and monitored to ensure the continuous implementation of good hygiene practices.

The law, and the industry guides to good hygiene practice, require every person working in food handling areas to maintain high standards of cleanliness and wear suitable, clean and, where appropriate, protective clothing. Personal cleanliness may be considered to include good hygiene practices necessary to prevent the contamination of food. This interpretation prohibits smoking, spitting, wearing excessive jewellery and requires cuts to be covered with waterproof dressings.

Food handlers should be in good health, have good eyesight and be able to read, especially if they are responsible for checking thermometers, instructions on labels, date coding and the condition of food. Furthermore, clean staff wearing clean protective clothing and exhibiting good practices will promote customer confidence and increased business.

All visitors to food premises, including managers from head office, enforcement officers and maintenance operatives, should observe the same hygiene rules as staff, and wear appropriate protective clothing. Major customers undertaking inspections will expect strict adherence to company rules.

STAFF SELECTION

The foundations of high standards of hygiene are built by the employment of the right calibre of staff and the provision of satisfactory training. The first stage in successful recruitment is the formulation of job specifications and descriptions that clearly define the level of responsibility and skill required. A reasonable salary commensurate with these requirements must then be offered. The following points should be considered when interviewing and appointing staff.

Personal appearance and attitude
Food handlers must have:
- a clean, neat and tidy appearance;
- an absence of skin infections;
- good dental hygiene;
- clean hands with short fingernails and no evidence of nail biting;
- an absence of excessive personal jewellery or make up;
- clean shoes, suitable for the work area; and

♦ a belief in the need for hygiene.

Persons who cannot take the trouble to present a good appearance at interviews will not respond to hygiene disciplines imposed in the food environment. They should show enthusiasm and have a suitable attitude relevant to the position for which they are applying. Hygiene awareness and a recognized qualification, especially for supervisors, is also advantageous.

Medical screening

It is advisable to ascertain some details of an applicant's medical history, particularly relating to foodborne illnesses, **food poisoning** or persistent diarrhoea. Applicants should have to complete a medical questionnaire (see Appendix II) and in large companies may be interviewed by a doctor or nurse. When necessary, persons should be asked to submit a faecal or blood specimen, although this is not normally recommended for routine screening of all applicants. Chest X-rays may also be required.

POST APPOINTMENT

The role of management

Before any staff are allowed to commence work they should be advised, in writing, of their personal obligations in respect of hygiene. They should be informed of the training they will be expected to undertake, including attendance on courses. It should be made clear that a breach of hygiene rules could result in disciplinary action. It is particularly important to stress that they have a legal obligation to notify their supervisor if they are suffering from any condition that exposes food to risk. Food handlers should be aware that if they are suspected of having, or have been in, close contact with a person who has had food poisoning, they would be expected to cooperate by providing faecal specimens. Interference with the personal freedom of the individual is likely to arouse resistance and possibly resentment, unless staff appreciate the reasons for particular restrictions.

After attending hygiene training, food handlers must be encouraged to implement what they have been taught. Effective instruction, supervision and competency-based testing will be essential to maintain high standards of hygiene. Continuous coaching is important to keep staff up to date and to reinforce the need for high standards.

Management must lead by example and provide a hygienic working environment together with adequate resources to facilitate the achievement of high standards, for example, suitable and sufficient protective clothing, constant replenishment of liquid soap, paper towels and **cleaning** materials with adequate cleaning time. A wash-hand basin at the entrance of food rooms, together with close supervision, is the best way of ensuring staff always wash their hands before handling food.

HAND HYGIENE

The role of the hands in transferring **pathogens** to high-risk food, especially from **carriers**, is probably underestimated in the UK. In the USA, infected food handlers are much more commonly implicated in outbreaks of foodborne illness, especially from campylobacter (45%), viral **gastroenteritis**, hepatitis, shigella (94%) and *Staphylococcus aureus* (71%). It is considered that up to 25% of foodborne illness in the USA may be due to improper handwashing. Barry Michaels of Georgia Pacific Corporation reported that his review of 260 outbreaks of foodborne illness attributed 40% to food handlers exhibiting symptoms of infectious intestinal disease or who were **asymptomatic**.

Furthermore, research undertaken in care homes demonstrated that when a programme of vigorous handwashing was implemented there was a 33% reduction in common colds and a 50% reduction in **cases** of diarrhoea. The US Centre for Disease Control and Prevention advise that "handwashing" is the single most important means of preventing the spread of infection. However, there is so much confusion and misinformation regarding handwashing, that even if it is undertaken, it may not be very effective.

The skin

To understand handwashing we need to understand the physiology of the skin. The skin consists of three layers, the epidermis, the dermis and the subcutaneous tissue. The epidermis is made up of millions of microscopic cells known as skin scales and contains many cracks and crevices that provide harbourage for resident and transient **micro-organisms**. A layer of dead skin cells is lost around every two days, although injuries and practices such as scrubbing with a hard nailbrush will result in a much more rapid loss of skin layers and eventually bleeding. A water repellent film produced by the sebaceous glands protects the outer layer of the skin. Removal of this layer by excessive handwashing or use of alcohol **disinfectants** is likely to cause irritation, dryness and roughness and increase the risk of dermatitis. All of these conditions deter people from washing their hands.

Resident micro-organisms such as *Staphylococcus epidermis, Staphylococcus saprophyticus, Enterobacter spp.* and **yeasts** are found throughout the epidermal layers and are impossible to remove, even with the most rigorous scrubbing. The presence of resident micro-organisms is important in

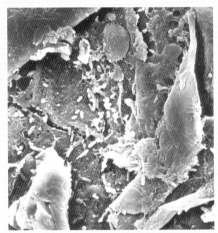

Skin magnification by around 3400 showing staphlococci, streptococci and bacilli, for example, *E. coli* or possibly salmonella.

preventing pathogens from colonizing the skin. We, therefore, would not want to **sterilize** the skin, even if it was possible. The only pathogen included as a resident micro-organism on healthy skin is *Staphylococcus aureus*, although it is only likely to be present in relatively low numbers. *Staphylococcus aureus* usually needs to grow in food to around 10^6 organisms per gram to cause illness by producing sufficient amounts of **toxin**.

Transient micro-organisms, including **bacteria**, viruses, **moulds, parasites** and yeasts, are those organisms that "hitch a temporary ride" on and in the epidermis. They do not usually live in the skin and the types present depend on our environment and what we touch. This means that food handlers touching contaminated poultry and raw meat are likely to have salmonella, campylobacter, *Staphylococcus aureus* and *E. coli* on their hands. High numbers of transient pathogens on the hands, fingertips and nails are also likely from faecal contamination after using the toilet, changing nappies or cleaning up after pets or ill people. High levels of

Staphylococcus aureus are likely after touching infected cuts or boils or changing a dressing. Food handlers are also much more likely to have *Listeria spp.* on their hands than, for example, office workers. Transient micro-organisms may survive on the skin from a few minutes to several hours; so, effective, regular handwashing is essential.

Handwashing strategies

Probably the most important thing is to ensure food handlers wash their hands as often as necessary. This involves explaining the hazards and demonstrating the correct procedure, which should be documented in the food safety policy. Staff will require supervision and need to be motivated to follow the procedure every time they wash their hands. Managers should lead by example.

Strategies to secure compliance will need to be developed. Handwashing should be monitored and may even be documented and verified, for example, by using hand swabs, agar plates or a fluorescent dye and ultraviolet light. Posters and signage will aid reinforcement. Disciplinary action may be necessary against employees who fail to wash their hands.

The learning process to change behaviour goes through the following stages:

- unconscious incompetence;
- conscious incompetence;
- conscious competence; and
- unconscious competence.

(Abraham Maslow)

Requirements for handwashing

Handwashing is best carried out at a designated hand-wash station and will require:

Dirty and obstructed washhand basin in cloakroom.

- an accessible wash-hand basin or trough, preferably with non-hand operable taps;
- a supply of clean, warm (35°C to 45°C) running water (4 to 8 litres per minute);
- suitable soap, preferably liquid, non-irritating in a replaceable cartridge;
- paper towels, preferably using an infrared dispenser; and
- a foot-operated or open-topped bin for used towels, which is emptied as often as necessary.

An effective management system must be in place to ensure the facilities are kept clean and well maintained and soap and paper towels are always available.

Effective handwashing

The objective of most food handlers when washing their hands should be to reduce the number of transient pathogens to a safe level. The method used will therefore depend on the number of pathogens likely to be present and this will usually depend on the last activity of the food handler. Activities likely to result in large numbers of pathogens should be followed by a double wash.

These activities include:
- using the toilet when there is a risk of faecal contamination, especially on the fingertips (toilet paper is porous and tends to slip). It should also be remembered that several cultures do not use toilet paper;
- cleaning up, for example, vomit from an ill person or a baby's nappy;
- changing, putting on a dressing or touching an infective cut, wound or boil;
- after handling raw poultry, meat or vegetables before handling ready-to-eat food;
- cleaning up animal faeces, for example, a guide or guard dog; and
- entering the food room at the start of the day or after taking a break.

The double wash procedure

Effective handwashing relies on friction and dilution to remove transient pathogens to a safe level. The first stage of the double wash requires the use of a clean, soft-bristled nailbrush. Running warm water is required to wet the hands and the nailbrush. The temperature of the water is not critical, although a temperature of around 35°C to 45°C is recommended, as no one likes washing hands in cold water (<20°C) or hot water (>49°C).

Around 3 to 5ml of a good quality, unperfumed liquid soap should be applied to the nailbrush. Bactericidal soaps are not necessary, as in the majority of food handling situations, aseptic conditions are not achievable or required. However, because some bacteria, such as *Pseudomonas spp.*, may contaminate and multiply in liquid soap, only cartridge replacement containers should be used.

Brushing under running water should produce a good lather and will remove contaminants from the fingertips and under the fingernails. Continue brushing until the lather disappears; this should result in a reduction of bacteria on the fingertips of around 1000 to one. (Dr P. Snyder, Jr.) The nailbrush should then be stored bristles up. In practice, cross-contamination from one food handler to another via the nailbrush is unlikely. However, nailbrushes should be changed at least daily. Cleaning and disinfection of nailbrushes may be achieved in a dishwasher, or by the use of chemical disinfectants followed by drying.

The single wash procedure

The second stage of the double wash is also used as a single wash on all other occasions when washing the hands. Around 3 to 5ml of liquid soap is added to wet hands. A good lather should be produced by vigorously rubbing the hands and fingertips together. In between the fingers, around the thumbs, the forearms and wrists should be thoroughly lathered for around 20 seconds. This time may be longer if the hands are heavily soiled, for example, with blood.

Highfield Poster,
How to wash your hands.

The hands should be rinsed thoroughly in warm running water (around 4 to 8 litres per minute) until all lather, dirt and contaminants are removed. The rinsing process is critical as it provides dilution and friction. A further reduction of bacteria by around 100 to one will result. The hands and arms should be completely dried, preferably using paper towels. Drying is an integral part of effective handwashing as the friction removes more transient bacteria, around 100 to one. The paper towel should, if necessary, be used to turn off the tap. It should then be placed in a foot operated, or open, waste container. The container or lid must not be touched with clean hands.

A single wash procedure should normally be used in the following situations:

- after combing or touching the hair, nose, mouth or ears;
- after eating, smoking, coughing or blowing the nose;
- after handling external packaging or flowers;
- after handling waste food or refuse;
- after cleaning, or handling dirty cloths, crockery etc;
- after shaking hands;
- after handling money;
- after touching shoes, the floor or other dirty surfaces;
- before putting on and taking off gloves; and
- as often as necessary throughout the day to remove soil and prevent product contamination.

Warm air dryers

Warm air driers are not recommended for food handlers, as they often take too long to completely dry the hands. This results in food handlers walking away, shaking wet hands or drying their hands on their protective clothing. Driers are unlikely to reduce the levels of bacteria remaining on the hands and some driers may even increase the number of bacteria after washing. Residual moisture on the hands after "drying" plays a significant part in the transfer of bacteria and viruses. If roller towels are used they must not result in contamination of the hands. Only clean sections of the towel should be used and clean and dirty parts of the towel must be segregated. Care must be taken to ensure the towel is changed before it runs out.

A non-hand contact paper towel dispenser and a knee operated handwashing station. (Courtesy of Holchem Laboratories.)

HYGIENE *for* MANAGEMENT *Personal hygiene*

Surfaces that have been found to be contaminated with coliforms and *E. coli* include the washhand basin, the tap and the paper towel dispenser. Occasionally the door handle may be contaminated. Disinfection of these surfaces is therefore necessary during cleaning. Non-hand operable taps and paper towel dispensers are recommended.

Common reasons for failure in handwashing:
- failure to train and motivate food handlers to wash their hands;
- failure to provide adequate facilities, soap or drying materials;
- poor accessibility of wash-hand basins;
- inadequate supervision, reinforcement or management commitment;
- poor quality soap, which does not remove fat from hands or is affected by hard water salts and fails to lather;
- lack of notices/posters regarding handwashing;
- lack of an effective system for handwashing;
- inadequate time allowed;
- extremely high handwashing frequency disrupting the skin barrier properties resulting in skin cracking and sores, especially in winter;
- high staff turnover;
- an unpleasant experience, including use of scalding or cold water, or the use of a poor quality soap with no emollient, which dries up the skin and increases the risk of dermatitis. Some bactericidal soaps also increase the risk of dermatitis;

AREAS COMMONLY MISSED DURING HANDWASHING

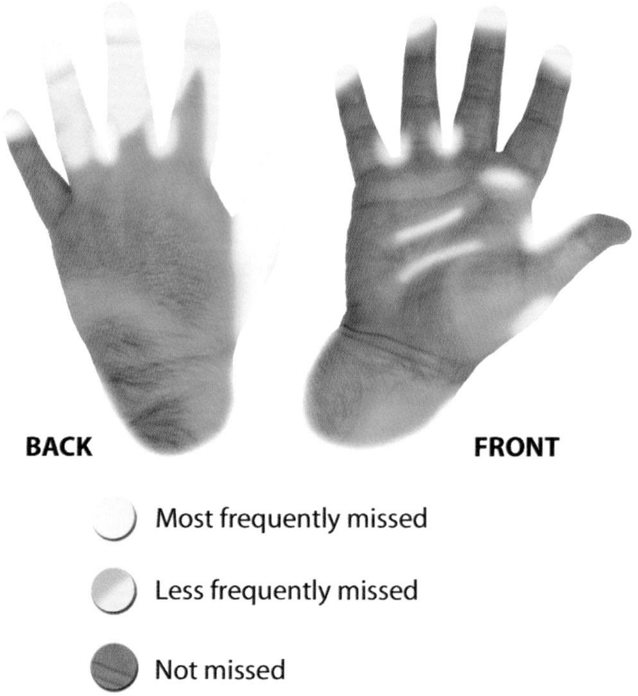

BACK **FRONT**

○ Most frequently missed

◐ Less frequently missed

● Not missed

- failure to remove all soap residue from the hands during rinsing, resulting in skin irritation;
- overuse of a stiff nailbrush resulting in damaged skin and/or inflammation; and
- long or false fingernails.

Hand contact during high-risk food service should be minimized and several procedures are used including tongs, utensils, plastic bags and gloves. However, cross-contamination can still occur if good practices are not observed.

The use of gloves for handling high-risk food

The use of disposable vinyl, latex and non-latex gloves has not been proved to be a safer method of handling food compared to food handlers who use effective handwashing techniques. Opponents of gloves claim they give a false sense of security, become contaminated if hands are not washed prior to putting them on and can result in cross-contamination from raw to high-risk food in the same way as hands. Furthermore, a significant number of gloves have defects and pinholes, which enable bacteria from the hands to pass through the gloves. Latex gloves may produce an **allergic** reaction in some people. The hand environment created by wearing gloves provides ideal conditions for the growth of pathogens such as *Staphylococcus aureus*. Punctures in the gloves during use may result in contamination of the high-risk food with large numbers of pathogens.

Proponents of gloves claim that provided hands are washed before putting on the gloves and they are disposed of frequently, they minimize the risk of contamination of high-risk food, especially if effective handwashing is not being achieved. It is also claimed that food handlers with gloves are more aware that they are handling high-risk food and are less likely to scratch their head, pick their nose or carry out other bad hygiene practices. Hands should also be washed when gloves are removed as significant multiplication of pathogens may have occurred. The use of suitable gloves is recommended, for example, when handling high-risk foods for vulnerable groups or when effective handwashing is not possible.

If gloves are used a glove policy should be provided. Food handlers should be trained on how and when to put on gloves and how they should be used to prevent cross-contamination. Clear instructions regarding the changing and disposal of damaged and contaminated gloves should be given. Effective monitoring and supervision of glove use, including checking for dermatitis, is essential.

Hand disinfection

The use of bactericidal soap or disinfectants such as alcohol and chlorhexidine solutions rarely produces significant improvements in the removal of transient pathogens compared to the double wash technique. Furthermore, they tend to be less pleasant to use, dry and irritate the skin and increase the risk of dermatitis. However, formulations are improving all the time and the use of hand disinfectants, with emollients to prevent drying of the skin, may be appropriate if handwashing by the double wash method cannot be undertaken and/or aseptic food production is required. It is always preferable to wash hands prior to using a hand disinfectant although some users claim excellent results from some alcohol base gels on hands that are not visibly soiled.

Skin lotions

Skin lotions are not usually recommended, as some support bacterial growth, and result in product contamination. However, in certain situations such as very frequent handwashing, an antimicrobial-containing skin lotion may assist in maintaining skin health and minimizing drying and/or cracking.

For more information on effective handwashing check www.hi-tm.com and www.gphealthsmart.com

Cuts, boils, septic spots and skin infections

Employees with boils and septic cuts should be excluded from ready-to-eat food handling areas, as such lesions contain *Staphylococcus aureus*. Uninfected wounds should be completely protected by a conspicuously coloured waterproof dressing. Cuts on hands may need the extra protection of waterproof fingerstalls. Waterproof dressings are necessary to prevent blood and bacteria from the cut contaminating the food and also to prevent bacteria from food, especially raw meat or fish, making the cut septic. Furthermore, waterproof dressings do not collect grease and dirt. It is preferable that green- or blue-coloured detectable plasters be used, to improve their visibility in food should they become detached. Metal strips incorporated in dressings assist detection where metal detectors are in use. Loss of dressings must be reported immediately. Perforated plasters are not recommended. Staff who report for work wearing unacceptable dressings must have them changed before they enter a food room or commence food handling duties.

Brightly coloured plasters should be used to cover cuts

The hair

Hair is constantly falling out and, along with dandruff, can result in contamination of food. As the scalp often contains pathogenic organisms such as *Staphylococcus aureus*, steps must be taken to prevent contamination from this source. The hair should be shampooed regularly and completely enclosed by suitable head covering. Hairnets (blue) worn under turbans, helmets and hats are recommended in very high-risk food handling areas, especially manufacturing. Combing of hair and adjustment of head covering should only take place in cloakrooms and should not be carried out whilst wearing protective clothing, as hairs may end up on the shoulders and then in the product. Hairgrips and clips must not be worn. After having a haircut staff should wash their hair before handling food or working in food rooms.

The nose, mouth and ears

Up to 40% of adults carry *Staphylococcus aureus* in the nose and 15% on their hands. Coughs and sneezes can carry droplet infection for a considerable distance and persons with bad colds should not handle open food. The hands should be washed after blowing the nose and soiled handkerchiefs should not be used; single-use paper handkerchiefs are preferable. The mouth may harbour staphylococci and food handlers should not eat sweets whilst working. Spitting is aesthetically unacceptable, can result in food contamination from staphylococci or streptococci and would result in enforcement action. Staff with discharges from the ears or eyes must not handle food.

Smoking

Smoking must be prohibited in rooms containing open food or whilst handling open food. This is not only because ash or cigarette ends may find their way into open food but also because:
- it encourages coughing;
- it may result in an unsatisfactory working atmosphere for non-smokers;
- there is a risk of contaminating food from fingers touching the lips while smoking; and
- cigarette ends, contaminated with saliva, are placed on work benches.

Legible notices should be clearly displayed, emphasizing the requirement not to smoke in food rooms.

Protective clothing

The proprietor must ensure that food handlers, visitors and maintenance personnel wear clean, washable (withstand up to 85°C), light-coloured, durable protective clothing, without external pockets. Protective garments should be appropriate for the work being carried out, should completely cover ordinary clothing and should not be removed whilst food handling. Jumper and shirt sleeves must not protrude and if short-sleeved overalls are worn, only clean forearms must be visible. Press studs are preferred to buttons as they are less likely to become detached. Clean overalls must be available as needed and stored away from contaminants. New

Don't play games in protective clothing during breaks.

products and practices may require a reappraisal of protective clothing provided.

Staff must be aware that protective clothing is worn primarily to protect the food from risk of contamination and not to keep their own clothes clean. Dust, pet hairs and woollen fibres are just a few of the contaminants carried on ordinary clothing. Protective clothing must not be worn outside food premises or for travelling to and from work.

When selecting protective clothing the manager must consider:
- the duties of the wearer and which parts of the body should be covered;
- how the garments are fastened;
- the colour and type of material;
- the smartness and fit (to improve customer image and generate pride in the wearer); and
- whether to purchase or use a laundry rental service.

Outdoor clothing and personal effects must not be brought into food rooms, unless stored in suitable lockers. Lockers should be adequately ventilated to keep them dry, eliminate odours and prevent mould growth. Cloakrooms which are well-ventilated and cleaned regularly are preferred.

Aprons, if worn, should be suitable for the particular operation and capable of being thoroughly cleansed. Disposable aprons may be obtained and should be disposed of after each use. The use of coloured aprons, to aid detection if damaged, is recommended. When used in factories, aprons are normally impervious. Aprons that are torn or have badly worn surfaces, which render them incapable of being cleaned, should be discarded. Facilities should be provided for cleaning aprons at various times during production and at the end of each working day. Cleansable hooks should also be provided for hanging up clean aprons.

Boots may be provided for wearing in wet areas. They should be anti-slip, unlined and easy to clean. Suitable facilities should be provided for cleaning boots, and storing cleaned boots.

Heavy duty rubber gloves may be worn for various jobs, for example, on inspection belts in factories or for cleaning operations. Gloves should be maintained in a clean condition and torn gloves should be discarded. If in contact with food, gloves should be of a different colour so that any detached pieces can be easily detected. After use, gloves should be thoroughly cleaned and dried. Only plain, unperfumed talc should be provided for use with gloves. The inside of all gloves provides a warm, moist environment ideal for the multiplication of bacteria, and gloves should not be left on food-contact surfaces during, for example, breaks.

Jewellery and perfume

Food handlers should not wear earrings, watches, jewelled rings or brooches, which harbour dirt and bacteria. Furthermore, stones and small pieces of metal may end up in the food and result in a customer complaint. Food handlers should not wear strong-smelling perfume or aftershave as they may taint foods, especially food with a high fat content.

Practices

Bad habits are not easily broken and if they are exhibited by a food handler, and present a risk of contaminating food, careful and conscientious control is required. Common bad practices include:
- wetting fingers to open bags or to pick up sheets of wrapping paper;
- picking the nose;
- scratching the head or spots;

- tasting food with an unwashed spoon;
- coughing and sneezing on to hands and handling food without first washing;
- using a food sink for handwashing;
- using a wash-hand basin to rinse utensils;
- handling the inner parts of crockery or glasses; and
- chewing gum, eating food or sweets in food rooms other than dining areas.

Managers must ensure that staff are not only familiar with the dos and don'ts of food hygiene but that they are always observed.

Exclusion of food handlers

Management should be aware that even with the strictest application of health standards, a healthy, symptom-free employee may be excreting pathogenic bacteria or viruses. Furthermore, laboratory tests cannot be relied on to detect small numbers of pathogens even though they are being excreted. Consequently, it must be emphasized that high standards of hygiene are essential to prevent the contamination of food, rather than a laboratory report indicating the absence of pathogens from a faecal specimen.

Food handlers must advise management if they know or suspect they are suffering from, or are a carrier of, a foodborne illness, for example, if they have diarrhoea or vomiting or because of close contact with a confirmed case or have consumed a meal known to have caused illness. They must also report infected wounds, skin infections and sores. Because absenteeism is undesirable and management appreciates continuous attendance, staff must be well motivated to report ailments that they believe may result in a salary reduction or even a loss of their job. All such reports should be noted on medical records. Proprietors have a legal responsibility to take action to avoid any risk of food contamination and this is likely to result in exclusion of ill food handlers from work until medical clearance is obtained. New staff should be given written instructions of their responsibilities prior to commencing work. Persons who are subsequently confirmed as excreting food poisoning organisms must not be allowed to engage in food handling until they present no risk to food safety.

Persons returning from holidays abroad, particularly from countries with warm climates and suspect sanitation, should complete a short medical questionnaire. Even if they have recovered from symptoms of diarrhoea or vomiting experienced on holiday, they should be excluded from food handling until they have provided at least one negative faecal specimen.

Persons suspected of suffering from food poisoning, typhoid, paratyphoid or dysentery can be excluded from food handling by the local authority and compensation paid to cover loss of wages. Several other conditions such as eczema and psoriasis, which are often associated with secondary infection, boils and septic cuts, respiratory tract infections from heavy colds to chronic bronchitis, infection of the eyes, recurrent discharge from the ears and dental sepsis or purulent gingivitis may also require the suspension of food handlers until successfully treated.

In 1995 guidance from the Department of Health, "Food handlers: Fitness to work", stressed that the greatest risk of pathogenic contamination of food is from persons with diarrhoea and vomiting and that good hygienic practices of food handlers are essential to prevent such contamination. The guidance confirmed that food handlers should be considered fit to return to work, following most gastrointestinal infections, if they had not suffered with diarrhoea or vomiting for 48 hours, once any treatment has ceased, and good hygiene, particularly handwashing, is observed.

Unfortunately, a significant number of food handlers do not wash their hands and a larger number do not wash their hands properly. Given this situation a risk assessment may suggest that it is unsafe to allow known excretors of foodborne pathogens to handle ready-to-eat food, even if they have been symptom free for 48 hours. Managers should also take into account the likely reaction of customers finding out that, for example, salmonella carriers are involved in the preparation of high-risk food. Furthermore, in the event of a food poisoning outbreak, the due-diligence defence may be more difficult to argue, if known carriers are employed.

The advice regarding typhoid carriers is even more worrying given that there may be an 8% risk that after six consecutive negative specimens the next specimen will be positive. Once again a risk assessment, based on current scientific knowledge, may suggest that a typhoid carrier should never be allowed to work in a high-risk food handling situation.

EXCLUSION OF FOOD HANDLERS PREPARING/SERVING UNWRAPPED READY-TO-EAT FOODS

(Recommendations of the PHLS Salmonella Sub-committee 1995)*

	Criteria for clearance to return to work	
	Case when symptom-free	Symptomless contact
Aeromonas spp.	48 hours after first normal stool providing adequate hygiene is practised (48 HAFNS)	None
Bacillus spp.		None
Campylobacter spp.		Clinical surveillance
Clostridium perfringens		None
Cryptosporidium spp.		Clinical surveillance
Cyclosporiasis		None
Escherichia coli (not VTEC)		Clinical surveillance
Giardia lamblia		Screen microbiologically
Norovirus/rotavirus		Clinical surveillance
Salmonella spp. (excluding typhoid & paratyphoid, *Vibrio parahaemolyticus*		Clinical surveillance
Yersinia spp.		Clinical surveillance
Staphylococcus aureus	Exposed septic lesion treated 48 HAFNS	Clinical surveillance
Shigella spp. (not *S. sonnei*)	Microbiological clearance	Screen microbiologically
S. sonnei		Screen microbiologically
Clostridium botulinum	None	Treatment of those at risk
Escherichia coli (VTEC)	Two negative specimens	Screen microbiologically
Hepatitis A	Seven days after onset of jaundice and/or other symptoms	Consider vaccination or passive immunization
Entamoeba histolytica	One negative specimen taken at least one week after the end of treatment	Screen microbiologically
Typhoid and paratyphoid	Six consecutive negative stool specimens, one week apart starting three weeks after the completion of antibiotic treatment	Two negative consecutive stools 48 hrs apart starting after case has commenced treatment
Taenia solium and thread worm	Until treated	Until treated

Faecal specimens should be taken at intervals of not less than 24 hours.

*Updated 2004. Communicable Disease and Public Health Vol 17 No 4 (Also available on www.hpa.org.uk)

11 Training & education of food handlers

Training is intended to modify or develop knowledge, skills and attitude through learning experience and to achieve effective performance in an activity or range of activities. Although ignorance may be a factor, most **food poisoning** incidents are caused by a failure of managers and **food handlers** to implement good **hygiene** practice. Consequently, **food** becomes **contaminated**, **pathogenic** bacteria are provided with the opportunity to multiply or they survive inadequate cooking or processing.

Training of managers, supervisors and all people who can influence the safety of food is essential to reduce the unacceptable high levels of food poisoning. However, the emphasis on classroom-based, foundation courses and certification in isolation has not been particularly successful in reducing levels of food poisoning. The main reason for this is the belief that ignorance is the main reason for food poisoning. Furthermore, it is assumed that if food handlers are aware of their legal obligations and they are provided with knowledge and understanding on the prevention of food poisoning, this will automatically result in the implementation of good hygiene practices. As Foster and Kaferstein stated in 1985 "human beings are not empty vessels into which correct information can be poured, thereby eliminating undesirable actions".

This hypothesis can be tested by considering the number of people who break the speed limit when driving. They all know what the speed limit is but it is often ignored. Prior to breaking the speed limit, most people will undertake a **risk assessment**. They will consider:
- the importance of getting to their destination on time;
- the consequences of speeding, including the road conditions and the risk to themselves and others;
- whether or not they believe the legal speed limit on the particular stretch of the road is reasonable or necessary;
- the likelihood of detection; and
- the deterrent, i.e. the punishment if caught.

In the same way, food handlers can be provided with the knowledge of the importance of handwashing after using the toilet but many of them continue to ignore this simple, but essential, good hygiene practice. Reasons for this can include:
- a belief that it is not important and that it is unlikely to result in food poisoning;
- it is unlikely that their failure to wash their hands will be discovered;
- they will not be punished for not washing their hands;
- there are no handwashing facilities, no hot water, no soap or no towels;
- management don't appear to care whether or not staff wash their hands; or
- they have insufficient time to wash their hands.

The importance of training

Having recognized that the objective of food safety training is to provide competent food handlers who continuously produce **safe food**, management must be convinced of the importance and benefits of training to provide the necessary resources.

Training should be considered as an investment in the future and evidence is available to show that companies providing high levels of training usually have better growth and profits. Training is necessary to:
- enable staff to fulfill their potential by understanding their responsibilities and improving their skills;
- promote confidence, increase job satisfaction, improve performance/morale and develop team spirit; and
- reduce the amount of supervision required.

The benefits of hygiene training

Hygiene training of staff should stress the importance of hygiene to the commercial viability of the organization and how a food poisoning incident or serious food complaint is likely to affect them. Such training should contribute significantly to the profitability of a food business by:
- assisting in the production of safe food;
- safeguarding the quality of the product and reducing food wastage;
- reducing complaints;
- generating a pride in appearance and practices, increasing job satisfaction and probably reducing staff turnover;
- contributing to increased productivity;
- ensuring that all the correct procedures, including cleaning, are followed;
- complying with any legal provisions or the requirements of industry guides or codes of practice;
- promoting a good company image, which should result in increased business; and
- improving the supervisory skills of managers.

The food industry will benefit by having available a pool of trained food handlers. Furthermore, if good food hygiene practices are also implemented at home, the level of food poisoning should reduce even more.

Effective food safety training

Effective food safety training involves two stages, firstly the provision of knowledge in a way that develops understanding and a positive attitude, for example, the importance of handwashing and the knowledge of when to wash the hands. The second stage involves the implementation of the knowledge by washing the hands properly when required. Practice, motivation and effective supervision, especially coaching, should result in the objective being achieved, i.e. the competency of the food handler and implementation of good practice at all times.

KNOWLEDGE + PRACTICE = TRAINING

UNDERSTANDING (LEARNING) ➝ ATTITUDE ➝ BEHAVIOUR ➝ COMPETENCY

To change behaviour, learning must be interesting, challenging, realistic, credible, relevant to the job and take account of existing knowledge.

LEARNING METHODS

Hygiene education revolves around some simple but important principles:

- if all I do is *hear* – I will *forget*;
- if I *hear* and *see* – I will *remember*;
- if I *hear, see* and *do* – I will *understand*.

Understanding helps *remembering* which leads to *learning*.

The first step in training is to motivate the employees to learn. They must appreciate that they could be a vital link in a chain of events leading to food poisoning and that suitable training will give them the knowledge to provide safe food.

The second stage in training involves explanation and demonstration to provide knowledge and skill respectively.

The third stage involves the trainee practising the task and the fourth stage should be to test that the task can be performed satisfactorily and that competency has been achieved.

Successful intervention (training)

There are many reasons why staff who have attended food safety courses fail to implement good hygiene practices, including:

- ignorance;
- poor working conditions;
- low pay;
- lack of job prospects;
- high turnover of staff;
- staff shortages; and
- lack of facilities, procedures and effective supervision.

To be successful, intervention must convince a person that:
- the problem can affect him/her;
- the outcome could be serious; and
- the proposed preventive measure is effective, practical and will benefit him/her.

Selling food safety solely on the basis of preventing a possible food poisoning outbreak is likely to be as successful as getting someone to stop smoking because of the long-term health effects to himself/herself or others. In other words some staff may respond, others won't. There are approximately 370,000 catering businesses and around 100 reported food poisoning outbreaks per year. Therefore, the chance of getting an identified outbreak in a catering premises is around one in 3,700. It is essential therefore to motivate food handlers to implement the knowledge gained, by using several factors in addition to "preventing food poisoning".

The motivation of food handlers to apply knowledge

The owners of the business must have a genuine commitment to achieving and maintaining high standards of food safety. Consequently they will:

- provide good working conditions and facilities for securing high standards of hygiene and personal hygiene;

- ensure managers and supervisors are committed and trained to provide the necessary coaching, instruction, motivating and training of staff;
- develop a culture of high standards, which ensures all staff believe that the right way is the hygienic way;
- provide the necessary resources and organization including finance, suitable food safety expertise and sufficient time (adequate staff levels); and
- communicate their views to staff to engender a positive attitude towards safety. Staff must know what is expected of them.

Staff may be motivated to implement high standards because they believe:
- it results in improved professionalism;
- it is a legal requirement;
- customers demand them;
- they will not cause food poisoning;
- it will assist their promotion (or other incentives);
- failure to do so may result in disciplinary action (deterrent); and
- their colleagues will object if they don't (peer pressure).

Management responsibility

Although staff cannot avoid their individual responsibility, especially when supervisors are absent, it is important to remember that most food poisoning results from management failure. In particular managers should:
- communicate the company policy on hygiene to staff and demonstrate full management support for the policy;
- lead by example. Managers must wear the appropriate protective clothing, wash their hands on entering a food room and always follow correct procedures;
- ensure staff are competent to provide safe food;
- remove any barriers to good hygiene practice;
- ensure staff have sufficient time to undertake good hygiene practices;
- take appropriate action against staff who exhibit unhygienic practices; and
- appreciate the potential effect of bad hygiene on the profitability of the business.

Management knowledge and skills

Managers of food businesses must themselves be trained to a level appropriate to their position and should:
- be able to prepare and implement a satisfactory food safety policy;
- know the requirements of legislation relating to their operation;
- have sufficient food hygiene knowledge to enable them to analyze hazards, determine risks, implement effective control and monitoring procedures at all points in the food business operation that are critical to food safety and, when necessary, take appropriate corrective action and implement product recall; and
- have the skill to select, motivate, train and supervise staff who have the necessary attributes to become competent food handlers.

The legal requirement for training

Section 16(1) and para 5(3) of Schedule 1 of the Food Safety Act, 1990 empower Ministers to make regulations requiring the hygiene training of persons engaged in food businesses. Section 23 of the Act empowers food authorities to provide food hygiene training courses for such persons. Chapter XII of Regulation EC No 852/2004 states, "Food business operators shall ensure:

(1) that food handlers are supervised **and** instructed **and/or** trained in food hygiene matters commensurate with their work activity"; and
(2) that those responsible for the development and maintenance of the HACCP system or for the operation of relevant guides have received adequate training in the application of the HACCP principles.

The level of training, instruction or supervision of food handlers is a responsibility placed on the food business to determine, having regard to the nature of the business and the role played by food handlers within it, and should be assessed as part of the HACCP system. Authorized officers should take into account any relevant UK or EC industry guides to good hygiene practice when assessing training requirements, but it is recommended that persons preparing high-risk open food require the level of training equivalent to that contained in the level 2 food safety courses accredited by the Chartered Institute of Environmental Health, the Royal Environmental Health Institute of Scotland, the Royal Society for the Promotion of Health, the Royal Institute of Public Health, the Society of Food Hygiene Technology and other similar training organizations.

It has also been suggested that food businesses may not successfully utilize the defence of due diligence, unless staff are properly trained and adequate records of training are kept.

Unfortunately the UK legislation, the codes of practice and the industry guides do not define "training" and fail to stress the need for the competency of food handlers.

Although the majority of food handlers engaged in high-risk food handling will benefit from attending an accredited course, it must be emphasized that obtaining an examination certificate relates to knowledge not competency. Consequently, a food handler who has a certificate but fails to implement the knowledge will not comply with the law.

It should also be made clear that food handlers involved solely with the handling and preparation of raw food, or staff handling only packaged high-risk food, will benefit from attending much shorter courses designed specifically for their needs. Two/three hours relevant on-site training will probably be more beneficial, and certainly more cost-effective than attending a classroom-based accredited course.

An alternative approach to training

The Food Safety Authority of Ireland are to be commended for their enlightened approach to food safety training. The FSA (Ireland) state in their "Guides to Food Safety Training" that food safety training is the key to food safety. It does not have to be certified. Training is only effective when training standards are being demonstrated in the food operation – effective training must be continuous.

Standards are expressed in terms of food safety skills, i.e. what the employee must be able to demonstrate in the area of food safety commensurate with their activity. For example:

Food safety skills	What the employee must do to demonstrate this skill	Employers, supporting activities
Monitor and record the temperature of foods.	Use a temperature probe correctly.	Provide appropriate, calibrated and maintained temperature probes.
	Know when to probe food products.	Provide appropriate rules for using and cleaning probes.
	Know the temperature limits (company/legal).	Provide relevant time/temperature recording sheets.
	Record the temperature.	Reinforce the need for time/temperature control using coaching, notices etc.
	Take corrective action and/or advise the supervisor if temperature outside specification.	Provide effective supervision.

A checklist may be designed to assess the knowledge and skills of food handlers during auditing or enforcement inspections. For example, food handlers should:
- know and adhere to their legal responsibilities regarding food safety;
- know the hazards and implement controls associated with their activities including cross-contamination and prevention of contamination;
- know high-risk and low-risk foods;
- avoid unnecessary handling of food/food-contact surfaces;
- avoid unhygienic practices that expose food to risk; and
- observe high standards of personal hygiene.

(If applicable):
- use a probe thermometer correctly;
- demonstrate effective checking of deliveries;
- demonstrate effective monitoring of controls; and
- know what appropriate corrective action to take if something goes wrong.

PLANNING AND IMPLEMENTING A TRAINING PROGRAMME

A structured approach to training will enable organizations to obtain the maximum benefits. Managers should consider future skill requirements, identify training objectives and set targets to achieve by specific dates. An appropriate budget allocation will be required. The first step is to determine whether sufficient expertise is available to plan and implement a training programme in-house. Many organizations will probably benefit from the use of experienced hygiene training consultants. Successful training programmes will usually involve the following stages.

Management of the training programme

One person should be responsible for the programme and continually monitor its progress and effectiveness. Management support will be essential and regular reports should be made to directors or the proprietor. A food safety policy statement should be signed by the chairman or proprietor to demonstrate his/her support for, and the commitment of the company to, training and high standards of hygiene. He/she must ensure, as far as practicable, that appropriate managers are appointed and systems established to ensure this policy is implemented.

Training needs analysis

The range of tasks and the risk to food safety of each task will need to be established, together with the existing level of skill and knowledge of all operatives, including managers, supervisors, cleaners and maintenance staff. (The definition of a food handler suggested by the Catering Industry Guide is "any person in a food business who handles food whether open or packaged [food includes drink and ice]".) The academic ability (the ability to assimilate knowledge, study, sit examinations and read and write) and competency of each person will need to be determined from personal records interviews, and observations. This will need to be carried out sensitively in the case of staff with literacy problems to avoid possible ridicule.

Assessment must also be made of the hygiene awareness, skill and behaviour of new employees, who should not be allowed to work in high-risk situations without appropriate induction training and close supervision. A recognized hygiene qualification is an indication of hygiene awareness but not a guarantee of competency.

Establishing a training committee

A training programme will probably be more successful if the views of the workforce are taken into account and it is not considered to be imposed.

In large organizations it may be beneficial to establish a training committee with representatives from appropriate departments and/or disciplines. Members may include the hygiene training/personnel manager, the production/catering/retail manager, an engineer, a supervisor and a first tier worker. All groups of staff should be given the opportunity to contribute, for example, by questionnaire or through discussion.

The committee will be responsible for considering the information collected and comparing the actual standards of hygiene with the required standards to determine the training needs. Objectives should be determined and targets set for the development and implementation of various parts of the programme. As far as practicable, the committee should ensure that the training undertaken is of benefit to individuals and the company as a whole.

The content of training programmes

The content of the programme will depend on the shortfall in standards/levels of skill, the time to be allocated and the finance available. These factors will also influence how, when, where and by whom the training will be undertaken. Competence assessment of operatives must be built into the programme at the beginning to ensure it is effective and provides value for money. The following questions will have to be answered:

Is the training to be provided in-house or are external consultants to be used?
To be effective, hygiene training must be continuous, accurate and up-to-date. It should include:
- pretraining assessment;
- induction training;

- on-the-job training (reinforcement, instruction and supervision);
- attendance of an appropriate food hygiene session/course (courses must meet the specific needs of the food business and be practical);
- continuous refresher training; and
- evaluation of training.

Trainers must have the ability to educate and motivate, and the necessary skill and knowledge to effect a change in behaviour of those persons being trained. In large organizations, managers with training skills who have attended an appropriate intermediate or advanced hygiene course may be able to provide level 2 hygiene training specifically designed for the company. Alternatively, an external trainer could provide a modified course to suit the requirements of the particular food business. One major advantage of in-house training is that it can be linked with training for hazard analysis and the monitoring of critical control points. The supervisor should be responsible for on-the-job coaching, effective supervision and for continuous refresher training. In the event of a food handler failing to carry out an important food safety procedure, the supervisor should remedy the problem immediately. On-site computer-based programmes (e-learning) are ideal as retraining can be provided immediately when it is required. The need for, or value of, recertification of food handlers every three years has not been established in terms of the benefit to food safety. However, provision must be made to update training when, for example, new legislation or new food safety systems, such as **HACCP**, are introduced.

Food Safety Foundation
Highfield.co.uk ltd e-learning.

Who will be trained and to what level?

The industry guides to good hygiene practice identify the appropriate level of training for food handlers. All staff should understand the essentials of food hygiene as part of their induction training, prior to starting work for the first time. Most food handlers will then require hygiene awareness instruction, within four weeks, which may include:
- an overview of the company's food safety policy, confirmation of the company's commitment to hygiene and the consequences of a breakdown in hygiene;
- the potential of bacteria to cause illness;
- personal health and hygiene, including the legal responsibilities of food handlers and the need to report illness to their supervisor;
- preventing contamination, including foreign body contamination;
- cleaning and disinfection procedures, materials, methods and storage;
- food storage (protection and temperature control);
- recognition of **pests** and signs of pests; and
- waste disposal.

The actual content of, and requirement for, training will depend on the type of food being handled and the risk to food safety posed by the tasks undertaken, together with the level of supervision. Organizations with a high turnover of staff will require more emphasis on instruction and supervision, although training should still be provided. It is recommended that high-risk food handlers attend an accredited level 2 food safety course or an equivalent in-house course. Alternatively e-learning, distance learning or on the job training may be utilized to provide the relevant level of knowledge. In order to comply with the law this knowledge must be implemented by staff who follow good hygiene practice and demonstrate competency to produce safe food.

Supervisors and managers will benefit from attending either a level 3 (around 20 hours) or level 4 (around 40 hours) food safety course. If they are responsible for the HACCP system they will also require training in the application of HACCP principles. HACCP training may be incorporated within the level 3 or level 4 course.

What provision has been made for pretraining assessment and induction training?

Pretraining assessment of new employees can considerably reduce the amount of time spent on training. Courses that include a significant amount of information already known to food handlers are often counter-productive and demotivating. Pretraining assessment is just as important for new employees with a food safety qualification as for those who have attended in-house training with their previous employers, or those who have received no training. It is essential that all new starters are properly supervised/instructed and receive essential induction training prior to carrying out any **food handling** duties. All companies should provide new employees with a basic induction book, which states the company food safety policy and details the hygiene rules that they will be expected to follow. Operatives should complete a short test to confirm they have read and understood the content of the book. The booklet could be used as a basis for short induction training sessions, which will be necessary to emphasize the dangers of bad hygiene, what is expected of him/her and the consequences for the employee failing to adhere to essential hygiene rules. On-the-job training is critical in developing the correct attitude of new food handlers. Staff must be shown the most hygienic way of performing their job as bad habits, once cultivated, are difficult to break. As well as explaining how to carry out a task, staff must be told the reason for carrying it out in a particular way.

What training and examination techniques will be used?

Most food handlers involved with the preparation of high-risk food attend a formal course of lectures provided by an internal or external trainer. However, certified courses presented in a classroom style with written examinations may be inappropriate for some operatives and other methods of training designed to improve hygiene standards may be necessary. It is estimated that around 25% of people in the UK have literacy problems. National vocational qualifications (NVQs) and other competence-based routes to attaining national food hygiene qualifications should also be considered. Open learning programmes with workbooks may be of use in certain circumstances.

Computer-based training (e-learning)

A considerable increase in food safety training using computers has taken place in the last few years. Computer-based training, especially for induction, level 2, level 3 and refresher training, has several advantages compared to classroom-based courses.

Advantages of high quality computer-based training
- *Consistent* – the same high quality interactive information is available 24 hours a day;
- *Flexible and accessible* – training is provided at a convenient time for a realistic period of time, training missed through illness or unexpected busy periods is easily rescheduled;
- *Convenient* – all training can be carried out on site;
- *Value* – very cost-effective, no travel, no subsistence, reduced training time because pretesting ensures you only provide necessary training;
- *Complete solution* – some packages incorporate induction, level 2 and level 3 and facilitate continuous refresher training;
- *Monitoring* – an accurate overview of training progress will be available;
- *Records* – comprehensive information of the performance of each food handler can be printed out and may be used to support a due-diligence defence;
- *Understandable* – can be used successfully by people with reading difficulties or those whose second language is English, or those with learning difficulties including dyslexia; and
- *Certificated* – on successful completion of the programme a recognized certificate of training may be provided. Furthermore, it may be possible for food handlers to sit an accredited examination if this is required.

However, computer-based training should not be used to cut training budgets or make trainers redundant. As stated earlier the major reason for food poisoning and food complaints is the failure to implement good hygiene practice. Trainers should be involved with the development and auditing of systems that ensure the implementation of good hygiene practice at all levels. This will also involve additional training of supervisors and managers to ensure effective and continuous training, through coaching and reinforcement, of all food handlers.

Examinations and testing

Given the considerable number of food handlers with literacy problems, or whose second language is English, the use and type of examination should be considered carefully, especially at level 2. Examinations which include out-of-date, confusing or irrelevant questions may cause more harm than good. It is also essential, at level 2, that food handlers should be advised which questions they get wrong. At no time should the examination or qualification at level 2 ever assume more importance than the training. Examinations should be practically based and relevant to the job of the food handler to avoid trainers stating "in practice this is the correct procedure, but the correct answer in the examination is different", or "you won't need this information for your job but you may need it for the examination".

It should be remembered that the information provided at level 2 is essential for ensuring food safety, it is not to test the academic ability of food handlers. There is no need to provide unusual, complicated or obscure questions.

Testing by the use of projects and assignments relevant to the specific food business is receiving increased support in preference to the theoretical examination of level 3 and level 4 courses.

Have arrangements been made to continually assess the competency of the trained food handlers?

Increased knowledge, obtained by food handlers attending hygiene courses, may soon disappear. Prior to sending staff on courses a record should be made of their understanding and implementation of good hygiene practices. A further assessment should be made on completion of the course to ensure it provided value for money. In addition, competence assessment should be incorporated within the day-to-day role of supervisors to ensure hygiene practices are observed and to provide continuous refresher training. Interactive hygiene software will also be useful to provide immediate refresher training.

Accurate records of task assessment, standards achieved, courses attended and progress through training of each employee should be maintained. Staff appraisals can be linked to staff development to determine future training needs.

It is unfortunate that some organizations and auditors put more emphasis on paper records and certification than they do on the competency of food handlers at foundation level.

Competency assessment cards should be used to:

- ensure staff implement good hygiene practice;
- ensure supervisors accept responsibility for the competency of staff;
- demonstrate compliance with the legal requirement "that food handlers are supervised and instructed and/or trained in food hygiene matters commensurate with their work activity"; and
- assist in a due-diligence defence as they will form an essential part of the training records.

Highfield.co.uk ltd competency assessment cards.

Training sessions and methods

When planning training it is important to consider the background of the food handlers including their age, previous experience, ability, language difficulties and interests. Objectives for training must be established and, more importantly, achieved. Finally, the training method and content must be decided and visual aids selected. Food handlers must be encouraged to participate, especially if in groups, and this may be accomplished by asking questions beginning with the words, how, why, what, when, who, where and which. E-learning programmes must be interactive if they are to be successful.

There are many different ways of providing knowledge and training, however, they should all have regard to the key phases of information processing, namely: attention; comprehension; retention; and recall. Training can be provided to a single food handler, a group of food handlers, in each case on or off the job.

TRAINING METHODS

Individual	Group	On-the-job
Distance learning	Classroom/lectures	Practical instruction
Revision	Meetings/discussions	Coaching (supervision)
Research	Workshops/case studies	Experience/practice
E-learning (computer-based)	Seminars	Management

In large companies hygiene committees, similar to safety committees, could be established to discuss ways in which hygiene can be improved and customer complaints reduced.

Whichever method is adopted, it should be designed to ensure that the food handler retains as many important facts as possible. The length of training sessions depends to a large extent on the complexity of the material being presented, relative to the ability of the food handlers to concentrate and comprehend. However, for the majority of food handlers, sessions should be held frequently and kept short and simple.

It is often advantageous to invite visiting lecturers, such as environmental health practitioners/officers, to reinforce hygiene training, as they usually present a slightly different approach, which may stimulate renewed interest, even if it is only to highlight the legal obligations placed upon food handlers. The use of guest lecturers also demonstrates the importance the company places on hygiene training.

Communication skills

To get the message across, the hygiene trainer must take account of factors that interfere with effective communication and these may be considered as factors affecting either the trainer or the trainee.

Factors affecting the trainer
- *Voice* – should not be monotonous, too quiet or too loud.
- *Appearance* – should be smart and suitably dressed.
- *Habits* – should not be distracting, for example, rattling loose change, walking up and down, frequent repetition of words or phrases such as "ok", frequent touching of the hair or face.
- *Eye contact* – should be maintained with as many of the group as is possible. Each trainee should feel that he/she is being addressed.
- *Enthusiasm* – should be conveyed to the trainees, who must be given the impression that the trainer believes in the importance of what is being said. Questions should be encouraged.
- *Knowledge of the subject* – should be thorough as reasoned arguments are more likely to succeed than dogma.
- *Presentation* – must be at the appropriate level. Trainers must not talk down to people or over their heads. Sessions must be well-planned as disjointed, unrehearsed sessions are not effective.
- *Vocabulary* – must be appropriate. Long difficult speeches, words and jargon will not be appreciated.

Factors affecting the trainee
- *Motivation* – are the trainees aware of the importance of the subject and how it affects them either directly or indirectly?
- *Comprehension* – if trainees do not understand what is said they will soon lose interest.
- *Time of day* – large lunches, too much drink or training sessions after a hard day's work may result in student indifference.
- *Length of session* – as for as practicable training sessions at induction and foundation level should be kept to about an hour.
- *Design of training room/area* – can everyone see and hear the trainer/visual aids?
- *Visual aids* – professional, legible, informative, interesting and, sometimes, amusing.

The location and timing of training courses

Provided adequate training facilities are available, it is preferable for training courses to be organized on the food premises. Although six-hour level 2 courses can be completed in a day, improved results should be obtained by organizing sessions of 30 minutes or an hour over several days or weeks. A frequent compromise is three x two hour or two x three hour sessions. Level 4 courses will usually be organized at training centres, hotels or colleges where appropriate facilities are available to allow use of videos and other hygiene aids as well as any necessary demonstrations.

Training should be undertaken at a time that is convenient to the business operation but also acceptable to the food handlers.

Visual aids equipment

Videos, PowerPoint™ presentations, overhead transparencies, desktop presenters, hygiene games, photographs, posters and exhibits such as pests and food complaints are useful to break up training sessions and maintain interest. They are also useful for highlighting actual situations that would be difficult to recreate in a training room. Care must be taken to ensure the information presented does not become dated. Presentations should be carefully prepared and used to emphasize, clarify or supplement talks. Interesting or amusing coloured illustrations that maintain attention are preferable to black and white text, and essential at foundation level. Presentations on screens should be clear and legible to everyone in the audience and not be crammed with too much information. It is better to use two or more PowerPoint™ slides or transparencies rather than cramming information. Graphs and illustrations are preferable to tables, and a good contrast between text and background is essential. Handwritten scrawl on transparencies is unacceptable. The number of words should be limited to a maximum of around 40 with adequate spaces between lines. Presentations should be rehearsed to ensure familiarization with sequences.

Multimedia and overhead projectors

The main advantage of the projectors are that they enable trainers to maintain eye contact with trainees and lectures must not be presented to the screen. The overhead projector should be switched off when when not in use – blank screens or the presence of information that is no longer being discussed is distracting. Information on transparencies should be revealed a little at a time, as and when it is relevant. PowerPoint™ presentations overcome this problem and are much more flexible. They are also very easy to amend and customize.

Trainers should be familiar with equipment and should be able to carry out minor repairs such as changing bulbs or fuses. Focusing and volume control should be attended to before trainees assemble.

Piloting of the programme

Prior to the introduction of the programme, it should be piloted to a cross-section of employees. Close supervision and monitoring is essential to obtain accurate feedback from those trained to enable the programme to be evaluated. Competence assessments following training should be compared with assessments prior to training.

Implementation of the training programme

Provided the evaluation of the pilot exercise is satisfactory and any necessary corrective action is approved, the programme may be introduced to all parts of the organization. Consultation with the local environmental health department during the planning and implementation of the training programme may be beneficial to confirm you are satisfying any legal requirement.

The programme must be flexible and monitored and evaluated continually to ensure the objectives are being achieved and, if necessary, modifications can be made.

Evaluation of the programme

Evaluation of the whole programme and each training session/course is essential. Evaluation of staff, supervisors and managers should be undertaken to determine:
- whether objectives have been achieved;
- if knowledge and awareness are greater;
- if attitude has changed;
- if behaviour has changed (improved skills, practices and standards);
- if staff are competent to provide safe food; and
- further training needs.

Various techniques of evaluation have been used following training and attendance at courses, including:
- pre- and post-training inspections;
- pre- and post-training interviews to assess knowledge gained;
- pre- and post-training testing, including practical assessment of competence and written assignments;
- the use of pre- and post-training questionnaires;
- bacteriological swabs taken of hand and food-contact surfaces;
- bacteriological food sampling (especially for the presence of indicators);
- trends in the number of hazards identified during audits; and
- trends in food complaints.

Evaluating specific courses

If possible, a control group of food handlers who have not attended the course should be compared to ensure that improvements in hygiene resulted from attendance on the course and not, for example, from improved supervision or on-the-job training. Managers who have attended advanced hygiene courses could be asked to report on hygiene improvements or achievements implemented three and six months after completing the course.

HYGIENE for MANAGEMENT Training & education of food handlers

Highfield.co.uk ltd Handwashing When? Poster.

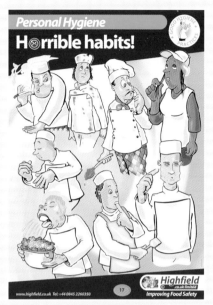

Highfield.co.uk ltd Personal Hygiene - Horrible Habits! Poster.

Reinforcement of training and staff motivation to ensure that hygiene standards are maintained

Effective supervision, reinforcement, coaching enforcement and continuous refresher training will be essential to ensure behavioural changes are maintained after attending training sessions. Incentives should be used to encourage high standards. In some countries deterrents have been used to maintain standards, for example, frequent hand swabbing, closed circuit television and fines.

Posters and notices may be used to encourage good practices and to reinforce the legal obligation of the individual. Posters may shock, amuse or instruct and should be prominently positioned and changed as often as necessary to maintain attention. Hygiene displays may be useful and free literature may be available from the local environmental health department.

Small groups of staff participating in hygiene sessions, for example, discussing case studies, can stimulate significant interest. External speakers may be brought in to chair such discussions.

Recognition may be given to employees who obtain hygiene qualifications and especially those who operate hygienically. This may take the form of increased remuneration or an award.

Many companies use some form of appraisal to determine whether or not supervisors and managers justify receiving larger increases in salary. Hygiene should be included as one of the categories on which this decision is made. Once managers realize that poor hygiene affects their salary, improved standards should be assured.

12 The design & construction of food premises

A considerable amount of thought must always be given to the design, construction and location of food premises, to ensure a cost-effective and hygienic operation. Newly built premises should comply fully with the requirements of the appropriate hygiene and safety legislation and the elimination of potential hazards and risks should be a key objective. Plans should be discussed with enforcement officers as it is much easier and cheaper to provide satisfactory finishes and facilities such as sanitary accommodation during construction, than it is to be told to provide alternative finishes or additional sanitary accommodation when the building is completed.

The use of satisfactory building materials and a well-planned layout are essential to achieve high standards of hygiene. The size of the premises must facilitate efficient operation and the site must be large enough to accommodate possible future expansion.

Selection of a suitable site

Choosing a suitable site for new food premises is of paramount importance. During the selection procedure, consideration must be given to the provision and availability of services, i.e. electricity and gas, water supply and effluent disposal and the accessibility for delivery and waste disposal. The site should not be liable to flooding or unacceptable contamination from chemicals, dust, odour or pests.

Close proximity of residential properties may cause problems for some premises, especially factories. Most food businesses generate noise, to a greater or lesser extent, and complaints from nearby properties can be expected, especially if noise is generated before 8.00 a.m. or after 6.00 p.m. Shift working, unloading or loading vehicles, extractor fans and refrigerated vehicles parked overnight are just a few examples of commonly occurring complaints. When the site has been selected, and outline planning permission obtained, time and effort must be devoted to the design of the premises.

General principles of design

Achieving a satisfactory design, which provides a linear workflow and so eliminates cross-contamination, facilitates effective cleaning and can be operated profitably, is far from simple. Full consultation between all involved parties, including architects, environmental health practitioners/officers and users is recommended. It should always be borne in mind that an increase in capital expenditure may well result in both increased life of the premises and a reduction in operating and cleaning costs. When designing food premises the following principles should be considered:

- the risk of contamination should be minimized under the proposed working conditions; clean and dirty processes must be separated. Where possible the work areas in large-scale manufacture should be segregated into precook and post-cook, and staff should not be allowed to move from one area to the other. If practicable, a separate area should be provided for de-boxing and unwrapping raw materials, especially when there is a risk of contaminating food with staples, paper, cardboard and string. Accumulations of refuse should not be permitted;

HYGIENE for MANAGEMENT *The design & construction of food premises*

- workflow should be linear and progress in a uniform direction from raw material to finished product. Food should not be kept at ambient temperatures for longer than is absolutely necessary. Distances travelled by raw materials, utensils, food containers, waste food, packaging materials and staff should be minimized;
- facilities for **personal hygiene** and disinfection of small items of equipment should relate to working areas and process risks;
- where appropriate, suitable facilities must be provided for temperature, humidity and other controls;
- the premises should be designed to avoid accumulation of dirt in inaccessible places and the risk of contamination from condensation and must be capable of being thoroughly cleaned and, if necessary, **disinfected** at the end of production;
- insects, rodents and birds must be denied access and harbourage;
- yard surfaces and roads within the boundary of the premises must have a suitable impervious surface with adequate drainage, and provision made for refuse storage. Refuse should not have to be taken through food rooms when collected. Facilities for cleaning yards must be provided, for example, a stand-pipe; and
- suitable provision must be made for staff welfare, including cloakroom and, if necessary, canteen and first aid facilities.

THE CONSTRUCTION OF FOOD PREMISES

If the design of a food premises establishes the method of operation, the construction and structure determine how easily it can be cleaned. It is essential that the correct materials are chosen for all ceiling, wall and floor finishes and even more important that they are properly fixed or applied. Insufficient attention to fixing generally results in a potentially satisfactory material becoming totally unsuitable and advice should be sought from specialists. Materials should be non-toxic, suitably durable and easy to maintain. Before a material is selected, the

Unsatisfactory food premises, staff toilets and washing facilities.

supplier should be requested to provide details of similar premises using the particular product and existing applications should be visited. Whichever surface is chosen, it should be kept in good repair and be capable of withstanding the frequent cleaning to which it will be subjected.

Ceilings

Ceilings may be either solid or suspended. The latter are advantageous as horizontal pipework and services are concealed. Access for inspection, pest control and maintenance must be built in. Structural walkways are often necessary and should always be provided in large premises. Suspended ceilings are normally made of metal lattice incorporating cleansable panels. Aluminium backed and faced fibre-board has proved successful in many food factories. Flush-fitting ventilation grilles should normally be provided.

Solid ceilings give less scope for a hygienic finish, as pipework and ventilation trunking normally protrude. They should be well insulated to avoid condensation and mould growth. Generally, ceilings should be smooth, fire resistant, light-coloured, coved at wall joints and easy to clean. Plasterboard with taped joints, skimmed with plaster and finished with a washable emulsion is common in smaller premises, although in factory areas with little steam, corrugated sheeting treated with suitable anti-fungal paint has proved satisfactory. Special attention must be paid to ceiling finishes above heat and/or steam-producing appliances such as ovens, sinks and retorts.

Ceiling height will vary depending on the type of operations being carried out but should only be high enough to provide satisfactory working conditions and allow the installation of equipment. Maintenance and cleaning is much easier if the ceiling is around 3.5m from the floor.

Walls

Smooth, impervious, non-flaking, light-coloured wall surfaces are required. Whichever finish is chosen it must be capable of being thoroughly cleaned and, if necessary, disinfected by the methods employed in the premises. Internal solid walls are preferable to those with cavities, which may harbour pests. Crevices and ledges that cause cleaning problems should be eliminated. False panelling must not be used.

Regard must be had to the operations carried out adjacent to the wall, as the surface may need to be resistant to spillages, chemicals, grease, heat and impact. Wall surfaces in use include resin-bonded fibre glass, ceramic-faced blocks, epoxy resin, glazed tiles with water-resistant grouting and rubberized paint on hard plaster or sealed brickwork, which is used in some food factories. Some paints incorporate a fungicidal additive but absorbent emulsion paint should not be used on walls.

Galvanized steel, aluminium and stainless steel

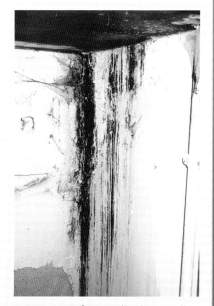

An unsatisfactory kitchen wall.

are used, although they may warp and strain their fixings. In addition, these metals are prone to corrosion and attack by some cleaning chemicals. Plastic sheeting is becoming more popular, although standards and suitability vary considerably. Some sheets can be heat formed to fit corners and angles. Fire resistance of materials must always be considered. Stainless steel splash backs are recommended behind sinks and working surfaces that may be exposed to impact damage.

Wall or floor stops are needed to prevent doors damaging wall surfaces, and wall corners should be protected by non-corrosive metal or PVC angles. Stainless steel angles bedded in mastic with countersunk screw fixings are recommended. Protection of wall surfaces from trolleys may be required. Crash rails or specially designed **coving** have proved successful.

Pipework and ducting should be bracketed at least 150mm from walls to facilitate cleaning. All lagging to pipes must be smooth and impervious. Pipes passing through external walls must be effectively sealed to prevent the ingress of pests.

Precautions for painting within food premises

- Select the correct paint. The paint chosen should:
    - adhere firmly so that flaking is minimized;
    - be of low toxicity; the minimum standard should be in compliance with BS 4310: 1968 for low-lead paints;
    - be cleansable; the higher the gloss the better the washability. If vigorous cleaning is used, for example, using hoses, specialist paints will be required;
    - dry rapidly. It may be possible to use a rapid-drying, emulsion-based primer and undercoat system followed by an eggshell. This would allow a two-coat system to be applied within a working day and so minimize production losses.
- Cease production during redecoration. This requires careful planning. It may be possible to decorate out of normal production times, for example, at weekends.
- Ensure tainting of food does not occur. Solvent vapour may spread to other sections not being decorated and is especially likely to contaminate food rich in fat or oil, such as butter.
- All fixed equipment must be well covered.
- All paint must be kept in properly marked containers that are not made of glass.
- Ensure the contractor is aware of hygiene and safety rules.
- On completion of painting ensure that the area is clean, free from odour and painting materials before processing starts.

Further information may be obtained on specialist paints for the food industry, and precautions to be taken, by contacting paint manufacturing firms.

Doors, windows, stairs and platforms

Windows in factories are often omitted as they take up valuable wall space, and if used for ventilation purposes will allow the entry of insects and dust. If windows are provided they should be fixed on north-facing walls to reduce glare and solar heat gain. Where this is unavoidable solar film will assist in counteracting heat gain. Cleansable, well-fitting fly-screens must, where necessary, be fitted to opening windows. Windows should be constructed to facilitate cleaning. If present, internal window sills should be sloped to prevent their use as shelves. The design of frames, of both windows and doors, should avoid acute angles. Right-angled joints between frames and walls should be beaded or filled to form continuous surfaces. Architraves should be avoided and reveals tile-surfaced rather than wood carcased. Woodwork should be well-seasoned, properly knotted, stopped, primed and given three coats of polyurethane paint.

A filthy window in a canteen.

Doors should have smooth, non-absorbent surfaces capable of being thoroughly cleaned. They should be tight-fitting and self-closing. Door handles and finger-plates should be capable of disinfection. Swing doors with kick-plates are preferable to handles. Many food factories use polypropylene or toughened rubber doors as they require little maintenance and are easy to keep clean. Clear plastic strips are also used. External doorways should, where necessary, be proofed against the entry of insects, and metal kick-plates should be provided to prevent gnawing by rodents.

Doorways must be large enough to allow for the movement of mobile equipment and possible replacement of fixed equipment. Swing doors that open both ways should be fitted with sight panels. Stairs, ladders and platforms must be capable of being thoroughly cleaned and should not expose food to risk of contamination.

Floors

When selecting a floor covering, the following criteria should be considered:
- the volume and nature of traffic, for example, fork-lifts;
- whether the area is wet or dry;
- how the area will be cleaned, particularly if steam is used;
- what chemical resistance will be necessary;

Area of floor beneath kitchen sink, main unit.

Dirt and flour build up on the floor and oven sides.

Failure to clean under equipment.

- whether production will need to be curtailed to effect repairs to the floor; and
- the type of sub-floor.

Ucrete flooring withstanding contamination from hot oil and potato starch, as well as heavy traffic and cleaning at high temperatures. (Courtesy of MBT Ucrete.)

To assess the cost-effectiveness of a finish, regard must be had to initial cost, durability, performance and safety. In food premises, floors should be durable, non-absorbent, anti-slip, without crevices and capable of being effectively cleaned. Where appropriate they must be resistant to **acids**, grease and salts and should slope sufficiently for liquids to drain to trapped gullies; a slope of one in 60 is the minimum recommended. The angle between walls and floors should be coved. Regard must be had to the possible odour taint of food during installation or repair. The base of H section vertical girders should be filled with concrete to avoid difficulties in cleaning.

Several types of floor covering may be used. For example, epoxy resin, granolithic (concrete incorporating granite chippings), welded anti-slip, vinyl sheet and slip-resistant ceramic or quarry tiles. Wooden floors are unacceptable. Whichever type of floor covering is chosen, it must be laid in accordance with the manufacturer's recommendations. Special attention must be paid to the jointing of quarry tiles and the sub-floor must be very smooth if anti-slip, vinyl sheeting is laid. If concrete is used, it must be steel-float finished and sealed with, for example, an epoxy sealant, to ensure that it is dust free and impervious.

Services

These include gas, electricity, water supplies, drainage, lighting and ventilation. Proper provision of services is essential to the hygienic and effective functioning of all food businesses.

Gas supplies

Specific hygiene implications are covered by more general safety and equipment fixing requirements. Supply pipes should always be mounted clear of the floor and never so close to other pipes as to restrict access for cleaning. Flexible connections, to facilitate removal of equipment for cleaning purposes, are recommended.

Electrical supplies

Adequate numbers of power points should be available for all electrical equipment, without the use of adaptors or the need for lengthy flexes. Provision should be made for maintenance, repair and cleaning operations. Cut-out switches for power circuits should be accessible and separate from lighting and ventilation supplies, so that cleaning can take place in safety. Separate cut-out switches should be provided for refrigeration equipment.

Controls should be fixed clear of equipment to avoid becoming dirty or wet during cleaning. Removable electrical components are advantageous. Electrical wiring should be protected by waterproof conduits. All switches should be flush-fitting and waterproof (especially in production areas).

Water supplies

All cold water supplies for use with food, for cleaning equipment or surfaces or for personal hygiene must be potable and should be mains supplied and not fed via an intermediate tank, unless chlorinated. The distribution system must not provide opportunities for contamination or the multiplication of micro-organisms. Water supplies should be subject to control and monitoring procedures determined by hazard analysis and, if necessary, include periodic microbiological and chemical sampling.

Hot water should have a target water discharge temperature of 60ºC. In hard-water areas, hot water supplies should be softened, otherwise scale build-up will cause cleaning and operational problems and add significantly to detergent usage. Water softeners and filters must be maintained in a good condition so that they do not contaminate the water.

Ice used in food, drinks or buffet displays must be made from potable water. Ice machines must not be exposed to risk of contamination and must be regularly cleaned and disinfected. Utensils must not present a foreign body hazard; glassware should not be used to "shovel ice". Ice for drinks should not be handled with bare hands.

Non-potable water used for steam generation or fire control must be conveyed in identifiable systems, which have no connection with, nor any possible reflux into, the potable water system. An external water supply for washing refuse areas and loading bays should always be available.

Drainage

Premises should have an efficient, smooth-bore drainage system, which must be kept clean and in good order and repair. Drains and sewers should be adequate to remove peak loads quickly without flooding. Sufficient drains should be installed to facilitate effective cleaning of rooms by pressure jet cleaners or other means.

Shallow, glazed, half-round floor channels within food rooms are best left uncovered, provided that they are not a safety hazard. If covers are necessary, these should be non-corrosive, continuous, of suitable strength and easily removed for cleaning. In certain circumstances trapped gullies are preferable.

To avoid fat solidifying and causing blockages, drainage systems must be well designed and constructed, with the minimum number of bends, and have an efficient self-cleaning velocity. Cleaning and flushing, at least six monthly, is essential. Grease traps, if fitted, should be large enough to allow adequate time for fat to separate. They should be emptied as frequently as necessary and, as the contents may be foul-smelling and obnoxious, traps should be positioned outside food rooms.

Inspection chambers should be placed outside food rooms but if interior location is unavoidable they must be airtight. Manhole covers should be double sealed, bedded in silicon grease and screwed down with brass screws. All drainage systems must be provided with sufficient access points to allow rodding in the event of blockages. Petrol interceptors may be required for yard drains.

Items of machinery such as potato peelers and dishwashers, connected directly into the drainage system, should be trapped to avoid waste pipes acting as vents for sewers. Waste pipes to fittings should be plastic with screw or push-fit connections to enable easy dismantling in case of blockage.

Drains should be constructed to inhibit the harbourage and movement of vermin. Defective drains may result in effluent, foul odours and rodents entering food rooms. They must be repaired as quickly as possible. All external rainwater fall-pipes should be fitted with balloon guards to prevent rodent access. Circumference guards should be fitted around all vertical pipes fastened to walls, to prevent rodents climbing up them. Waste pipes exposed to constant high temperatures should be constructed of Alkathene. The direction of flow should be from clean areas to dirty areas. Toilets should feed into the system after food rooms.

Ventilation

Suitable and sufficient ventilation must be provided to produce a satisfactory, safe working environment and to reduce humidities and temperatures that would assist the rapid multiplication of bacteria. Condensation encourages mould growth and drips onto food. Ambient temperatures should be below 25°C. The ventilation system, which should always flow from a clean to a dirty area, must prevent excessive heat, condensation, dust and steam, and remove odours and contaminated air. Good ventilation will assist in reducing grease and staining of ceilings, so reducing the need for frequent decoration.

When planning a ventilation system, expert advice should be sought to ensure that food rooms have the recommended number of air changes, for example, in kitchens, the minimum ventilation rate should not be less than 17.5 litres per second per square metre of floor area and not less than 30 air changes per hour. Inlets must be suitably filtered to prevent dust, dirt and insects being brought into the food room. The input capacity should be approximately 85% of the extract capacity. Natural ventilation through screened windows normally needs supplementing by mechanical ventilation to ensure effective air circulation.

Ducting should be as short as possible with man-sized access points at 3 metre intervals to

Filthy extraction canopies in kitchens.

facilitate cleaning and removal of dirt, grease and pests. Fan motors should be located outside the kitchen otherwise the noise produced may result in staff switching them off. Fans should be of a low-noise type but silencing may result in cleaning difficulty and fire risk.

Steam-producing equipment such as cookers, boilers and blanchers should be provided with adequately sized canopies constructed of anodized aluminium or other suitable material. Canopies should overhang by around 250mm. Filters, if fitted, should be cleaned frequently to eliminate fire hazards and maintain efficiency. Drainage gutters at the base of canopies should be provided to collect condensate.

Much of the equipment used in food rooms, especially for heat processing and cooking, emits radiant heat, which is not directly affected by air flow. Provision of lower heat-emitting equipment such as pressure vessels and microwave ovens, and upgrading insulation on ovens should be considered to reduce heat production.

Workroom temperature where food is handled

Generally, food hygiene law regulates food temperatures, and health and safety law regulates air temperature. Health and safety requirements can usually be met by:
- maintaining a "reasonable" temperature of 16°C (13°C if work involves serious physical effort); or if this is not practical
- providing a warm work station; or if this is not practical
- providing suitable protective clothing, suitable heated rest facilities and minimizing the time in uncomfortable temperatures.

Hygiene of ventilation and water systems

Clean air and safe water are vital for a pleasant working environment and hygienic food production, and failure to carry out routine maintenance and cleaning of ductwork, pipes and cooling towers can result in increased risk of infection.

Water in cooling towers for industrial processes and air conditioning systems is often contaminated with legionella bacteria. This organism develops in neglected water and air

Sampling a header tank for harmful bacteria.
(Courtesy of Rentokil Initial PLC.)

conditioning systems and is spread by aerosols, for example, from showers and cooling towers. Poorly maintained water systems result in a build-up of slime and dirt, which support many bacteria and algae and can result in product contamination.

Those organizations without the necessary equipment or in-house experience should consider employing a reputable contractor to carry out regular surveys and maintenance. Filters, header tanks, cooling tower sumps, packing and drift eliminators, overflows and plant rooms should be cleaned and, where necessary, disinfected. Such servicing enables water treatment programmes to operate efficiently, reduces the risk of infection and prolongs the life of pipes.

Lighting

Suitable and sufficient lighting must be provided throughout food premises, including store rooms, passageways and stairways, so that employees can identify hazards and carry out tasks correctly. High standards of lighting will produce an environment conducive to clean and safe working without eye strain. Other advantages include the enhanced appearance of food in service and retail areas together with the fact that insects and rodents generally shun well-lit areas.

Artificial lighting is often preferred to natural lighting because of problems of solar heat gain, glare, shadows, and insects entering open windows. Fluorescent tubes, fitted with diffusers to prevent glare and product contamination in the event of breakage, are recommended. Suspended ceilings can incorporate flush light fittings, which are also satisfactory, but in certain premises, particularly bakehouses, panel lights incorporated into the ceiling may buckle due to excessive heat.

Standards of lighting may be identified in the industry guides. The Catering Guide suggests 150 lux in storerooms and 500 lux in preparation areas. As a general guide, in areas where decoration is bright and ceiling heights do not exceed 3 metres, 20 watts from a fluorescent fitting per square metre of floor space will give 400 lux.

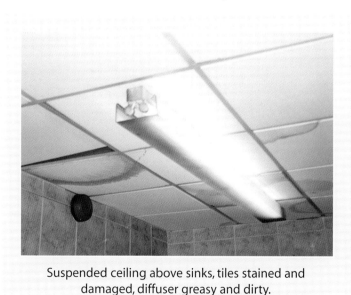

Suspended ceiling above sinks, tiles stained and damaged, diffuser greasy and dirty.

HYGIENE for MANAGEMENT *The design & construction of food premises*

Untidy, dirty pot wash and food preparation area.

Wash-up area, sink not sealed to adjacent wall surfaces.

Handwashing facilities

Adequate facilities for handwashing and drying should be provided wherever the process demands. In particular, a suitable number of basins or troughs should be sited at the entrance of food rooms to ensure all persons entering wash their hands. Wash-hand basins must be easily accessible and should not be obstructed. They must be kept clean and maintained in good condition. Non-hand operated spray taps providing warm water at 35ºC to 45ºC are essential to minimize the risk of cross-contamination. Washing facilities provided in food rooms should be additional to those used in conjunction with sanitary accommodation. All basins and troughs, preferably made of stainless steel, should be connected to drains by properly trapped waste pipes.

Cleaning and disinfection facilities

Where appropriate, adequate facilities for the cleaning and disinfection of utensils, crockery, cutlery, glasses and equipment should be provided. These facilities should be constructed of corrosion-resistant materials, normally stainless steel, capable of being easily cleaned. Twin sinks are preferable to facilitate washing and disinfecting/rinsing. Sinks should be provided with adequate supplies of hot and cold water. Mixer taps are acceptable. Taps should be wall-mounted with no direct connection to the sinks. It is good practice to provide facilities for draining and air drying equipment close to the area where it is washed. Waste pipes should be plastic push-fit ware. Sinks should be freestanding so that they can be removed easily after unscrewing the lower trap joint freeing the waste pipe. "Sterilizing" sinks and units should be capable of operating at 82ºC, as should the rinse cycle of dishwashing and tray-cleaning machines. It is recommended that if any of the above are positioned against the wall, a suitable stainless steel upstand of 300mm, or a splashback, be provided. If glasswashers are used they should clean and disinfect. Hoses and other equipment for cleaning and disinfecting fixed equipment may be required.

Separate sinks must be provided for food preparation and equipment washing if the volume demands it. In small operations the same sink may be used if there is no risk to food safety. Exclusive food sinks may be provided with cold water only. It is good practice to have signs above sinks indicating their use.

Sanitary conveniences and washing facilities

All new premises should be provided with adequate staff sanitary accommodation in accordance with the Workplace (Health, Safety and Welfare) Regulations, 1992 and the approved Code of Practice and Guidance L24. Suitable and sufficient sanitary conveniences must be provided at readily accessible places, in adequately ventilated and lit rooms that are kept clean and tidy. Rooms containing sanitary conveniences must not communicate directly with a room where food is processed, prepared or eaten. Internal wall and floor surfaces should permit wet cleaning.

Non-hand operated flushing devices are recommended. Doors to intervening spaces and sanitary accommodation should be self-closing and clearly illustrate the sex of the user. It is good practice to display a conspicuous notice requesting users of the toilet to wash their hands before leaving. Suitable and sufficient washing facilities must be provided at readily accessible places. In particular, facilities must be provided in the immediate vicinity of every sanitary convenience and supplied with clean hot and cold or warm water, liquid soap, nailbrushes and appropriate drying facilities.

Highfield.co.uk ltd self-adhesive hygiene notices.

Misused and dirty handwashing trough.

Cloakrooms

Adequate accommodation for outdoor clothing and footwear, not worn by the staff during normal working hours, must be available. Such articles must not be stored in a food room unless in suitable cupboards or lockers provided only for this purpose. Adequate facilities for drying wet clothing should also be provided. Cloakrooms must be kept clean and tidy and food scraps should not be allowed to accumulate under benches and behind lockers as this may result in cockroach or rodent infestations.

STAFF SANITARY CONVENIENCES AND WASH STATIONS

Males or females at work			Alternative, males only		
Males/Females	W.C.s	Washstations	Males	W.C.s	Urinals
1 – 5	1	1	1 – 15	1	1
		16 – 30	2	1	
6 – 25	2	2	31 – 45	2	2
26 – 50	3	3	46 – 60	3	2
51 – 75	4	4	61 – 75	3	3
			76 – 90	4	3
76 – 100	5	5	91 – 100	4	4

Public sanitary accommodation

Section 20 of The Local Government (Miscellaneous Provisions) Act, 1976 enables a local authority to require the provision of sanitary appliances at places of entertainment and places selling food or drink to the public for consumption on the premises. The number of sanitary appliances is not stipulated, although hot water can be required. It is good practice for public toilets not to be used by food handlers.

PUBLIC SANITARY ACCOMMODATION RECOMMENDED FOR RESTAURANTS, PUBLIC HOUSES AND CANTEENS

	Males	Females
W.C.s	1 per 100	1 per 50
Urinals	1 per 25	
Wash-hand basins	1 per W.C. + 1 per 5 urinals	1 per 2 W.C.s

Wash-hand basins to be provided with hot and cold water supplies.
Provision should also be made for at least one cleaner's sink.

Sanitary disposal

Wherever women are employed or catered for, the law requires that suitable and hygienic provision is made for the disposal of sanitary dressings. Customer welfare should also be considered when selecting a disposal method.

Problems may arise from soiled dressings if the disposal of sanitary towels and tampons is inefficient or inadequate, and there is also a risk of blocked drains if attempts are made to flush such items down the toilet Pedal bins are inevitably unhygienic and present cleaners with the unpleasant task of emptying and cleaning them. Mechanical macerators and incinerators may produce unpleasant smells and may not completely destroy used sanitary dressings – especially where dressings contain synthetic cellulose rather than natural fibres.

An alternative method of disposal is a regular contract service using specially designed containers placed within W.C. cubicles to afford privacy and which are regularly collected and replaced by fresh, sterilized ones. The contractor is responsible for the safe disposal of the contents away from the customer's premises.

Some containers use a **bactericide** that works as a vapour as well as a liquid to ensure adequate protection and elimination of unpleasant smells. The bactericide should be capable of keeping the contents sterile for periods well in excess of the normal frequency of exchange.

Contractors should be asked to undertake a survey to establish the optimum number of sanitary towel dispensers, rate of replenishment and the number, and siting, of disposal units required. Some contractors may also provide other hygiene services or amenities such as air fresheners, coin-operated dispensers of sanitary dressings and a disposal service for such things as hypodermic needles.

The storage and disposal of waste

Waste disposal systems must be planned, along with other services, when food premises are designed. Refuse must not be allowed to accumulate in food rooms and should not be left overnight. Waste generated within the premises may be stored in polythene bags which are removed when full and at the end of each working day. Stands for such bags, must be maintained in a clean condition. Employees must be educated to "clean as they go", to replace lids and wash their hands after receptacles are used. Sacks should not be overfilled and should be tied to prevent problems from insects. Refuse collectors should not have to enter food rooms or dining areas.

Waste food should be kept separate from paper and cardboard packaging. In some instances, waste may be stored under refrigeration pending collection, for example, bones in butchers' shops. It is preferable for all waste food to be removed from food premises at least daily and general refuse to be removed at least twice a week.

Suitable facilities must be provided for the storage of waste externally, prior to removal from the establishment. The number and type of receptacles used will depend on the type and quantity of waste, the frequency of collection and the access available for the refuse vehicle. Wheelie bins are commonly used, although skips and compactors are more appropriate for large food factories. All receptacles should be durable, impervious, capable of being cleaned and provided with suitable tight-fitting lids or covers to prevent insects, birds and rodents gaining access.

The refuse area must have a well-drained, impervious surface, which is capable of being kept clean. Standpipes, hoses and, possibly, high-pressure sprayers should be provided for cleaning purposes. The receptacles, and the refuse area, should be thoroughly cleaned after each emptying. Covered areas to protect refuse from the sun and rain are recommended. Satisfactory provision should be made for the disposal of liquid food waste, such as oil. It should not be flushed down the drain.

Accumulations of rubbish in room adjacent to wash-up area.

Unsatisfactory refuse storage.

Refuse areas should not be too far from food rooms to discourage their use but they should not be too close to encourage flies to enter the food rooms. They should not be sited next to the main food delivery entrance. Receptacles used for the storage or collection of food must not be used for refuse.

Refuse compaction

Charges for refuse collection are continually rising and, since levies are usually based on volume of waste produced, often a price per bin, it is advantageous to use a refuse compaction system. Compactors vary from units similar to a large dustbin to refuse-sack compactors and skip rams. A well-run compaction system improves hygiene, as flies and other pests are less likely to be attracted. Cleaning of refuse areas is also easier as spillages are considerably reduced. Capital costs for purchasing compactors should be compared against collection charges, cleaning and pest control. However, before ordering a compactor the local authority cleansing department should be consulted, as special arrangements may have to be made for emptying.

Perimeter areas

A concrete path, at least 675mm wide, abutting the external walls should be provided around all food buildings. This removes cover for rodents and enables early signs of pests to be discovered, for example, rodent droppings. Paths should be kept clean, free of vegetation and inspected regularly. A smooth band of rendering, around 450mm, at the base of external walls will discourage rodents from climbing.

Whenever possible, a perimeter fence should be constructed around food premises to deter unauthorized entry. Areas within perimeter fences must be kept clean and tidy.

Unsatisfactory drainage.

Cluttered rear yard offering harbourage for pests.

Rubbish, old equipment and weeds must not be allowed to accumulate or provide harbourage for insects or rodents.

Kitchen design

Kitchen design to operate cost effectively, meet the needs of users and satisfy essential safety requirements and enforcement officers, demands considerable expertise. The layout of a well-designed commercial kitchen has three main characteristics:
- clearly identified and separated workflows;
- defined accommodation, and services specific to the purposes allocated; and
- economy of space provision (commensurate with good hygiene practice).

The unit will also be so arranged that it can be easily managed. The layout will afford management personnel easy access to the areas under their control and good visibility in the areas that have to be supervised. Space will be allocated for the management function and for equipment such as telephones and computers.

Workflows

Four separate workflows need to be considered from delivery to service: product; personnel; containers, utensils and equipment; and refuse. Cleaning of premises, equipment, utensils, crockery and cutlery will also need to be considered.

Product flows should be subdivided into **high-risk** and ready-to-eat foods, and contaminated (raw food) sections. Clear segregation should be maintained between the two. As far as practicable, workflows should be unidirectional, without backtracking or crossover.

Accommodation and services

Accommodation should be sized according to operational need. Essentially, the working areas, stores, the equipment and its relative spacing should all be determined and laid out to suit the operation.

Areas should be allocated according to environmental compatibility; hot functions with hot, dirty with dirty, wet with wet, dry with dry, defined overall by segregation between high-risk and contaminated food handling.

Adequate working space and good staff changing facilities are important to create good working conditions. The siting of electricity, gas, water and drainage must be considered at an early stage. Sufficient connections must be provided for mobile equipment. Specially designed heavy-duty connections will be required.

Size

There should be a minimum of circulation and dead space, commensurate with the efficient functioning of the unit. Size should be neither too small nor too large; there are penalties in the over-provision of space as much as there are problems with too small a provision. The use of standard-size equipment will be beneficial on space requirements.

The size of the kitchen can only be determined when its exact purpose and function have been defined. A unit designed for a fast food operation will not work effectively for à la carte service. When a catering operation is planned, the kitchen should be designed first and the rest designed around it. Haute cuisine, for example, will require ample storage space for fresh vegetables, a larder and pastry area. Fast food will require none of these but will need ample space for freezer storage. Other items to take into consideration include:

- the state of raw materials, for example, ready prepared or not;
- the extent of the menu, for example, à la carte or set menus;
- the length of meal times and the number of sittings;
- the scope of the operation. Does the business support multiple outlets such as banquets, or does it operate as a wholesaler;
- is the kitchen used purely for residential or non-residential;
- the type of equipment used, for example, cook-chill, microwave oven etc.;
- the use of modular equipment;
- the type and amount of dishwashing and cleaning. Are disposable plates and utensils to be used;
- if ready prepared food, such as bread and pastries are to be used; or produced in the kitchen; and
- if such functions such as dish-wash and vegetable preparation are to be included in the kitchen area or ancillary rooms.

Siting of equipment

Great care is needed to ensure that the operation works as a single system, with equipment required for specific functions grouped and accessible, to avoid excessive walking and the temptation to take shortcuts. For instance, refrigeration should be close to the working areas it serves, otherwise excessive amounts of food may be brought out to save the trips to and from the unit. Mis-en-place refrigeration is recommended. Wash-hand basins should be strategically located to ensure that operatives entering clean areas wash their hands. Personnel should also be encouraged to wash their hands when leaving the dirty area. Depending on the size of operations, additional wash-hand basins should be provided close to work stations. Minimizing the distance that staff need to walk to undertake their duties improves safety (the easy way becomes the safe way) and reduces costs, as it saves time.

A non-hand operable tap.
(Courtesy of Marks & Spencer.)

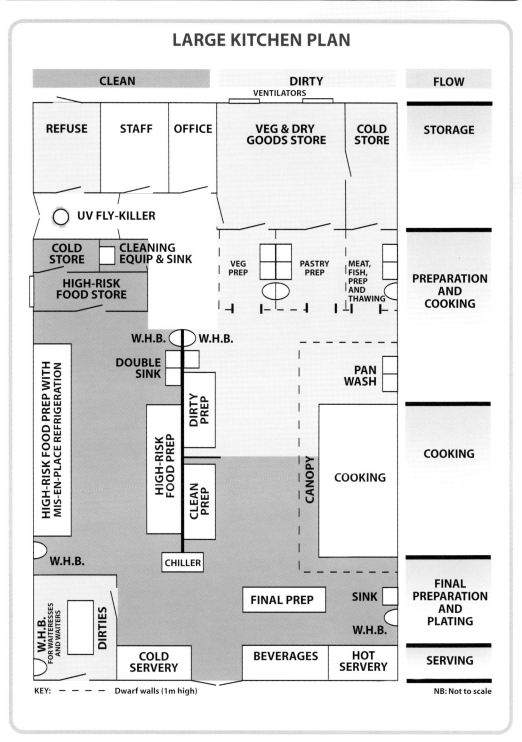

*The photographs used to show unsatisfactory premises in this chapter were provided by Kingston-upon-Hull City Council, Environmental Health (Food Team).

13 The design of equipment

Hygienic design of equipment is necessary to comply with legislative requirements, avoid product contamination and facilitate cost-effective cleaning and, if necessary, disinfection (downtime for cleaning will be minimized). Good design necessitates the consideration of the complete process, the environment and methods of cleaning and maintenance and not just the equipment in isolation.

Poorly designed equipment, which cannot be dismantled, may be uncleanable, incapable of being chemically disinfected and may result in product contamination by pathogenic bacteria, which will be disastrous in the case of high-risk food. Even if equipment can be dismantled, unhygienic design may make cleaning and disinfection prohibitively expensive. The ultimate test of whether or not a machine is hygienic is the standard of the product.

The legal requirements

Hygiene legislation requires all articles, fittings and equipment, with which food comes into contact, to be kept clean and be so constructed, of such materials and maintained in such condition and repair as to minimize risk of contamination and enable thorough cleaning and, where necessary, disinfection. Furthermore, equipment must be installed in a way that allows the surrounding area to be cleaned. Proprietors of businesses governed by specific regulations will need to comply with similar requirements and all businesses should be aware of the advice contained within any relevant UK or EC industry guides.

Failure to keep equipment clean and in good condition is likely to result in prosecution.

HYGIENE for MANAGEMENT *The design of equipment*

Dirty blade of can opener.

Dirty food production equipment.

Filthy exterior of gas hob/oven and deep fat fryer.

Unsatisfactory raw meat preparation area.

The cleanliness of, and risk of contamination from, equipment will need to be considered as part of the identification of steps in the activities of the businesses that are critical to ensuring food safety.

In June 1989 the Council Directive relating to machinery (89/392/EEC) as amended by Directive 91/368/EEC, was published. Although mainly concerned with safety, an important section deals with the hygienic design of agri-foodstuffs machinery.

In the UK the requirements of the Directive are enforced using the Supply of Machinery (Safety) Regulations, 1992 (as amended in 1994). The responsibility for compliance rests primarily with the manufacturers and suppliers (including importers), although the user must ensure that the machine is properly installed and maintained and used for the purpose for which it is intended. The Directive requires that new machinery used for preparing and processing foodstuffs carries a CE marking and must be designed and constructed to avoid health risks and in particular:

♦ contact materials of foodstuffs must satisfy the conditions set down in the relevant directives. Machinery must be designed and constructed to facilitate cleaning;

HYGIENE for MANAGEMENT *The design of equipment*

Equipment must be kept clean and well maintained.

- all surfaces and joints must be smooth, without ridges or crevices that could harbour organic materials;
- projections, edges and recesses should be minimal. Continuous welding is preferable. Screws and rivets should not be used unless technically unavoidable;
- contact surfaces must be easily cleaned and disinfected. The design of internal surfaces, angles, etc. must allow thorough cleaning;
- cleaning residues must drain from equipment surfaces, pipework, etc., there must be no retention in voids;
- the design should prevent organic accumulations or insect infestation in uncleanable areas, for example, by the use of castors or alternatively, sealed bases; and
- lubricants must not come into contact with product.

In addition, equipment manufacturers must provide information on recommended products and methods of cleaning, disinfecting and rinsing. This requirement is particularly relevant as, no matter how well designed, equipment will lose the potential to be hygienic if clear instructions for installation, operating, cleaning and maintenance are not provided. The simpler the instructions the better, as workers often attempt to short cut complicated and time-consuming systems, sometimes with disastrous consequences.

The European Committee for Standardization (CEN)

The European Committee for Standardization has been charged with the task of preparing new European standards and harmonizing documents and promoting the implementation of international standards in relation to machinery safety and hygiene.

A series of technical committees has been established. CEN/TC 153 is concerned with food processing machinery, safety and hygiene specifications. Another working group is considering the principles of hygiene design that apply to all equipment. In addition, there are nine groups working on standards for specific machines: bakery; meat; catering; slicers; edible oils; pasta; bulk milk coolers; cereals; and dairy.

Construction materials

Materials in contact with food must be non-toxic, non-tainting and constituents from their surfaces must not migrate into the food or be absorbed by the food in quantities that could endanger health. Materials must have adequate strength over a wide temperature range, a reasonable life, be corrosion and abrasion resistant and be easily cleaned/disinfected.

The Machinery Regulations state that materials in contact with foodstuffs must satisfy conditions set down in the relevant EU legislation. Regulation (EC) No 852/2004 on the hygiene of foodstuffs requires surfaces of food contact equipment to be smooth, washable, corrosion resistant non toxic and easy-to-clean and disinfect. Equipment must be constructed, be of such materials and be kept in good condition so as to minimize any risk of contamination. Equipment must be installed to allow adequate cleaning of the equipment and the surrounding area.

The most widely used material is food grade stainless steel, either type 304 containing 18% chromium and 8% nickel or the more corrosion resistant, and expensive, type 316 with a 10% nickel content and 2% molybdenum. Handles of knives, brushes and other equipment should all be made from cleansable materials such as stainless steel or high-density polypropylene.

Aluminium is attacked by sodium hydroxide, which forms the basis of some detergents, and sodium hypochlorite. Even hard anodizing fails to protect against severe pitting corrosion. Copper and alloys of copper hasten oxidation if in contact with oil or fat. Several metals, for example, copper, zinc and cadmium, are unsuitable as they are absorbed by acid food and may cause illness.

Some plastics may be suitable, but must be approved for food use. A useful indication can be obtained if materials comply with safety standards of the US Food and Drug Administration or the German Bundsgesundheitsamt. In the case of uncertainty, advice should be sought from the manufacturers, particularly with regard to high temperatures achieved during cleaning and disinfection.

Surface finish

Hygiene legislation requires surfaces to be in a sound condition and easy to clean and, where necessary, disinfect. This requires the use of smooth, washable, non-toxic materials unless food business operators can satisfy the competent authority that other materials used are appropriate.

Surfaces should be continuous, non-porous, non-flaking and free from cracks, crevices and pits. Surfaces will need to retain a satisfactory finish throughout their life including anticipated abuse and normal wear and tear. Instruments are available for measuring surface texture and a profile graph can be made of the shape, height and spacing of surface irregularities. The American National Standard for Food, Drug and Beverage Equipment suggests a value of no greater than the equivalent of 0.8mµRa for surface roughness of food contact surfaces.

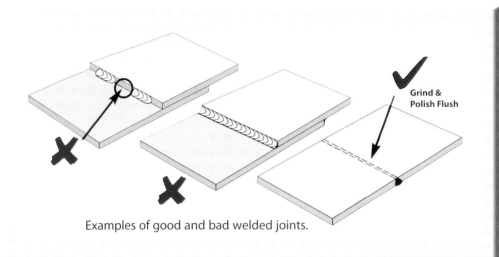

Examples of good and bad welded joints.

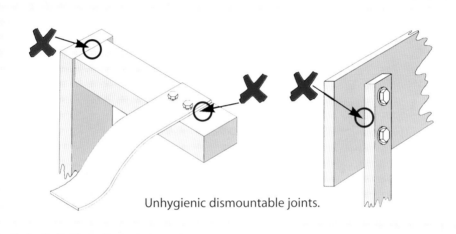

Unhygienic dismountable joints.

Unpolished cold rolled sheet from steel mills is well below this. (Roughness average Ra is the arithmetic average value of the department of the profile above and below the mean line throughout the specified sampling length.)

A rough surface provides a better mechanical "key" for soil. Furthermore, soiling is more apparent on bright surfaces, which is why it is cost-effective to electropolish stainless steel. (Care is needed with siting to avoid problems with sunlight reflecting.)

Joints

Joints should be made by welding or continuous bonding to reduce projections, edges and recesses to a minimum. Butt welding is preferred to lap joints provided the welds are ground and polished to the same standard as the rest of the surface, unless it can be demonstrated that the untreated weld is cleanable. Solders used in the construction of **food-contact surfaces** must also be non-toxic and should not contain cadmium or antimony. Silver soldering may be used to seal joints but must not contain cadmium.

Dismountable joints, for example, those relying on bolts, may contain product residue and high bacterial counts when disassembled. If such joints are not cleaned daily they should be sealed against the ingress of product and **micro-organisms** by means of a gasket. (Controlled compression of gaskets is essential.) Metal to metal joints, even if leak tight, may still permit the ingress and egress of micro-organisms. Flanged joints must present a smooth, continuous internal surface and be sealed with an appropriate gasket. (Some gaskets are made of materials such as rubber, which harden and crack over time.)

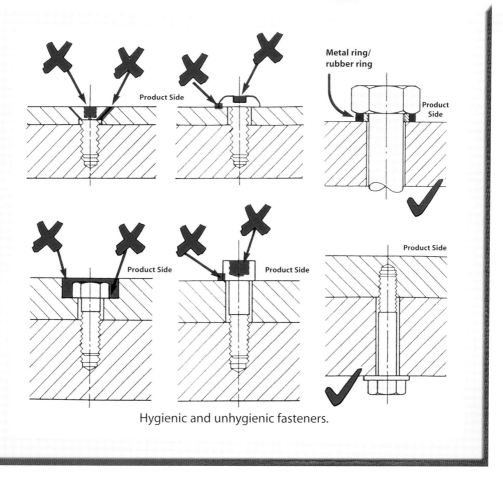

Hygienic and unhygienic fasteners.

Fasteners

Exposed screw threads, nuts, bolts and rivets should be absent from food-contact surfaces, although if removed daily for cleaning purposes are less likely to cause a problem. Annular recesses and screwdriver slots retain product residues. Nuts and bolts, etc. should be located on the non-product side where this can be achieved.

The hygienic design of equipment also includes avoiding physical contamination of the product. The accidental ingress of foreign matter such as bolts, nuts, washers and gasket materials must be prevented.

HYGIENE for MANAGEMENT *The design of equipment*

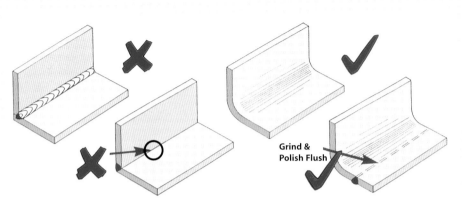

Examples of good and bad internal angles.

Internal angles and corners

The Machinery Regulations require inside surfaces to have curves of sufficient radius to allow thorough cleaning. When components are bolted or screwed together, sharp corners are usually unavoidable and metal that is bent or machined is preferred. Routine dismantling and cleaning is the only way to deal with unsatisfactory internal angles.

Dead spaces

Dead spaces allow product retention, bacterial growth and product contamination. They must be avoided in the design of equipment and not introduced during installation or as a result of modifications. The Machinery Regulations require the design and construction of equipment to prevent organic matter accumulating in, or liquids or insects entering, areas that cannot be cleaned. In effect, this means that gaskets of material approved for food use or contact must be used, or the design must enable regular dismantling for the space to be cleaned and disinfected.

Bearings

Bearings should, wherever practicable, be mounted outside product areas to avoid contamination by lubricants. Where this is not possible edible lubricants, or if liquid the product itself (as in the case of foot bearings), should be used.

Equipment exterior

The external surfaces of equipment must avoid ledges and dust traps, for example, round stretchers are preferred to rectangular. It is important to avoid recessed corners, sharp edges, unfilled seams, uneven surfaces and hollows and projecting bolt heads, threads, screws or rivets that cannot be cleaned. Inaccessible spaces, pockets and crevices where product may accumulate must be absent.

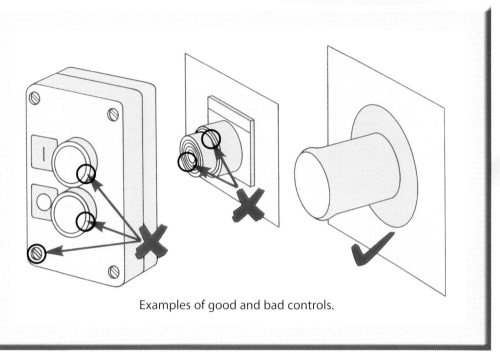
Examples of good and bad controls.

Instrumentation

Instrumentation must be constructed from appropriate materials and any transmitting fluids should be approved for food contact. Hygienic installation of instruments is essential. Controls such as push buttons must also be of hygienic design so they can be maintained in a clean condition and be capable of being cleaned by whatever system is in use. Dials that are fitted to machines must have adequate clearance to allow cleaning.

Fixing and siting of equipment

Equipment must be sited so that there is sufficient space to facilitate access to all external and internal surfaces and, where required, to allow for rapid dismantling and reassembly. Stationary equipment should be at least 500mm from walls and have a clearance of at least 250mm between the floor and underside of the equipment, to allow thorough cleaning of wall and floor surfaces. Alternatively, machines should be fixed firmly to the floor and sealed. Narrow areas or angled contact with the floor should be avoided. Machinery may be mounted on coved, raised platforms of concrete to facilitate cleaning. Where necessary, additional space may need to be provided. The bases and lower parts of machines, including motors and gears, may be difficult to clean and consequently collect dust and spillages, which make ideal breeding sites for insects. Skirting or cover plates tend to trap dust.

Where practicable, and with due regard for safety, equipment can be mobile to facilitate its removal for cleaning. This is particularly important if sited close to walls. Tubular metal storage racks, refrigerators and ovens should be castor-mounted with brakes on all wheels. Tables and benches should not be fitted with castors as even when braked, slight movement can still occur, posing a hazard to knife users.

When pipework is installed, future cleaning must be considered and sufficient space left between the pipes and floors or walls to allow easy access for cleaning. Alternatively, pipes may be built into the walls. Lagging of pipes must have an impervious cleansable finish. Pipework and equipment directly above open food must not expose food to risk of contamination from condensation, rust or flaking paintwork.

Conveyors can be designed to enable automatic cleaning. Manufacturers must be pressured to ensure that if dismantling is necessary it is as simple as possible.

Guards, required under health and safety legislation, should be capable of being cleaned and, where necessary, removed quickly and easily, even if special tools are required. Sheet-metal and wired-perspex guards are preferable to painted mesh.

Gas and electrical connections

The presence of electrical equipment such as motors and switches poses problems for cleaners, as even routine cleaning may expose users to some danger of electrocution. Such equipment must always be waterproof and/or capable of removal without tools, prior to cleaning the main equipment. If practicable, electrics should be sited outside food rooms, to reduce cleaning problems. It also enables servicing and maintenance to be carried out by electricians who do not need to enter food rooms.

Gas and electricity supply pipes to ovens should be flexible and capable of being disconnected. In the case of gas, the coupling to the mains must incorporate a self-sealing valve and a shut-off cock immediately upstream of the fitting. The flexible tube must conform to gas suppliers requirements and the appliance must be secured to a fixed point by a strong, detachable wire or chain, shorter in length than the tube, to avoid accidental stretching. Trailing wires and cables must be avoided.

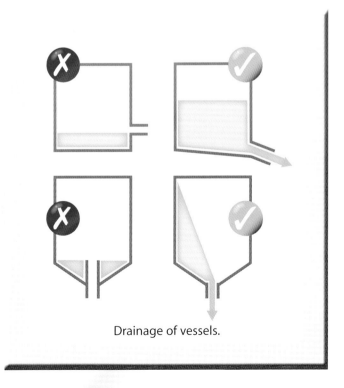

Drainage of vessels.

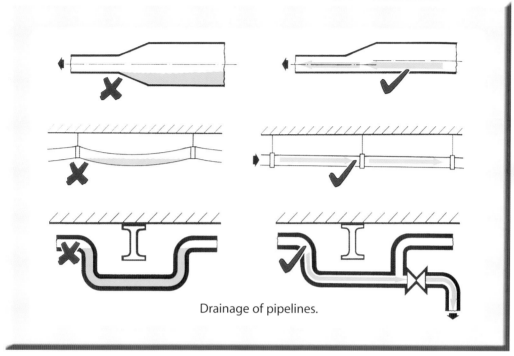
Drainage of pipelines.

Drainage
All pipelines, vessels and equipment should be self-draining, not only to enable liquid deriving from foodstuffs to be discharged but also for cleaning and rinsing fluids. It is particularly important that pipelines and vessels cleaned by CIP systems are designed to facilitate complete drainage. Valves need to be accessible for cleaning.

Preparation surfaces
Preparation surfaces should be jointless, durable, impervious, the correct height and provide a firm base on which to work. If materials other than stainless steel are used, for example, food grade plastic, care should be taken to seal the edges and gaps, which may harbour food scraps. They must be able to withstand repeated cleaning and disinfection without premature deterioration, pitting or corrosion. Flanged-lip designs for tables and shelves, which harbour food residues and are difficult to clean, should be avoided. Abutting surfaces that can be separated for cleaning may be used.

Chopping boards
A variety of non-absorbent, plastic chopping boards is now being produced, for example, good quality polypropylene. However, some plastics are unsatisfactory and hardwood boards in good condition can be more hygienic. Soft wood and open jointed boards are unacceptable. When selecting a chopping board regard should be had to its:
- water absorbency;
- resistance to stains, cleaning chemicals, heat and food **acids**;
- toxicity and odour;
- durability; and
- cleanability.

Microwave ovens have been used to successfully disinfect wooden chopping boards but not plastic boards. Temperatures of 95ºC can be achieved in 4 minutes in 800W ovens. Longer times may result in smoking. Increased moisture (wet boards) enhances killing efficiency of micro-organisms in wood. Furthermore, a good plastic board should not split or warp when passed through a dishwasher, which is one of the best ways of disinfecting it.

The use of boards containing the bactericide triclosan is increasing. Triclosan is used in toothpaste, mouthwash and deodorants and should remain active for the life of the board. Concerns over the use of bactericides in chopping boards, equipment and wall finishes have been expressed because of the risk of developing resistance in pathogenic micro-organisms. As yet no ideal replacement has been found for hard wood chopping blocks used by butchers. However, these should be maintained in good condition and used solely for chopping or sawing raw meat.

An unacceptable cutting board.

Contamination

To avoid cross-contamination, it is important that the same equipment is not used for handling raw and high-risk products without being disinfected. To prevent the inadvertent use of equipment for high-risk and raw food it is recommended that, where possible, different colours and/or shapes are used. Colour coding may be extended from knives and chopping boards to include washing facilities, trolleys, protective clothing, cloths and packaging material. A common colour-coding system involves using: red (raw meat), green (salad), blue (fish), yellow (high-risk), white (dairy) and brown (raw vegetables).

Deterioration

Equipment deteriorates with age and eventually becomes incapable of being thoroughly cleaned. Routine maintenance is essential. Chipped, broken and badly pitted equipment allows the harbourage of dirt and bacteria and should be replaced. String or tape must not be used to repair defective equipment, including knife handles. Equipment that is no longer used should be removed from food premises.

The cleaning of equipment

All operating instructions and procedures must be clearly communicated to equipment users and cleaners. The equipment should be capable of being cleaned and, if necessary, disinfected safely, thoroughly and rapidly without the need for skilled fitters and specialized tools. If dismantling is necessary this must be achieved relatively easily, as should reassembly.

Sharp edges are a serious hazard for cleaners. A reluctance to clean equipment because of poor design will result in a lowering of hygienic standards. Hinges should be capable of being taken apart for cleaning. Angle iron is difficult to clean and tubular construction is preferred, open ends to tubular legs must be sealed.

Storage of clean equipment

Having spent a considerable amount of time cleaning and disinfecting equipment and utensils, it makes sense to ensure that subsequent storage does not result in contamination. The USA Food and Drug Administration Food Service Sanitation Manual states that clean equipment and utensils should be stored at least 450mm above the floor in a clean, dry location, in a way that protects them from contamination by splash, dust and other means.

Maintenance

Equipment must be maintained in efficient working order and good repair. This involves routine inspection and preventive maintenance as recommended by the manufacturer and determined by experience. Regular effective cleaning and waste removal will assist in extending the life of equipment. Maintenance must be carried out by competent persons who have received adequate information and training. Pressure systems must be inspected by engineering inspection companies.

Build up of flour dust and debris, leading to failure to keep equipment clean and in good condition is likely to result in prosecution.

*The photographs used to show unsatisfactory premises in this chapter were provided by Kingston-upon-Hull City Council, Environmental Health (Food Team), and the illustrations by Campden & Chorleywood Food Research Association.

14 Cleaning and disinfection edited by Peter Bird & Steve Bagshaw

Even in the most highly automated food business, operating under the most controlled conditions, soiling of both surfaces and equipment is unavoidable. The type and extent of soiling will vary considerably but, whatever the operation, it is essential that such residues are not allowed to accumulate to levels that expose food to the risk of contamination. Removal of these residues is the process of cleaning.

The objective of the clean varies depending on the standard required. A physical clean may be used to prevent physical contamination of the product. However, operations involving high-risk foods require a microbiological clean, which involves cleaning to remove soiling, and disinfection to destroy micro-organisms. A clean surface has no soiling or objectional odour, is not greasy to touch and will not discolour a white paper tissue wiped over it.

The law and cleaning

Food Safety legislation requires food premises and equipment to be kept clean. The standards of cleanliness and frequency of cleaning required will vary depending on the type of operation and the type of food being prepared. This means that, for example, the standard of cleanliness in a cook-chill kitchen will be higher than a room used solely for peeling potatoes. Food-contact surfaces in high-risk areas will require periodic disinfection, whereas thorough cleaning of areas used for low-risk food preparation may suffice.

The frequency of cleaning and the removal of waste and other refuse, must ensure that there are no accumulations in food rooms, except so far as is unavoidable for the proper functioning of business. Adequate facilities must be provided for cleaning and disinfecting work tools and equipment and, where necessary, adequate provision for any washing of food.

All food must be protected from contamination that would render it unfit for human consumption, injurious to health or unreasonable to consume. This means that food must not be exposed to risk from cleaning. Several prosecutions have resulted from the contamination of food with cleaning chemicals, or with bits of metal, plastic or wood from cleaning equipment.

The benefits of cleaning

Cleaning is an essential and integral part of a profitable food business. In addition to satisfying legal requirements, cleanliness will:
- disrupt routes of contamination, e.g. cleaning and disinfecting tables used for raw and high-risk foods;
- ensure a pleasant, safe and attractive working environment, which will encourage effective working and reduce the risk of accidents to both staff and customers;
- promote a favourable image to the customer and assist in marketing the business;
- remove matter conducive to the growth of micro-organisms so facilitating effective disinfection and reducing the risk of food poisoning, spoilage and wastage;
- remove materials that would provide food or harbourage for pests and prevent early discovery of infestations;
- reduce the risk of foreign matter contamination and thereby obviate customer complaints; and

♦ prevent damage to, or a reduction in, the efficiency of equipment and services, and reduce maintenance costs.

HACCP and cleaning

Cleaning must be carried out at every stage in the production of food, from the delivery bay to the service point and is often included in **HACCP** plans. However, in most food businesses it is preferable for effective cleaning and disinfection to be considered as an essential prerequisite to the introduction of HACCP.

Hazards and problems from cleaning

Although cleaning is intended to eliminate **hazards** or reduce them to a safe level, negligent or ineffective cleaning can result in many serious hazards and other problems.

Hazards from cleaning

Microbiological contamination may result from:
- cleaning from raw to high-risk areas;
- using contaminated cleaning equipment especially brushes and cloths;
- using the wrong chemicals or the right chemicals at the incorrect dilution;
- dust created during sweeping and dry cleaning; and
- aerosols created by pressure spraying, floor scrubbing and boot washing.

Microbiological multiplication may occur when:
- cleaning and disinfection are not carried out properly or at the correct frequency; and
- there is a failure to remove food debris thoroughly and frequently.

Microbiological survival may occur when:
- surfaces are not adequately cleaned prior to disinfection; and
- the type, concentration or **contact time** of a **disinfectant** is inadequate.

Chemical contamination may result from:
- using the wrong chemicals, which produce **tainting**, for example, phenolic **disinfectants**;
- using chemicals at the wrong concentration, for example, hypochlorite for disinfecting lettuce;
- a failure to rinse properly, for example, clean-in-place systems;
- storing chemicals in food containers or bottles; and
- storing chemicals with food.

Physical contamination may result from:
- using inappropriate, worn and defective cleaning equipment;
- using inappropriate cleaning substances, for example, abrasives;
- using defective **protective clothing**, especially torn rubber gloves; and
- a failure to remove food debris, waste and packaging material.

Problems from cleaning

In addition to food safety hazards, negligent cleaning results in many other problems:

- financial consequences arising from using the wrong chemicals, the wrong dilution, the wrong temperature and inadequate contact time. This not only applies to direct costs relating to product, equipment and premises, but also the cost of effluent treatment;
- a poor quality product, which may lead to a reduction in shelf life, customer complaints, loss of reputation, court proceedings, food poisoning, redress from suppliers and loss of sales;
- wastage of food and production re-runs;
- corrosion and premature replacement of equipment;
- production breakdowns, for example, following the incorrect use of caustic soda, which has removed grease from bearings;
- unacceptable deterioration of floor surfaces and drainage systems;
- fire hazards, for example, because of build up of grease in ventilation ducting from extract hoods; and
- accidents due to slipping on wet floors, mixing chemicals or not using appropriate protective clothing.

CLEANING TERMS

Technical terms commonly used in cleaning science include:

Abrasives

Abrasives include hard abrasives, such as pumice and fine sand, which may be compounded with soaps to form pastes, and mild abrasives, such as lime and chalk, which are used in cream cleaners. Not generally approved of because of the risk of foreign body contamination.

Amphoteric surfactants

The active ions of amphoterics can be either positively or negatively charged, depending on the pH (acidity or alkalinity) of the solution. Some amphoteric surfactants are very mild in nature and are used in shampoos; others have excellent biocidal properties and are used as disinfectants. They have low human toxicity but high unit cost.

Anionic surfactants

Anionic surfactants separate in solution to form ions. The active ion is negatively charged. They are the most commonly used surfactants and are often mixed with non-ionic surfactants to form basic commercial detergents such as washing-up liquid. They are generally biodegradable, non-toxic and have good wetting properties.

Biodegradability

Biodegradability refers to the property of a surfactant, which allows it to be broken up by bacteria in sewage works, so reducing pollution and toxic hazard to fish. All modern surfactants tend to be biodegradable, although some form intermediate toxic compounds and have been banned in some countries.

Cationic surfactants

Cationic surfactants ionize in solution, the active ion being positively charged. Mixed with non-ionics, they are very commonly used as disinfectants or **sanitizers**. They are high foaming and have relatively poor wetting characteristics.

Compatibility

Detergents with different ionic charges should not be mixed as they are not compatible and become inactivated. Even mixing similar types of detergent may impair performance, as blends are carefully balanced to **optimize synergistic** effects.

Emulsion

An **emulsion** is a suspension of one liquid in another. Detergents have the ability to surround microdroplets of oil or fat and hold them in suspension as an "oil-in-water" emulsion that can be easily rinsed away.

Foaming activity

Foam is not an essential part of a detergent and can interfere with mechanical cleaning. In manual cleaning, however, foam increases the surface area of a solution, increasing the speed of activity and enables more dirt to be suspended. Too much foam can cause rinsing difficulties.

Non-ionic surfactants

Non-ionic surfactants do not ionize in solution and therefore do not carry a charge. They are compatible with anionic, cationic and amphoteric surfactants, commonly being used as components of blends to improve overall cleaning performance. They can also be mixed with **acids** to form acid cleaners. They are generally biodegradable, non-toxic, have good wetting properties and may be high or low foaming.

Saponification

Saponification is the process of making soap, usually by boiling vegetable and mineral fats and oils with an alkaline material such as caustic soda. This also occurs in cleaning when alkalis react with fatty or oily **organic** soils. The soap thus formed assists in the further cleaning action. Soap is anionic in character and forms a scum in hard water when it reacts with the calcium and magnesium ions.

Scouring powder

Scouring powders are cleaning powders combining abrasive and often minute amounts of bleaching agent. They will seriously damage enamel or stainless steel surfaces and should not be used for cleaning these materials.

Sequestrants

Sequestrants are chemicals that counteract the effect of **water hardness** salts, preventing the formation of scum, which would interfere with the cleaning action of chemicals.

Solvents

Solvents are chemicals that have the effect of dissolving specific types of soil. Solvents, such as glycol ether, are often used in small amounts as a part of a detergent formulation to improve its performance for dealing with oils and fats.

Synergism

Synergism is an enhanced performance achieved by the mixture of two or more chemicals, the sum of the activity being greater than the total effect of the components acting separately. This effect occurs with ionic and non-ionic detergent blends and can also be important in disinfectant formulations.

MANAGEMENT FUNCTIONS OF CLEANING

Cleaning is essentially a management function and the physical process of removing dirt is only the final stage. The prime responsibility is to ensure that premises and their contents are capable of being effectively cleaned. Not only does this make economic sense but it is also a legal requirement.

Securing the commitment

Paying lip service to standards is not enough. Everyone involved in a food business must be personally committed to ensuring that satisfactory standards are achieved. Requisite standards must be clearly defined, effectively communicated and reinforced by management. Staff must be motivated, instructed, supervised and controlled.

Providing the means

Sufficient numbers of adequately trained cleaning staff, properly supervised and supplied with appropriate materials and equipment, must be employed. Adequate supplies of hot water are essential. To avoid confusion the number of different cleaning chemicals used should be minimized.

Reducing unnecessary labour

Careful planning, assembly of materials and adopting the most suitable method will save considerable time and effort. Systematic wiping or spraying will avoid duplication of coverage and the risk of areas being missed. If "clean as you go" is in operation, cleaning will be easier as residues will not have hardened. Organizing work in teams is often highly efficient.

The use of contractors

It is often worth considering the use of specialist cleaning contractors to supplement in-house cleaning by intensive deep cleaning of the structure, ventilation trunking, drains and difficult-to-reach surfaces. Contractors may use steam, special solvents, degreasers and sophisticated equipment to establish and maintain standards of hygiene not otherwise readily attainable in inaccessible areas. However, contractors must be selected carefully and suitable references should always be obtained before allowing the use of potentially dangerous chemicals that could, in inexperienced hands, cause thousands of pounds worth of damage and extensive contamination of food.

Cleaning schedules

Cleaning schedules are a communication link between management and staff and are necessary to ensure that equipment and premises are effectively cleaned and, if necessary, disinfected as frequently and as economically as possible. When planning a cleaning schedule, the following items should be considered:

- the size, type and temperature of area to be cleaned, the structure of the building and the wall, floor and ceiling finishes;
- the type of equipment and the material from which it is made. The manufacturer's recommendations should always be followed;
- whether or not the equipment can be dismantled;
- the type of soiling and the water hardness;
- the presence of electrics;
- the available water pressure and drainage system;
- the time constraints, manpower and training needs;
- if cleaning is necessary during food preparation, or if food should be removed;
- whether or not disinfection is required;
- the risk of spreading contamination, for example, the role of pressure hoses in disseminating *Listeria monocytogenes;*
- the requirements of the Health and Safety at Work etc. Act, 1974, and all relevant health and safety legislation;
- the controls available to monitor the effectiveness of cleaning; and
- the overall cost, including labour, equipment, chemicals, water and heat. In large premises the cost of effluent treatment must also be considered.

The cleaning schedule itself must be clearly and concisely written, without ambiguity, to ensure that instructions to staff are easy to follow and result in the objective of the schedule being achieved.

Written schedules should specify:
- what is to be cleaned;
- the chemicals, materials and equipment to be used;
- the dilution and contact time of the chemical;
- how it is to be cleaned (method);
- when it is to be cleaned (frequency);
- the time necessary to clean it;
- who is to clean it;
- the protective clothing to be worn;
- the safety precautions to be taken; and
- who is responsible for monitoring and recording that it has been cleaned.

Control and monitoring of cleaning

A system for identifying problems must be established. Faults may be caused by poor management, administrative deficiencies, unsatisfactory staff, inadequate training or unreasonable expectations of individual performance.

Standards should be routinely monitored through inspections, which are carried out by management personnel not directly involved in cleaning operations. All findings should be recorded and utilized to rectify faults and improve the effectiveness of cleaning.

ADMINISTRATION

To implement management policy, certain administrative functions must be carried out and these include:

Selecting a chemical supplier

Price is obviously important, but cheap chemicals can result in expensive cleaning. Suppliers should have a knowledge of the type of operation, a satisfactory range of products, a knowledge of all appropriate health and safety legislation, the expertise to undertake a comprehensive audit to determine cleaning needs, the time to spend demonstrating how to use the product to best effect and easy to understand literature that does not mislead. They must be reliable and have the ability to respond to emergencies and preferably be quality assured.

Stock control

Provisions similar to those applicable to food stocks should apply, with proper stock rotation, delivery checking and inventory control. Comprehensive records should be kept of all transactions and minimum and maximum stock levels should also be established. Authorization to order should be restricted to named personnel.

Distribution

Transfer of materials from the store to the point of use should be properly organized to ensure that the right materials, in the right quantities, reach the right people, at the right time. Over supply, or storage of bulk stocks in working areas, should be avoided.

Financial control

Records of expenditure on each item, identifiable with the area of use, should be kept. Expenditure norms should be established and any deviations investigated, as this is often the first indication to management that procedures are amiss. Underspending is as significant as overspending.

TECHNOLOGY OF CLEANING

Management and administration apart, the remainder of the cleaning function is the application of technology. In this sense cleaning is defined as "the systematic application of energy to a surface or substance, with the intention of removing dirt". Central to the definition is the concept of energy.

Energy in cleaning

ENERGY IS AVAILABLE FOR CLEANING IN THREE DISTINCT FORMS

kinetic energy:	physical – manual labour;
	mechanical – machines;
	turbulence – liquids (clean-in-place);
thermal energy:	hot water; and
chemical energy:	detergents.

Normally, a combination of two or more energy forms is used. Manual labour is the most expensive and chemical energy the most economic, although adequate contact time is also important. The correct energy balance is essential for cost-effective cleaning.

Cleaning costs

CLEANING COSTS MAY BE APPORTIONED AS FOLLOWS			
labour	70%	chemicals	6%
equipment	12%	heating	4%
water/effluent	6%	corrosion	2%

The above figures should only be considered as a rough guide, for example, if only buckets and mops are used, equipment costs will be lower and labour costs increased. If the wrong chemicals are used, corrosion costs may be significantly greater than 2%. Water and effluent costs have increased significantly over the last ten years and water use should be minimized.

Chemical cleaning

The usual cleaning medium is water, which dissolves certain residues and forms a solution that can be rinsed away. Water, however, is not efficient in dissolving many of the soils that occur in the food industry. To improve the efficiency of water as a cleaning chemical, and to counteract the effect of the impurities in water (hardness salts), other chemicals, for example, detergents, are added.

Detergents

Detergents are chemicals, or mixtures of chemicals, made of soap or synthetic substitutes, with or without additives, which are used to remove grease or other soiling. They are available as powders, liquids, foams or gels.

Detergents can exhibit two types of action depending on their make-up and the soil they are acting upon. These are: chemical actions where a constituent of the detergent reacts chemically with the soil, for example, the reaction of mineral scale and acid detergents; and physical actions where a constituent of the detergent helps to penetrate, remove and disperse the soil from a surface, for example, the removal and *emulsification* of fats from a surface using a neutral detergent.

Manual detergents used in the food industry should be non-toxic, non-tainting, non-irritant, non-corrosive, free-rinsing, soluble in water and not form scum in hard water.

Surfactancy

This is the property of a detergent that enables it to increase the "wetting power" of water by reducing the surface tension. This increases the contact between the soil and the detergent solution, which is able to penetrate the minute irregularities of the dirt more effectively.

Dispersion

This is the ability of a detergent to break up large accumulations of matter into smaller particles.

Suspension

When dirt is broken up into particles, they become coated with a thin film of detergent, which keeps the particles apart and buoyant, i.e. in suspension, allowing them to be rinsed away. This characteristic is commonly referred to as the emulsifying action of detergents.

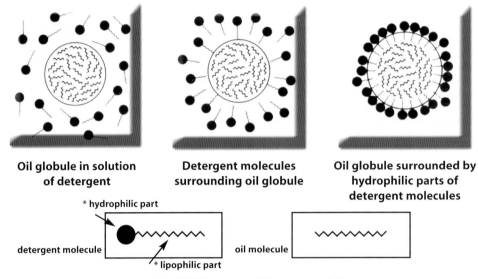

Emulsifying action (diagrammatic).
* hydrophilic means water loving and lipophilic means fat loving.

Classification of detergents
For convenience, detergents tend to be classified by their acidity or alkalinity.

Alkaline detergents

The most commonly used alkaline detergent is caustic soda (sodium hydroxide). It is corrosive to skin, aluminium and zinc, has poor wetting properties, but is effective for fat and protein solubilization and is relatively cheap. Sequestering agents (chelates), such as the amino carboxylic acids, EDTA and NTA, are added to prevent scale formation in hard water. Inorganic phosphates are effective sequestrants used in the UK, mainly in powder products, but are banned in some European countries because of the eutrophication of rivers.

Acid detergents

These are mainly used to remove mineral, protein and vegetable deposits and commonly contain phosphoric acid, which is one of the least corrosive acids. It must never be allowed to come into contact with chlorinated compounds because of the consequential release of toxic chlorine gas.

Neutral detergents

These are generally blends of surfactants used for manual dishwashing and manual cleaning. They are by their nature relatively safe to use, but gloves should still be worn to avoid defatting of the skin and to reduce the risk of dermatitis.

Detergent action
The reduction of surface tension enables detergent solutions to penetrate dirt and grease and lift them from the surface to form a suspension.

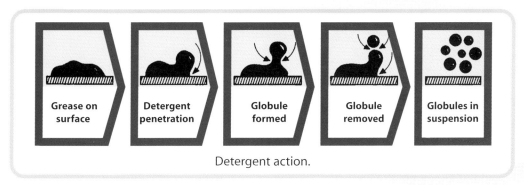

Detergent action.

Types of soil

In very general terms there are two types of soil: organic and inorganic.

Organic soils

These are derived from living matter and include animal fat, vegetable oils, starch, sugars and proteins from milk, egg, meat or blood. Normally they may be removed by using neutral or alkaline detergents. However, if heated, dried or allowed to remain for a prolonged period, then alkalis must be used to remove them.

Grease or oil will, if heated, form a tenacious, dark, sticky deposit known as polymerized grease. This may be removed by a mild alkali such as sodium carbonate (washing soda) or sodium metasilicate. Further heating of polymerized grease results in the formation of carbon, which must be removed by the more aggressive, caustic alkalis such as sodium hydroxide or potassium hydroxide (caustic potash).

Another type of organic soil is tannin, which is derived from tea, coffee or wine. It is normally removed by an oxidizing agent such as sodium hypochlorite (bleach) or sodium perborate. However, if the tannin is present with water hardness salts it may be better to use an acid detergent.

Inorganic soils

These include water hardness salts (scale), oxidized metals (rust), uric acid salts (urinal stains), beerstone and calcium salts deposited from milk (milkstone). Inorganic soils are normally removed by acids such as hydrochloric and phosphoric, although these should not be used on aluminium due to potential corrosion problems.

Soils are very rarely made up of one component, for example, a hard scale in a meat processing area may be a complex of a mineral scale and protein.

DISINFECTION

Disinfection is defined by British Standard 5283 as "the destruction of micro-organisms, but not usually bacterial spores; it may not kill all micro-organisms but reduces them to a level which is neither harmful to health nor the quality of perishable foods". Disinfection may be achieved by using heat, chemicals, irradiation or UV radiation. UV is usually effective for atmospheres and clear water but not for surfaces. The term sterilization relates to the destruction of all micro-organisms and spores and is normally unnecessary and impracticable to achieve within the food industry.

Disinfection is commonly used to describe one of the last stages of the clean, where any remaining micro-organisms are reduced to an acceptable level. However, it must be remembered that the most important stages in the overall regime are cleaning and rinsing. The disinfection stage on its own, or following poor cleaning, will give poor microbiological results.

Heat disinfection

The application of heat is the most reliable and effective means of destroying micro-organisms, although it may not be the most practicable, especially for surfaces. It is used in machines, such as dishwashing machines, with a water temperature of 88°C and a contact time varying from one to 15 seconds. It is also used in "sterilizing units" where articles may be fully immersed for a period of 30 seconds at 82°C.

Steam disinfection

Lances producing steam jets may be used in large food factories to disinfect machinery or surfaces that are difficult to reach. Steam-cleaned equipment is self-drying, although disadvantages may include:
- an adverse effect on some materials, for example, some plastics and certain wall and ceiling finishes, such as paint, may be stripped off;
- the removal of grease and lubricants from machinery;
- a safety hazard if steam lances are used by untrained personnel; and
- the formation of condensate on other equipment, especially overhead pipes and metal girders, leading to corrosion and potential product contamination.

Chemical disinfection

Disinfectants suitable for use in the food industry are limited to those that, when used correctly, will not have a deleterious effect on food, equipment or personnel. Types of chemical disinfectants available include:

Chlorine release agents

The commonest chemical **bactericides** are hypochlorites, which are salts of hypochlorous acid (HOCl). The most useful is sodium hypochlorite, which disinfects by oxidation of protein, an essential part of the structure of bacteria, **viruses**, **yeasts** and **fungi**.

They are normally supplied in solutions containing 6 to 14% available chlorine and should be diluted to between 100 and 2,000ppm of available chlorine, depending on the particular situation. A freshly made-up solution of hypochlorite containing 100ppm of available chlorine will satisfactorily disinfect a completely **clean surface**, provided a contact time of approximately three minutes is allowed. However, depending on the amount and type of soil, pH, contact time, temperature and organisms or spores present, a concentration of up to 2,000ppm of available chlorine may be necessary to ensure disinfection. In the event of doubt, the manufacturer's advice should be obtained. For general use, to minimize corrosion, it is recommended that hypochlorites are used below 40°C, at a maximum of 200ppm with a contact time of up to 20 minutes. Upon completion of disinfection, they should be rinsed off with clean, cold water. They must never be used in conjunction with acid cleaners due to the production of chlorine gas. Hypochlorites may react with phenols, present in some resins used in glass fibre vessels, to form chlorophenols, which result in taint problems.

Hypochlorites are relatively cheap and are active against most micro-organisms. They are unaffected by hard water. However, they have a pungent odour, are corrosive to man and many metals and are readily inactivated by organic soil. They have no wetting power and therefore a surfactant may be added. Liquid hypochlorite should be stored in a cool, dark place. The use of chlorinated powders containing, for example, sodium dichloro-isocyanurate, eliminates most of the above disadvantages.

Provided they are used on clean surfaces, allowed adequate contact time, residues are rinsed off and suitable safety precautions are taken during handling, the advantages of using hypochlorites far outweigh their disadvantages.

Quaternary ammonium compounds (QACs or **quats***)*

Cationic in characteristic, QACs are safe, non-corrosive, stable, and taint-free bactericides with some inherent detergent properties and little odour. In isolation, they have a narrow range of activity and are not as effective as hypochlorites against Gram-negative bacteria, viruses and fungi. They may be inactivated by hard water, organic material and some plastics. Effectiveness can be significantly improved by the incorporation of a sequestrant as part of a well-built formulation with non-ionic detergent and alkali ingredients. QAC-based disinfectants are the most widely used surface disinfectants in the food processing industry. Because of their inherent substantive properties, which can make it difficult to completely remove them by normal rinsing, they are not used in brewing as any residual QAC effects beer head retention.

Disinfectant soak.
(Courtesy of Holchem Laboratories Ltd.)

Amphoteric disinfectants

Amphoterics exhibit bactericidal properties. They are of low toxicity, relatively non-corrosive, tasteless and odourless. They are good disinfectants but are expensive. They are high foaming and are unsuitable for use with machines and high-velocity sprays.

Biguanides

These are cationic bactericides, similar in function to QACs but with better all-round performance. They have no wetting properties and are therefore compounded with non-ionic detergents. They are sometimes used in the licensed trade for glass washing and are non-foaming. They are unsuitable for use in highly alkaline conditions and should not be used after caustics.

Iodophors

Iodophors are expensive but effective bactericides consisting of iodine and non-ionic surfactant in an acid medium. They have both detergent and disinfecting properties and their acid characteristics make them suitable for the dairy industry, breweries and soft drink manufacturers. Advantages include the ability to kill a wide range of organisms, effectiveness at low temperatures (iodine may sublime above 40ºC), tolerance of soiling and hard water and brief contact time. Iodophors are stable, of low toxicity and virtually odourless if properly used, although they may combine with food to cause taint. Furthermore, if used negligently they may be corrosive.

Iodophors stain soiling matter such as milkstone and so indicate poor physical cleanliness. Plastics may be stained but metals are not. In use, the yellow-brown colour of the iodophor is absorbed by organic matter and the colour disappears, a useful indicator of potency. They should not be used on aluminium or copper.

Peroxy compounds

These fairly specialized products are increasingly used for terminal disinfection in dairies and breweries (per-acetic acid, replacing hydrogen peroxide) and as peroxy compounds within a detergent blend for agricultural and other applications.

Per-acetic acid destroys micro-organisms by an oxidative action and remains effective at temperatures as low as 0ºC. Once diluted, it has a limited life and is only suitable on glass or stainless steel surfaces.

Alcohols

Where there is a requirement for light cleaning and disinfection in an essentially "dry" area, for example, delicatessen point of sale or packing belts on a production line, the use of an alcohol-based spray/wipe product is particularly useful. This is typically a blend of alcohol, QAC, and possibly mild detergent additives, formulated to provide good disinfection in lightly soiled conditions and without the use of water. The product flashes dry after application thus removing the need for an undesirable wipe-dry operation. Care must be exercised because of the flammability of alcohols.

Ethanol and iso-propanol are two of the alcohols commonly used for surface and hand disinfection.

Aldehydes

Formaldehyde and glutaraldehyde have a wide spectrum of activity but are not used because of their high toxicity and irritancy characteristics.

Choosing a disinfectant

The choice of disinfectant depends on many factors including:
- amount of soiling;
- type of cleaning equipment;
- water hardness;
- contact time available;
- stability;
- type of micro-organisms that need to be destroyed;
- type, imperviousness and smoothness of surface to be disinfected;
- possibility of taint;
- temperature of application;
- toxicity of the disinfectant and the effect on personnel;
- ionic nature of the detergent used before disinfection; and
- the way the disinfectant is to be used/method of application.

Most suppliers of disinfectants to the food industry will have tested their products for effectiveness against a range of micro-organisms. Users should satisfy themselves that the product is suitable for the use intended.

Where to disinfect
Disinfection is not a universal process applicable to all surfaces. It should only be used on those surfaces where the presence of micro-organisms, at the levels found, will have an adverse effect on the safety or quality of the food handled. If disinfection is considered necessary, it should normally be restricted to:
- food-contact surfaces;
- hand-contact surfaces;
- cleaning materials and equipment;
- the surface of fruit and vegetables to be consumed raw; and
- hand disinfection (only when essential).

Disinfection of food and hand-contact surfaces
As micro-organisms are rarely mobile and need to be physically carried on to food, the disinfection of non-food contact surfaces, such as floors and walls, is rarely necessary. Surfaces not directly coming into contact with food, but frequently touched by food handlers, need disinfection to avoid the build-up of micro-organisms on hands.

Hand disinfection
Hand disinfection is only necessary in critical situations, for example, when the handling of high-risk food is unavoidable or to protect food handlers, such as fish filleters, from developing septic cuts. In the majority of food handling situations, washing the hands properly with liquid soap, rinsing in running water and thorough drying is quite satisfactory. If hand disinfection is considered necessary, it should be carried out after normal handwashing. Hand disinfectants should be fast-acting, rapid-drying and contain ingredients to protect the skin. Alcohol-based handrubs are becoming more popular.

Disinfection of cleaning equipment and materials
Cleaning equipment is often an important vehicle of contamination. It should be cleaned and disinfected frequently and always before storage. Normal machine laundering at 65°C or above, in the case of cloths and towels, will achieve this.

Incorrect storage of mop.
Courtesy of Vikan (UK) Ltd.)

Disinfecting fruit and vegetables
An increasing number of food poisoning and foodborne disease outbreaks are being attributed to the consumption of contaminated fruit and vegetables, such as cut melons, lettuce and tomatoes. Contamination can occur because of the use of animal fertilizer, sewage-polluted irrigation water or post-harvest washing with polluted water. Infected food handlers may occasionally be implicated, especially in the case of low-dose pathogens, such as *Shigella spp*, *Salmonella* Typhi, or Norwalk-like viruses and hepatitis A.

Blanching in hot water at around 70ºC for 15 seconds is probably the most effective way of destroying organisms, although hypochlorites (50ppm free chlorine at pH of 6.00 to 7.5 for around two minutes) ozone or acetic acid and peroxide (around 75ppm) will also reduce levels of transient pathogens. Vegetables should be washed prior to disinfection to remove dirt which would inactivate the chemicals. Agitation during washing provides the friction to remove bacteria from the food surface. Each wash will reduce the numbers of bacteria by about ten to one. Some experts suggest that a double wash may avoid the need for disinfection.

It is recommended that the water used for washing vegetables and fruit should be around 10ºC higher than the food being washed to avoid contaminated water containing bacteria being sucked into the food.

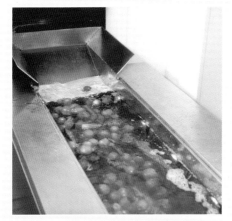

Disinfecting cucumbers and tomatoes. (Courtesy of Drywite Ltd.)

Airspace disinfection

Micro-organisms can be carried in the air and transferred to food or food-contact surfaces. This is only an issue where high-risk food is prepared. Disinfection of the air space can be carried out using a gas, such as ozone, or by aerosol misting using a liquid disinfectant. Airspace disinfection only takes place when no exposed food or personnel are in an area. It does not replace the need for surface disinfection.

Disinfection frequency

In most operations, contamination by micro-organisms takes time to build up to a

Disinfecting fog. (Courtesy of Holchem Laboratories Ltd.)

significant level unless extraneous contamination is introduced by poor operational practice. Therefore, under normal circumstances, disinfection can correspond with cleaning intervals dictated by visual soiling or with work cycles. However, intermediate cleaning and disinfection may be needed to counter the effects of inherent bad practice, such as the use of the same surface or equipment for cooked and raw meats.

Resistance

Some bacteria, under certain circumstances, develop resistance to disinfectants at concentrations that would normally be lethal. This problem can be avoided by:

- ensuring that a fresh solution of disinfectant is used, at the correct concentration;
- ensuring that the components of a system are compatible with the disinfectants used;
- using disinfectants of adequate performance against the bacteria requiring control;
- using only disinfected equipment and/or disposable or freshly laundered cloths; and
- applying disinfectants to clean surfaces.

Only in extreme cases, where the above conditions have not been observed, is it necessary to change the type of disinfectant used.

Inactivation

This phenomenon occurs when a disinfectant mixes, or comes into contact, with a substance that interferes, partially or completely, with its ability to kill micro-organisms. Such substances are called inactivators, the primary one being dirt, which affects all disinfectants to a greater or lesser extent. Hard water will inactivate uncompounded QACs, and plastics, cork, cellulose and certain organic materials will inactivate cationic disinfectants in free solution.

Control procedures

It is unfortunate that there is no visible difference between a physically clean but bacteriologically contaminated surface and one that has been effectively disinfected. Furthermore, even minor deviations from manufacturers' instructions may render the disinfection process ineffective.

If necessary, the effectiveness of disinfection should be monitored by bacteriological swabbing of surfaces and also examination of the food product. It is important to establish trends over a period of time and not to react to isolated unsatisfactory results.

Rapid cleanliness testing

Systems are now available for use by non-technical staff to ascertain whether or not physically clean surfaces are microbiologically satisfactory. Tests measure the amount of adenosine triphosphate (ATP), a chemical found in food debris, bacteria, moulds, yeasts and all living cells, remaining on the surface. When the ATP comes into contact with luciferin luciferase, the chemical that causes the firefly to glow, light is produced proportional to the amount of ATP present. Measurement of the light can be undertaken in seconds and provides an accurate standard of cleanliness. Another test involves using Vericleen strips.

CLEANING METHODS

The first step in effective cleaning is to determine the most appropriate method. Factors to consider include:
- accessibility, are surfaces accessible, for example, work surfaces, or conveyor belts? Is dismantling required or are the surfaces enclosed, for example, pipes and tanks?
- what standards are required to ensure safe food production, quality and shelf life;
- the type and mixture of soils;
- the type and level of microbial contamination and the required level of reduction;
- the risks to food, people or equipment;
- the resources available, labour, time, equipment and finance; and
- the impact on the effluent disposal system.

Manual clean

Hand cleaning utilizes the physical energy from an operative (elbow grease) coupled with application of a neutral or mildly alkaline detergent or detergent/disinfectant or sanitizer to a clean surface. A cloth, scouring pad or brush and detergent from a bucket or sprayer provides a very effective and flexible cleaning method. However, hygiene and foreign body contamination issues exist with all cleaning tools.

Manual clean.
Courtesy of Vikan (UK) Ltd.)

Dry clean

Dry cleaning methods include wiping, scraping, brushing and vacuuming of dry debris and are commonly used for dry ingredient areas, for example, in bakeries. The absence of water reduces the opportunity for microbial growth but, unfortunately, may not result in effective soil removal. Disposable cloths dampened with either a water- or alcohol-based detergent/disinfectant are often used to enhance soil removal where excessive water can be a problem, for example:

- on electrical or electronic equipment; and
- during product change clean on lightly soiled surfaces in high care environments, for example, in sandwich production.

Hand sprayers containing sanitizers are commonly used with disposable cloths in fast food outlets and cafeterias to wipe tables between customers.

Dishwashing

Unfortunately, many catering premises only use a single sink for dishwashing. Wherever possible this should be supplemented by a dishwasher. Where a single sink is used it is essential to use water as hot as possible, just below 60°C, and to replace the water frequently as it becomes cool or greasy. Suitable gloves will be required at these temperatures. Air drying may not occur and disposable paper towels should be used in these circumstances.

Although the use of a single sink is not as effective or hygienic as either the double sink or the dishwasher, provided the crockery and cutlery are clean, free from grease and dry, then there will be minimal risk of causing illness to customers.

Detergents must be compatible with the local water supply and correct amounts, in accordance with manufacturers' instructions, must be used. Measuring equipment must be cleaned frequently.

Soak clean

Soak cleaning is used for small items to extend contact time with the detergent. It is beneficial on heavy or tenuously soiled items. Highly carbonized trays, for example, pizza trays can be successfully cleaned in this manner. Detergents are often highly alkaline and may need to be formulated to be suitable for use with aluminium. Ultrasonic energy can be used to enhance soil removal.

Foam or gel cleaning

A foam or gel is simply a vehicle to increase contact time between soil and detergent. A wide variety of foam and gel products are available, which when applied to a soiled surface, will remain in contact for typically 10 minutes to 1 hour. This contact time enhances the chemical actions between soil and detergent, making it easier to remove the soil.

However, the loosened or reacted soil still requires some physical energy to remove it. This can be provided by a cleaner and a brush and/or by energy from a washdown system. In a large number of cases, some manual energy is necessary while the foam detergent is in contact with the soil because the physical energy from the washdown system is insufficient on its own.

Foams are applied by a variety of application systems, for example, Jetstream Foamer. They all have the benefit of keeping the cleaning operative away from direct contact with the foam, and also enable large areas to be quickly covered.

Foam or gel cleaning can improve performance. (Courtesy of Holchem Laboratories Ltd. and Perkins Chilled Foods Ltd., Barnsley.)

Foam detergents include, neutral, alkaline, caustic, chlorinated alkaline and acidic products.

- Alkaline products are the most commonly used foam formulations because of their good all-round performance on fats and proteins.
- The higher alkalinity the better the performance against carbonized and polymerized soils, such as those found in cooking and frying operations.
- Products safe for use on aluminium and other "soft metals" would typically be the neutral and alkaline products; but data sheets must be checked for compatibility.
- The chlorinated caustic products are highly effective at removing protein soils and staining. The chlorine donor in the formulation is primarily there to aid removal of proteins and not as a biocide.
- Acidic products are effective on mineral scale and also on proteins.

Automated washing machines

Trays, racks, bins and utensils can be automatically washed using purpose-built washing machines. The machines maximize energy through the washdown jets, high temperature and aggressive detergents, for example, highly alkaline, to enable quick cleaning. Typically a tray passing through a tunnel traywasher will be in the detergent zone for around ten to 20 seconds.

Highly alkaline or caustic-based detergents are used for washing plastic or stainless steel parts. These products are available in different formulations to suit site water hardness. A caustic detergent sequestered to deal with water hardness will keep the traywasher relatively free of mineral scale, although the rinse sections of the washer often need periodic descaling with an acid detergent.

Aluminium or tinned utensils are cleaned using either neutral or aluminium-safe alkaline products.

Disinfectants are typically low foam, quaternary ammonium compounds or amphoterics. An exception to this is in poultry processing, where live bird crate disinfection uses Department of Environment, Food, and Regions approved products, which are often per-acetic acid based.

Washdown systems

An increasing number of food premises are being designed to facilitate cleaning by the use of water hoses. Satisfactory fall and drainage of floors are essential and wall/floor finishes and equipment, must be resistant to water ingress and damage. Care must be taken to protect electrical components against the ingress of water.

A washdown system is used to provide energy for soil removal, to rinse away soil and detergent residues and may be used as a means of applying the detergent or disinfectant.

Pressures range from 20 to 196 bar (300 to 2,800 psi), although extreme care must be exercised at the higher pressures. Increasingly, lower pressures of around 20 bar are being used for many applications, particularly when combined with foam or gel detergents. The correct machine or system must be selected for a particular job. Inadequate pressure may result in poor cleaning, whereas excessive pressure will increase the risk of damage to surfaces and equipment. This may cause unnecessary splashing and misting which leads to potential cross-contamination. A volume of water of at least 9 litres per minute is required to wash away dislodged soils.

Nozzle design, pressure and flow rate are all important in achieving soil removal in each cleaning situation, without causing cross-contamination by overspray or aerosols. Various nozzle designs are available giving a spreading angle of between 15º and 40º.

Washdown systems may be fixed installations or mobile. The fixed systems are generally fed with hot water in the range of 55 to 65ºC. Mobile systems may be electric, diesel or petrol driven. The large electric washers require a three-phase supply. Safety must be considered in relation to mobile electric units, which should be protected by an approved earth-leakage circuit breaker. All electrical installations should satisfy the requirements of the Electrical Engineers Regulations and should be checked regularly. Fumes from petrol- and diesel-driven washers may make them unsuitable for use in food premises unless fixed outside food rooms.

Clean-in-place (CIP) equipment

Cleaning-in-place has replaced hand cleaning in dairies, breweries and potable liquid installations. Some food processing plants, including cook-chill systems, are now using CIP systems. It involves circulating non-foaming detergents and disinfectants through process

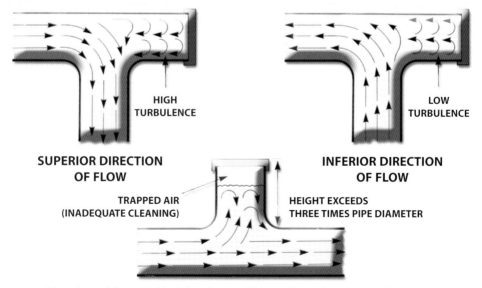

Direction of flow and height of tees affects effects cleaning performance.

equipment in the assembled state. The combined effects of solution turbulence, chemical energy and heat remove soil debris and micro-organisms from pipework and ancillary plant, without time-consuming dismantling and manual cleaning. Efficiency of cleaning depends on the type of soil, temperature, concentration of detergent, velocity, turbulence and the design of the CIP system.

The use of heat changes the physical state of the soil, for example, melts fat. It also accelerates chemical reaction. As a rough guide, an increase of 10°C may be considered to double the speed of chemical reaction. The most cost-effective results are usually obtained by using detergents at the highest temperature possible, taking into account equipment, venting, stressing and detergent manufacturers' recommendations. Allowances must also be made for the expansion of pipe lengths during CIP when hot solutions are used. It may not be possible to use hot solutions for cleaning refrigerated systems or tanks because of the risk of damaging heat-sensitive thermometers and thermostats. For light soiling, a concentration of detergent between 0.1 and 1.0% causticity is preferred. Acid products are more effective on some soils. Disinfection normally involves QACs or amphoterics, terminal disinfectants or per-acetic acid.

Careful design of CIP systems is essential to avoid unsatisfactory results. Inaccessible crevices and pockets, which can become foci for bacterial multiplication, must be eliminated. Valves and other components must be installed in the correct orientation and configuration to avoid uncleanable dead legs and tees that retain air or debris. The maximum length of dead legs and height of tees should be three times the pipe diameter to ensure satisfactory cleaning. Direction of flow into a pocket is superior to flow away from the same pocket.

All sections of the pipe-line, including valves and fittings, must be filled with solution without leaving any pockets of air. A fluid velocity of around 1.5m per second is satisfactory for most applications, although 2m per second may be required to remove stubborn soil with low-temperature solutions. Suitable fall must be built in to horizontal pipes as all parts of the system must drain completely.

Tanks, exceptionally large diameter pipes and large containers may be thoroughly cleaned using spray balls or rotating jet devices, as this avoids the unnecessary and uneconomical filling of vessels with solution. It also obviates the need for men to climb into vessels to carry out manual scrubbing.

Mains services required by CIP units may include: electricity, which may be single- or three-phase; a suitable clean, softened water supply; a clean, filtered, oil-free air supply; a saturated, filtered steam supply; and a suitable splash-proof drain or gully, resistant to temperatures of up to 90°C, capable of transporting CIP acids and alkalis and removing effluent. All systems include cleaning solutions, reservoirs, pumps, feeds and control equipment, which may vary from simple timers and manually operated switches and clocks to complex, fully integrated computer operations.

A typical basic CIP sequence consists of five stages:
- pre-rinse with cold water to remove gross soil;
- detergent circulation to remove residual adhering debris and scale;
- intermediate rinse with cold water to remove all traces of detergent;
- disinfectant circulation to destroy remaining micro-organisms; and
- final rinse with cold water to remove all traces of disinfectant.

The time allowed for each operation can be determined for each particular plant or circuit being cleaned.

EFFECTIVE CLEANING

Having selected the appropriate cleaning method, the effectiveness of the whole cleaning operation depends on:

- choosing the correct chemical;
- applying it at the optimum temperature and concentration;
- allowing it time to function; and
- using it with the correct equipment.

The correct mix of the above will reduce the work required.

Choosing cleaning chemicals

Cleaning product choice (the detergent or disinfectant) cannot be considered in isolation from the cleaning method or from the dispense and application method. A detergent or disinfectant is purely a tool in the overall cleaning procedure that assists in achieving the required end result; namely a clean facility that is safe for food production. For example, a neutral detergent applied with a bucket and brush can achieve an equally good cleaning result on a stainless steel tabletop as one cleaned using an alkaline foam detergent rinsed off with a medium pressure washdown system.

The primary choice of a chemical depends on the soil. However, other factors must be considered including: compatibility of surface being cleaned with detergent, for example, highly alkaline detergents may cause corrosion on soft metals; method of application; personal safety issues; suitability for use with water supply; time available; and compatibility with other chemicals used.

SELECTION MAY BE AIDED BY THE FOLLOWING CHECKLIST

- soil
- situation
- surface
- safety

The optimum temperature

Temperature can be critical to chemical performance, generally the higher the better. However, heat may denature the chemical or fix proteins such as egg, in which case warm or cold applications must be used initially.

Dosing aids and applicators

This classification applies to devices used to deliver or control chemicals. They may be manually operated or automatic. Dispensers may be of the plunger or proportioning or metering type. Applicators include aerosol, trigger spray, pump-up sprayer, backpack sprayer, air or electrically driven sprayer and foam or gel applicators.

Swan-neck or mushroom dispensers

These are spring-loaded, push-operated pumps, screwed vertically into concentrate containers.

Venturi proportioners

These draw detergent by water suction from a container. Consistency can be a problem with low or varying water pressure.

Proportioning pumps

These simple water driven-pumps are more expensive than venturi proportioners but give greater consistency of dilution.

Hand sprayers

Sprayers may be trigger-pump or pressure-operated and vary in size from 500ml to 25 litres. They can be adjusted to deliver an atomized mist or a fine, long-reaching jet.

Time to function

Chemicals must be given sufficient time to act. This period is normally referred to as the contact time. Vertical surfaces may not allow sufficient contact time due to run-off. Alternative cleaning methods, such as repeated applications, or the use of foams, gels, highly viscous liquids, or even soaking, can be used to extend contact times. Attempts to remove dirt before the chemical reaction is complete results in a considerable amount of extra work.

Equipment

A growing range of specialist equipment is available for cleaning operations and the correct choice will ensure that the most cost-effective cleaning is carried out. Cleaning equipment must be cleaned and disinfected prior to storage.

Selecting the right equipment

The quality of cleaning equipment varies considerably and is usually price related. Effectiveness and durability depend on the choice of materials used in manufacture, the design and quality of construction and the suitability for a particular task.

Manual cleaning aids

Consistent, high standards of cleanliness will only be achieved if the cleaning tools have been specifically manufactured for the stringent demands of the food industry. Correct choice is essential if operatives are to avoid recontamination of a cleaned surface with dirt or bacteria. The quality and cleanliness of tools that touch surfaces in direct contact with food is particularly important.

Cleaning equipment must be stored satisfactorily.
(Courtesy of Vikan (UK) Ltd.)

Design

The design must ensure that tools can be effectively cleaned and there are no hiding places for residues or bacteria to accumulate. If handles are hollow they must be properly sealed. The use of ergonomics in design minimizes operator fatigue, improves safety and consistently results in higher standards. Correct length and diameter of handles reduce the risk of muscle tension and provide the most comfortable grip. Handles should reach the chin of the user, when measured from the floor.

HYGIENE for MANAGEMENT *Cleaning & disinfection*

Materials

Modern materials such as high-density polypropylene, polyester and rilsan are resistant to acid and alkaline cleaning agents and will withstand repeated heat sterilization up to 130°C.

Colour coding

The use of colour coding for cleaning equipment, for example, handles of brushes, bristles and cloths used in high-risk situations, with different colours being used in potentially contaminated raw food areas, assists in reducing the risk of cross-contamination. It also reinforces hygiene training relating to the need to separate raw and high-risk food.

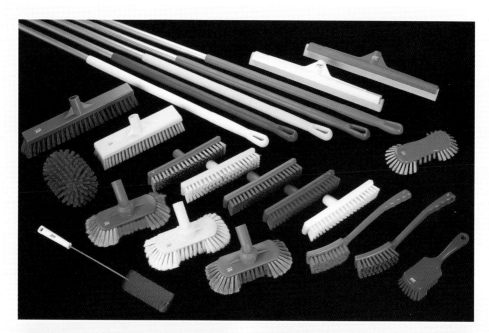

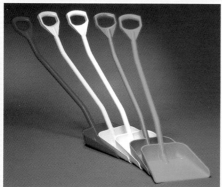
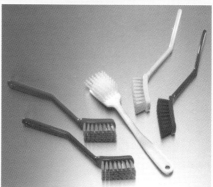

Colour-coded brushes and shovels can reduce the risk of cross-contamination.
(Courtesy of Vikan (UK) Ltd.)

Cloths

Cloths vary from the durable, high-strength textile to the semi-disposable, non-woven with high absorbency and relatively low mechanical strength. Disposable paper is commonly used in preference to cloths.

Brushes

Brushes constructed from modern materials such as high-density polypropylene for stocks with bristles of polyester or rilsan give improved performance and resistance to wear. They are capable of withstanding boiling water and the normal cleaning chemicals used throughout the food industry. Wood and natural bristles must be avoided and worn out brushes must be replaced. As brushes become worn they become less effective, they discolour and bristles are more likely to drop out. Blue coloured bristles are often preferred as they are more easily detected if they become loose. Nylon filaments are porous and quickly lose their stiffness in wet conditions, and inferior materials may even distort in hot water.

Care is necessary to ensure damage to surfaces does not occur because the bristles are harder than the surface being cleaned.

Mops

The common or socket mop should not be used in the food environment. The twist mop and the Kentucky mop, with its detachable head that can be washed and boiled, are much more appropriate. Mops must be thoroughly cleaned after use and then left to dry. The storage of mops in buckets, with or without disinfectants, should not take place as this may allow the multiplication of, or the development of resistant, bacteria.

Buckets

Buckets may be high-density nylon or polypropylene with steel handles. Domestic-grade buckets should be avoided. Buckets should not be placed on the floor and then on to food contact surfaces.

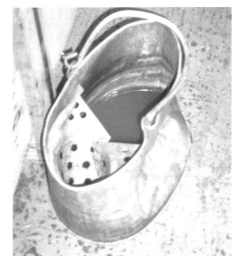

Unsatisfactory cleaning equipment. (Courtesy of Vikan (UK) Ltd.)

Sinks and tanks

Sinks are the basis of any manual utensil and small equipment cleaning and are essentially fixed vessels for the containment of a cleaning solution. Sizes may vary from 20 to 450 litres. Construction material should be food grade stainless steel.

Tanks are similar to sinks but are used for soaking. They tend to be under-exploited. A stainless steel tank containing an aggressive cleaning solution can be used to soak a variety of equipment, removing encrusted carbon with a minimal labour input.

Mechanical aids

These include floor scrubbers, rotating washers, power washers, air lines, steam cleaners, vacuum pick-ups, dishwashing and tray-cleaning machinery. Judicious use of mechanical equipment can significantly reduce labour requirements, but considerable care should be taken in selection to ensure that it is suitable for the use intended.

Vacuum cleaners

Suitable industrial vacuum cleaners, which suck up dirt and dust in a stream of air and then filter the air to trap the dirt, should be used to clean such areas as dry stores and carpeted public rooms. The suction must be powerful enough to ensure efficient operation and some models may be used to remove liquid spillages. Extension pipes are available to clean ceilings. The use of suction vacuum cleaners is preferable to sweeping as dust clouds are avoided.

A single disc scrubbing machine and a suction vacuum cleaner.
(Courtesy of JohnsonDiversey.)

Pre-rinsing prior to dishwashing.
(Courtesy of Marks and Spencer)

Pre-clean
(Courtesy of Vikan (UK) Ltd.)

CLEANING PROCEDURES

Whatever the location, industry, soiling type or circumstances, cleaning and disinfection comprises six basic stages:

- pre-clean: sweeping, wiping or scraping off loose debris, pre-rinsing and/or pre-soaking;
- main clean: applying detergent and loosening of the main body of dirt;
- intermediate rinse: removal of loosened dirt and detergent residues;
- disinfection: destruction of remaining micro-organisms;
- final rinse: removal of disinfectant residues (if required); and
- drying: removal of final rinse water and storing to prevent contamination.

In light-soil conditions, the pre-clean may be combined with the main clean. Disinfection can be omitted on non-critical surfaces. Drying can either be natural, as in air drying, or physical, using disposable paper towels, hot air or a clean dry cloth.

Disinfection can take place in combination with the main clean using specific chemicals known as sanitizers, although this is not as effective.

Preliminary and post-cleaning actions

In the case of dangerous or electrical equipment, preliminary safety procedures and final safety checks will be required. Several types of equipment will need partial or complete dismantling to ensure effective cleaning.

Double-sink washing

This procedure is recommended for washing-up in catering premises, public houses and retail outlets selling high-risk foods, when suitable dishwashing machines are not available. It also applies to food-processing, packing and distribution plants where small items are hand washed. Thermal disinfection is usually more effective if double sinks are used. However, the evaporation of water often results in poor visibility and condensation, leading to mould growth on ceilings and walls. Furthermore, hot water at 82°C may pose an unacceptable health and safety risk. Because of these problems, a suitable chemical disinfectant such as a QAC, an amphoteric-based disinfectant or hypochlorite in a tablet form may be used. However, the rinse water should still be hot enough to allow air drying. The requisite temperature will depend on the particular piece of equipment/utensil and the material from which it is made, the temperature of the item prior to immersion and the time the item is left in the second sink. Melamine plates were found to air dry rapidly after immersion for 30 seconds at 65°C.

The full, six-stage procedure should always be followed:
- remove any heavy or loose soil by scraping and rinsing in cold water;
- place articles in the first sink in detergent solution at 53°C to 55°C, scrub with a suitable brush and/or wipe with a clean cloth to loosen dirt residues. Higher temperatures bake on proteins and are too hot for the hands;
- re-immerse in the first sink to wash off loosened dirt;
- place articles in the second sink to rinse off chemical residues;
- leave for sufficient time at a high enough temperature to ensure rapid air drying. Dilution (large volumes of water in sinks) also aids disinfection. Baskets for disinfecting purposes should be maintained in good condition and inspected regularly. They should be loaded so that all surfaces of crockery and equipment are fully exposed to the rinse water. Hollow items such as cups should be placed on their side; and
- remove articles, allow to drain and evaporate dry on a clean, disinfected surface.

The order of washing-up should be planned so that glasses and lightly soiled articles are cleaned first. For safety reasons, glasses should be washed one at a time. Some articles may need pre-soaking and treatment with abrasive pads. The temperature of the wash water should be so high as to be uncomfortable, necessitating the wearing of rubber gloves, but should never exceed 60°C to avoid fixing proteins. A rinse aid may be added to the rinse water to promote smear-free drying. Disposable paper towelling placed on the draining surface helps absorb water and reduces noise. Clean pans, cups etc. should be inverted and protected from contamination during storage. Gloves must be washed and dried after use, as must the sinks, drainers and any other surfaces involved in the process.

Dishcloths should be semi-disposable, discarded daily or, if woven fabric, should be withdrawn daily and laundered; normal laundry procedures are sufficient to secure disinfection. Cloths should never be left wet or soaked in disinfectant overnight as this can result in a build-up of resistant bacteria. Bleaching, if desirable, may take place prior to laundering but should not be considered as a substitute. High-density polypropylene brushes are preferable to cloths as they withstand autoclaving temperatures and can be cleaned and disinfected frequently.

Mechanical dishwashing

Mechanical dishwashing is preferable to, and often more economical than, manual washing, provided the machine is used according to the manufacturer's instructions. Machines, in addition to cleaning, are also a highly efficient means of disinfecting small items of equipment. They should be used for articles such as the removable parts of slicing machines, polypropylene chopping boards, nailbrushes and other items that come into contact with high-risk foods, provided that no damage to the item will result. The sequence is as follows:

HYGIENE for MANAGEMENT Cleaning & disinfection

Mechanical dishwashers. (Courtesy of JohnsonDiversey.)

- remove excess food into suitable waste bins; if necessary pre-soak or spray, unless the machine is fitted with a pre-wash cycle;
- pack articles in a neat, orderly fashion so that items do not overlap (racks may be used);
- articles pass through the wash cycle of hot detergent solution (49°C to 60°C);
- articles are rinsed at 82°C to 88°C, with the injection of rinse aid;
- the dishwashing cycles are complete and disinfected articles may be accessed; and
- allow cleaned items to drain and evaporate dry before storage (inverted).

To ensure that the best results are obtained, it is important that machines are serviced regularly, operated in accordance with manufacturers' instructions and that:

- utensils are washed as soon as possible after becoming soiled;
- the correct detergent for the level of water hardness is used. It is often more economic and efficient to install a water softener;
- the detergent dosing equipment is working properly and detergent rinse levels are properly adjusted;
- sprayer arm jets are clear from obstruction and strainers are in place. Both should be cleaned daily; and

♦ track speed, where appropriate, is properly adjusted to give the contact time required for the detergents used.

As a rule-of-thumb guide to the efficiency of a machine, if items coming out are too hot to handle and dry rapidly to a clean, smear-free finish, then the machine is operating correctly.

Cleaning a work surface

The procedure for work surfaces will vary according to the finish and the types of cleaning chemicals used. For stainless steel and similar surfaces the routine is as follows:
- remove loose debris with a clean, loosely folded cloth (pre-clean);
- wash with hot detergent solution and a clean cloth or brush (main clean);
- rinse with hot water and a clean cloth (intermediate rinse);
- apply a suitable disinfectant and allow sufficient contact time;
- rinse off with fresh water and a disposable paper towel (final rinse); and
- allow to evaporate dry (drying).

Cleaning a slicing machine

Pre-clean
- Switch off power socket and remove the plug.
- Set the slice thickness control to zero.
- Dismantle the machine and pass removed parts through the dishwashing machine.

Where the machine is of a type that has a removable blade, a blade guard must be fitted before the blade is removed. Cleaning may then commence, with the proviso that no person may clean a slicer or other dangerous machine unless they have reached their eighteenth birthday and have been properly trained.

If a dishwashing machine is unavailable, the blade may be cleaned and disinfected in situ or adjacent to the slicer. Alternatively the two-sink method may be used.

Post-clean
- Reassemble the machine.
- Redisinfect parts handled.
- Check the guards are properly fitted, reconnect the power and switch on the machine. Test run to check safe working. This procedure is vital because accidents have been caused by guards that were improperly fitted after cleaning. If any adjustments have to be made, the machine should be switched off and disconnected and the test run repeated.
- Switch off the machine, disconnect the plug and cover with a freshly laundered tea towel or other suitable covering.
- Supervisor to check.

Cleaning a soft ice-cream machine
- Remove all ice-cream from the machine.
- Rinse thoroughly with cold water.
- Switch off, disconnect the plug and dismantle the machine.

- Wash the parts and machine interior with a suitable detergent, reassemble the machine.
- Clean the outside of the machine and reconnect.
- Fill the machine with a suitable disinfectant solution, circulate for two minutes and drain. Leave the machine empty.
- Prior to use the next day, fill with fresh clean water, circulate and drain before filling with ice-cream mix.

Beverage vending machines

These machines may present difficult cleaning problems. The complexity of some of the working parts, the time allocated for cleaning and the tenacious nature of the soil deposits has often resulted in machines being imperfectly cleaned, as well as accelerated surface damage from the use of abrasives.

While the prevalent and most troublesome soiling is often identified as tannin, it has been shown that the stubborn nature of the soil is caused by an intermediate layer of water hardness scale, bonding to machine parts. The scale and tannin deposits can be removed by soaking in a suitably formulated acid detergent for a few minutes. Specific tannin removers merely remove the tannin without affecting the scale.

Problems have also been identified with feed tubes. Conventional procedures do not remove all soil and when a brush is pushed down a tube and pulled out from the same end, an amount of the soil is redeposited. Brushes have therefore been adapted by crimping the end loops so that a whole brush can be passed through a tube in one direction. Complete soil recovery and disinfection are effected by inserting a small piece of disinfectant-soaked cloth in the crimped end-loop, which is pulled through in much the same way that a rifle barrel is cleaned.

Cleaning procedure
- Remove all contact parts from the machine, dismantle where necessary and soak in acid detergent solution. (The waste bucket, after emptying and cleaning, can be used.).
- Remove ingredient containers, clean auger discharges with a nylon scraper, refill as necessary, wipe down the exteriors and set aside.
- Wipe down the interior of the machine and dry thoroughly with a disposable paper towel.
- Remove soaked parts, wipe, brush and "pull through" as appropriate, disinfect and dry. Reassemble and replace.
- Replace ingredient containers.
- Empty, clean, disinfect and dry the waste bucket to prevent fungal growth. Replace bucket.
- Flush all lines with clean, hot water from the machine system, test-vend all drinks, close the cabinet and wipe down the exterior. Provided that high-risk beverages such as soups are closely monitored, the above procedure may be carried out every three days.

THE HEALTH AND SAFETY OF CLEANING

The main Act applying to the safety of cleaning is the Health and Safety at Work etc. Act, 1974. This Act requires employers to ensure the health and safety of all persons on site, including employees, visitors, members of the public and sub-contractors.

There are a large number of regulations made under this Act. The most common are known as the European Six Pack Regulations; namely:

1. Management of Health & Safety at Work Regulations, 1999

These Regulations cover a broad range of duties applicable to most work environments. They are general in nature and may overlap with specific regulations, for example, The Control of Substances Hazardous to Health Regulations. They include **risk assessment**, which is one of the main principles in Health & Safety. Risk assessment involves a careful examination of the workplace to identify hazards (anything that has the potential to cause harm), exposure (the likely contact with the hazard) and risk (the likelihood and seriousness of someone being harmed).

Stages in risk assessment

- Look for the hazards that may result in significant harm, for example, physical hazards from chemicals, slipping on wet floors, moving parts of machinery, work at height (steps), pressure systems (steam, water or air), electricity (contact with water), manual handling, noise, poor lighting and low or high temperatures.
- Consider who might be harmed and how.
- Evaluate the risks arising from the hazards. Establish the chance of exposure occurring, level, time and frequency of exposure. Conclude whether or not a significant risk exists.
- Is the risk adequately controlled? Control measures will involve providing information, training, instruction and setting up safe systems and procedures. Precautions must comply with legal requirements and reduce the risk as far as practicable.
  Preventive and protective measures may involve:
    - removing the risk completely;
    - using a lower risk option;
    - preventing access;
    - organizing the work or area to reduce exposure;
    - issuing personal protective equipment; and
    - welfare and health surveillance.
- Assessments should be suitable and sufficient and recorded.
- Periodic reviews should be undertaken to establish any significant change.

2. Provision & Use of Work Equipment Regulations, 1998

These Regulations refer to the provision and use of all work equipment. The definition of work equipment is very broad, covering almost all equipment used at work.

3. Manual Handling Operations Regulations, 1992

Manual handling refers to lifting, putting down, pushing, pulling, carrying and moving by hand or bodily force. In addition to weight they deal with problems caused, for example, by sharp edges, slippery surfaces and extreme temperatures.

4. Workplace (Health & Safety & Welfare) Regulations, 1992

General requirements in the workplace including: working environment (lighting, temperature etc.); safety (access, falls, etc.); facilities (toilets, eating etc.); and housekeeping.

5. Personal Protective Equipment (PPE) at Work Regulations, 1992

These Regulations require the employer to provide suitable personal protective equipment for employees exposed to health and safety risks, when such risks cannot be adequately controlled by other means. PPE may include eye protection, respiratory protection, protective overalls and hand protection.

Employees must be fully trained in the use of such equipment, must be aware of the risks that the equipment will prevent or limit, must be aware of purpose and the manner in which the PPE is to be used and the maintenance required. Furthermore, employees must wear the PPE provided, take care of it and report any faults or damage.

6. Health & Safety (Display Screen Equipment) Regulations, 1992

These cover the health & safety requirements for work with display screen equipment.

Additionally, the following regulations are applicable to the safety of cleaning.

The Control of Substances Hazardous to Health Regulations, 1999 (COSHH)

These Regulations are applicable to many cleaning chemicals. Hazardous chemicals should be identifiable by the presence of a warning label.

The COSHH Regulations require employers to assess the risks from substances that are hazardous to health. These substances are those classified as very toxic, toxic, harmful, corrosive, irritant or those with specified maximum exposure limits (MEL), occupational exposure limits (OEL) or have biological or dust hazards. The risk must then be removed or reduced to acceptable levels.

Chemicals (Hazards, Information & Packaging for Supply) Regulations, 1994 (CHIPS)

These Regulations are mainly associated with the requirements placed upon the supplier of the hazardous chemicals and include the label requirements.

Information required on label
- The name, address and telephone number of the supplier.
- The name of the substance or preparation.
- An indication of the danger and the relevant appropriate risk symbol and safety phrases.

HAZCHEM SYMBOLS

CORROSIVE · TOXIC · FLAMMABLE · EXPLOSIVE · OXIDIZING AGENT · IRRITANT/HARMFUL

A safety data sheet must also be supplied. This includes detailed information about the product, its hazard classification under the CHIPS Regulations and the information needed by the user to carry out a COSHH assessment.

Guide for the storage, handling and dispensing of hazardous products

Receipt

All containers should be checked for damage, labelling and actual delivery against the delivery note.

Chemical stores

Lockable, clearly identified storage, which is separate from foodstuffs, should be made available for cleaning materials. Access to the store should be restricted. Stores should be dry, cool, well-lit, and well-ventilated and out of direct sunlight. There must be adequate space for safe and tidy storage, with sufficient racking to allow systematic placement of stocks. Information on storage and safety should be displayed.

Stores should be kept clean and spillages should be cleared away promptly. Stores should not freely drain and chemicals should not be allowed to enter the drainage system in quantities that could cause problems. Chemicals should be stored in original containers and used before the expiry date, which is usually written on the label or the container. Incompatible products must be kept apart, for example, acidic and chlorinated products.

Stores must be specially designed if bulk containers of potentially harmful chemicals are broken down, and a sink, water supply and adequate working surfaces should be provided. Care must be taken when making up cleaning solutions, and manufacturer's instructions must always be followed. For example, caustic soda flakes must always be added to water. If water is added to flakes, problems are likely because of heat and effervescence.

Hazardous chemicals
- Chemicals should never be transferred into unmarked containers.
- Lids must be firmly screwed on containers, especially when being carried/transported.
- There must always be adequate washing facilities for both routine and emergency situations. They should include eyewash facilities.
- There should be no contact with concentrated products.
- When transporting, decanting or dispensing concentrated products, appropriate personal protective equipment must be worn including overalls, gloves, and eye protection to the appropriate European norm.
- Protective clothing must be kept in good condition, dry and not contaminated with chemicals.

Dispense

Where available, automatic equipment is preferable for dispensing. Products must not be mixed because they may neutralize each other, form a very corrosive chemical, produce poisonous gases, or heat up rapidly causing boiling or an explosion.

First aid

The aims of first aid are to preserve life, to limit the effects of the condition and promote recovery. The first aid for cleaning chemicals generally involves the dilution of the product in or on the body.

- Eye and skin contact: dilute and remove the chemical by flooding with water. If necessary, arrange for the removal of the casualty to hospital. (Always take material safety data sheets.).
- Ingestion: maintain airways, breathing and circulation, obtain medical assistance and identify the substance swallowed. Do not induce vomiting.
- Inhalation: remove the casualty from the affected area, restore adequate breathing and obtain medical aid if necessary.

ENVIRONMENTAL CONSIDERATIONS

The cleaning process can have a significant impact on the environment and careful consideration must be given to minimize this impact. Waste food should be disposed of through dry waste handling and not rinsed to drains; thereby reducing the load on the effluent treatment system. Cleaning products can also have a negative impact on treatment systems if used excessively or at too high a concentration; always follow manufacturers' instructions.

Accidental spillage

In the event of chemical spillage, there is an immediate risk to site personnel and surfaces, and if chemicals mix in the drains there will be a subsequent risk to persons working in sewers.

Procedure in the event of spillage:
- wear appropriate personal protective equipment;
- contain the spillage to stop it reaching the drain;
- pump/decant the spillage into suitable containers for disposal/re-use, or absorb the spillage using sand or an absorbent material such as attapulgite; and
- if disposed of, use a registered waste disposal company.

Spillages entering the drain

The sewers may become dangerous for personnel working in them and the effluent treatment plant may be damaged/overloaded so that the chemicals end up in a river. The Water Authority/Environment Agency must be contacted for advice, and prosecution may follow. (Make sure the emergency numbers for the Water Authority and the Environment Agency are readily available.)

Storage
- Store chemicals safely, locked away.
- Never transfer chemicals into other containers.

Use

Note: If used incorrectly chemicals can be dangerous.
- Follow manufacturers' instructions.
- The hazard of chemicals can vary; a washing-up liquid is relatively safe whereas an oven cleaner will cause burns and blindness if splashed in the eye.
- Do not mix chemicals.
- Always wear appropriate personal protective equipment.

COMMON CLEANING PRODUCTS

GENERIC OTHER NAMES	TYPICAL USE	CONSIDER
General purpose detergent (neutral or mildly alkaline) *Hard surface cleaner*	• Used at 1% in hand hot water to clean all surfaces. • When used on food contact surfaces must be thoroughly rinsed off before applying a disinfectant.	• If in same room as food or on food-contact surfaces then product must be unperfumed and of low taint. • Ensure that product will not damage sensitive materials such as aluminium.
Heavy duty cleaner (highly alkaline) *Degreaser*	• Used at 2% to 10% in hot water to clean burnt or very heavily soiled surfaces. • May be recommended to be used neat. • Cleaning burn-on from heated surfaces, such as hobs, ovens, grills and rotisseries. • Boiling out brat pans and fryers. • Must be thoroughly rinsed.	• Generally hazardous to personnel, causing burns. • May attack aluminium. • Personal protective equipment.
Oven cleaner (highly alkaline)	• Commonly used neat. • Cleaning burn-on from heated surfaces, such as hobs, ovens, grills and rotisseries. • Must be thoroughly rinsed.	• Generally hazardous to personnel, causing burns. • May attack aluminium. • Personal protective equipment.
Sanitizer (QAC based)	• Light duty, single stage cleaning and disinfecting of surfaces (typically 1 to 2%). • Must be thoroughly rinsed.	• Single stage cleaning using a sanitizer is **not** as effective as cleaning with a detergent, rinsing and then disinfecting.
Disinfectant (QAC based)	• Disinfect at recommended strength (typically 1%). • May be suitable to leave on food-contact surface without rinsing.	• All detergents used for cleaning prior to the application of a disinfectant must be thoroughly rinsed away.
Disinfectant (sodium hypochlorite based) *Bleach*	• Disinfect at recommended strength (typically 0.1%). • Must be thoroughly rinsed.	• Contact with acid products will produce toxic chlorine gas. • Will produce taint even at low levels on certain foods.

HYGIENE for MANAGEMENT *Cleaning & disinfection*

GENERIC *OTHER NAMES*	TYPICAL USE	CONSIDER
Disinfectant (Amphoteric)	♦ Disinfect at recommended strength (typically 1%).	♦ Not inactivated by detergents. ♦ Inactivated by soil. ♦ May be suitable to leave on food-contact surface without rinsing.
Acid descaler (phosphoric or hydrochloric acid based) *Limescale remover*	♦ Specific products for limescale removal. ♦ Often used neat in urinals or toilet bowls.	♦ Contact with chlorinated products will produce toxic chlorine gas. ♦ Should not be used on food contact surfaces.
Washing-up liquid (neutral) *Neutral detergent*	♦ Used at 1% in hand hot water to clean surfaces. ♦ Used at 0.5 % in utensil sink for utensil, crockery and cutlery cleaning.	♦ When used on food-contact surfaces, should be thoroughly rinsed and followed by application of a disinfectant and allowed to air dry.
Dishwash detergent	♦ May be in tablet, powder or liquid form. ♦ Follow manufacturers' instructions.	♦ Should only be used in automatic dishwasher. ♦ Contact with acid products may produce toxic chlorine gas.
Dishwash rinse aid	♦ Usually in liquid form. ♦ Follow manufacturers' instructions.	♦ Should only be used in automatic dishwasher.

Dilution
- ♦ A 1% solution of a chemical is made by adding 1 part chemical to 99 parts water.
- ♦ Never use food utensils (for example, cups, jugs) to measure out chemicals; always use specific chemical dosing or measuring equipment.

15 Pest control edited by Adrian Meyer

Pests are the direct cause of most of the statutory closures of food businesses, they are a major factor in the thousands of food complaints reported to Environmental Health Departments each year and they feature prominently in many of the prosecutions taken under food hygiene regulations. Furthermore, pest attack is responsible for a significant proportion of the unfit food surrendered each year. If one also takes into account the amount spent on pesticides, the damage to buildings and fittings, the distress to customers and the spread of disease, it is obvious that pest control is an inseparable part of profitable and hygienic food production. Pests are a source of foodborne pathogens and pest control is usually considered as a prerequisite for HACCP.

It should be the responsibility of a senior person to ensure effective pest control. Persons involved in pest control should be fully trained, and it is essential to ensure that the treatment itself does not expose food to risk of contamination.

The common pests found in the food industry include:
- rodents: rats and mice;
- insects: flies, wasps, cockroaches, psocids, silverfish, stored product insects and ants;
- birds: mainly feral pigeons and sparrows; seagulls and starlings also cause problems in some areas; and
- mites.

Pest infestation often result in the closure of food businesses.

Effective pest control necessitates rapid detection and identification of the species causing concern, a knowledge of its life cycle and the most economical, rapid and safe way of eliminating it. Pests require food, shelter, warmth and security. Denial of these environmental factors will prevent their survival. Access may be denied by inspection of raw materials, design, maintenance and proofing, and food and harbourage may be removed by good housekeeping. This form of control may be termed environmental control and is the first line of defence against possible infestations.

Environmental controls may not be entirely successful and other steps must be taken to destroy any pests that gain access to food premises. Eradication methods may be considered under two main headings:
- physical;
- chemical.

Usually, physical control methods are preferable as the pest is caught, either dead or alive, and consequently is not able to continue contaminating food. Examples of physical control include electronic fly killers and rodent traps. Unfortunately, physical control methods are rarely completely effective and pesticides have to be used.

When chemicals are used the pest is not killed immediately and may, therefore, drop or crawl into food if adequate precautions are not taken. Furthermore, the safety implications of using pesticides must not be overlooked.

Pest control programmes should therefore encompass an integrated approach to pest control, i.e. **integrated pest management** (IPM). IPM is the cost-effective implementation of prevention and eradication strategies based on the biology of pests, intended to ensure a pest free food operation. This should include the following elements:

- early detection and identification of the pest species;
- the application of environmental management – improvements in **hygiene**, exclusion and proofing;
- the application of chemical and physical control. Undertaken with due regard to safety and the environment, i.e. the least toxic treatments; and
- the development of an effective monitoring strategy that will monitor the progress of any control operations.

Reasons for pest control
To prevent the spread of disease

Rodents, insects and birds are all hazards and can spread diseases which affect man and other animals. Rodents, sparrows, flies and cockroaches are all capable of transmitting **food poisoning** organisms and a range of additional **viral**, **bacterial**, **protozoal** and **endoparasitic** diseases, either by direct contact with food from their contaminated bodies or legs, by faecal deposits, or in the case of rodents, by urine. Furthermore, disease may be spread by:

- consuming food contaminated by rodent urine or droppings;
- contact with rat urine, which may result in Weil's disease (leptospiral jaundice). In the food industry, fish filleters and slaughterhouse operatives are most at risk;
- eating undercooked pork affected by Trichina cysts, which may infect pigs that have eaten dead rats;
- ectoparasites that live on rats; and
- rat bites.

To prevent wastage of food

Considerable financial loss is incurred by pest infestations in food and packaging materials. The presence of insects either dead or alive, rodents, droppings or hair, bird feathers or droppings found in food results in loss of production, recall of contaminated foods and the destruction of large quantities of food. Bagged foodstuffs under long-term storage can collapse due to heavy rodent attack and the cost of rebagging and **cleaning** can be considerable. Furthermore, birds and insects, but particularly rodents, eat food in fields, warehouses, commercial and domestic premises.

To prevent damage

As a consequence of the specially constructed incisors that grow throughout their lives, rodents are able to gnaw continuously; woodwork, soft metal pipes and electric cables are common targets. The damage caused by rodents from fire, flooding due to burst pipes, and

Rat damage to stored maize and rat bites on the neck of a young girl. (Courtesy of Acheta)

subsidence caused by burrowing results in considerable financial loss each year. Furthermore, at least one death has been recorded as a result of rodents gnawing gas pipes.

To comply with the law

Food safety legislation requires that all food premises must have adequate procedures in place, to ensure that pests are controlled. Food must be so placed and/or protected, as to minimize any risk of contamination. Food premises must be maintained in good repair and condition and designed and constructed to prevent contamination by pests.

Food safety legislation makes it an offence to sell food that is unfit or contains foreign bodies. Food contaminated with pests, parts of pests or droppings, etc. can be dealt with under the legislation. Furthermore, food premises with serious infestations of rodents, insects, particularly cockroaches, or birds that are a danger to the health of customers, could be the subject of closure procedure under the Food Hygiene (England) (Scotland) (Wales) (NI) Regulations 2006. The loss of business resulting from the prosecution of a food premises for offences involving pests can be considerable and may even result in bankruptcy. In 2006, a large supermarket was find £30,000 for selling a packet of crisps that had been nibbled by rodents. In 2002, an unlicensed butcher in Coventry was fined £40,000 for having dirty premises with a mouse and cockroach infestation.

The Prevention of Damage by Pests Act, 1949 requires the occupier of any land or buildings to notify the local authority of any rodent infestation (not applicable to agricultural land). The authority can insist that the occupier carries out any necessary treatment, including the removal of harbourage and repair of buildings. This Act also requires local authorities to take steps to ensure that their district is kept free from rats and mice.

Under the Health and Safety at Work, etc. Act, 1974 employers have a legal obligation to ensure, as far as reasonably practicable, the health, safety and welfare of employees. The presence of an infestation of certain pests could result in unsafe working conditions.

To avoid losing business, staff and profit

Most customers object to shopping in pest-infested premises or purchasing food contaminated by the droppings of pests or their bodies. The adverse publicity arising from prosecution may result in similar effect. Good staff are unlikely to work in badly infested premises.

The design, maintenance and proofing of buildings

Harbourage in food premises is not only provided in dark, undisturbed areas but also within the very structure of some buildings. For this reason, false ceilings must always have access points to enable inspection and treatment to be carried out. Boxing or ducting of pipes creates ideal conditions for harbourage and should normally be avoided. Where ducting is installed, it should be fitted with access plates at two metre intervals and should never finish in an open end. Surface panels and finishes that are not properly fixed and sealed to walls often provide ideal harbourage. Cavities within internal walls should also be avoided. All parts of the structure should be capable of being easily cleaned. High ledges, pits for elevators and ovens must be fully accessible. Elevators and conveyor intakes must have tight-fitting doors at the delivery end. Shutter boxes for roller doors must be checked to ensure they are not used as nesting sites for rodents or birds. The use of cupboards should be minimized and no gaps should exist around pipework passing into cupboards.

The design and installation of cables, electrical trunking and motors should eliminate harbourage. Motor housings for refrigerator compressors make ideal nesting sites for mice. All structural damage such as holes in walls, broken windows, loose tiles and damaged insulation should be repaired immediately to obviate the potential for insect harbourage. Silicon mastics are particularly useful for sealing small gaps.

All buildings should be adequately proofed; a small mouse can pass through 9mm wire mesh and so all access points greater than 6mm should be proofed. Rats can jump to a height of 90cm and so all possible points of entry below this height should be closed.

Doors should be close-fitting and provided with metal kick plates. Bristle strip is often used as a method of proofing the bottom of doors and other openings. Such strips are not rodent proof and can be accessed by rodents. In the absence of more thorough exclusion, they do however, act as a potential barrier. Gaps where pipes and girders pass through walls should be adequately proofed.

Defective drains, both above and below ground, must be made good. Tight-fitting inspection chamber covers must always be provided and replaced immediately if they become broken. Chamber walls should be kept in good condition with the mortar joints intact, otherwise rats may break through the walls. Disused drains should be properly sealed, especially foul drain inlets from toilets, which should be filled in using concrete mixed with broken glass. Water seals in gullies, sinks and W.C. pans must be maintained.

All external ventilation stacks must be provided with wire balloons fixed in the top of each pipe. Access to roof tops, via the outside of vertical pipes close to walls, can be avoided by fixing 20 gauge metal pipe guards to the pipe by an adjustable metal collar and projecting about 22cm. Cone guards must fit tightly against the wall, whilst square guards are best built into a brick joint and should have their edges turned down by about 5cm.

Insects and rodents thrive in warm conditions. Effective ventilation is required to keep food rooms as cool as possible. All ventilation openings, including opening windows, must be adequately proofed to avoid pests gaining access, for example, ventilation grids and air bricks should be proofed externally either with 6mm mesh, 24 gauge expanded metal or equivalent materials. It should be noted that if a pencil can pass through a gap, so can a young mouse.

Rats require a free water source and therefore denying access to water will assist control. Dripping taps, defective gutters, leaking roofs and puddles are all examples of common sources, which must be removed.

Good housekeeping

Despite all proofing precautions, pests will inevitably get into a building at some time. There is a difference, however, between the occasional invader and the establishment of a stable population. To reduce the risk of an infestation, it is important to deny the lone invader the conditions it likes and in particular to ensure that:

- premises are kept in a **clean** and tidy condition to reduce sources of food and harbourage. Attention must be paid to staff locker rooms, changing, dining and washroom areas. The consumption of food should be restricted to dining areas. Lift shafts must be regularly inspected to remove debris and food deposits. Adequate cleaning and dust extraction equipment is essential to avoid dust build-up, especially when handling dry powders such as flour. Fixtures and fittings should be at least 250mm above the floor to facilitate cleaning. Cooperation between cleaners and pest control contractors is essential to ensure baits are not removed, repositioned or washed away;
- spillages are cleared away promptly and food is not left outside;
- food is kept in rodent-proof containers and lids are always replaced; used ice-cream tubs are ideal for use in catering operations;
- **stock rotation** is carried out and all stock is stored correctly;
- unused equipment, packaging material and similar articles are rotated and checked frequently as rodents prefer living in undisturbed areas;
- special attention is paid to waste disposal. Receptacles should be of adequate capacity to avoid overflowing and external containers should be provided with tight-fitting lids or covers. Waste must be removed promptly and efficiently and refuse areas should be hosed down after waste is collected. Receptacles must be cleaned after emptying to prevent deposits providing breeding sites for flies. Incinerators must not be allowed to cause problems and food premises should not be built on or adjacent to refuse tips;
- vegetation, old equipment, rubbish and other cover or harbourage must be removed from the immediate vicinity of the site. It may be appropriate to undertake joint action with neighbouring premises to, as far as practicable, keep adjacent areas pest free; and
- all raw materials, including food, packaging, equipment and laundry must be checked to ensure their freedom from infestation.

Correct storage

The correct storage of goods is essential to reduce pest incidence. The following principles must be adhered to:

- all areas must remain accessible for cleaning and inspection, which should be carried out at frequent and regular intervals;
- damage to containers must be minimized to reduce spillage;
- all goods must be kept clear of the walls, windows and ventilators (at least 500mm);
- adequate gangways must be left for inspection between stacks;
- all goods must be kept off the floor, for example, on pallets or low stands, taking care that enough room is left to clear spillages;
- all areas must be well-ventilated and lighted;
- storage areas must be in good repair and effectively proofed against pest entry;

- storage space should be cleaned and inspected before new stock arrives; and
- goods that are infested or susceptible to infestation must be segregated from those that are not; raw materials, packaging and finished products should be stored separately.

PEST CONTROL STRATEGIES

The risk of infestation will always remain, however effective the environmental management and the storage practice. It is therefore essential to have a pest control strategy in place. There are two options for developing such a strategy, either the use of a professional pest control contractor or the development of an in-house pest control competency.

The use of contractors

Most food businesses rely on the expertise of a pest control company, or the local authority, to ensure their freedom from infestations. The final decision on whom to choose normally depends on the type of pest and the methods required for its control. However, the destruction of pests that are observed in the premises is not sufficient; regular inspections should be carried out to ensure the complete absence of pests from the immediate surrounding area.

It should be noted that the use of contractors does not absolve managers from their responsibility of keeping premises pest free. Furthermore, their use in isolation, is not a defence, should legal proceedings be instituted for a complaint regarding food contaminated with insects or parts of rodents or their droppings. However, food authorities should consider the attitude of, and precautions taken by, companies when deciding whether or not to institute proceedings, and the court will also take these factors into account when considering a due-diligence defence or the level of fine.

Selecting a contractor

The following matters should be considered when selecting a contractor:
- the ability of the contractor to undertake a complete survey and provide a clear report of recommendations and action required. The contract should detail pests covered, frequency of visits and reports, arrangements for additional treatments, including emergency response, preventive measures and include an unambiguous quotation;
- the experience of the contractor of pest control in the food industry and provision of appropriate references from current clients;
- the adequacy of appropriate insurance cover with regard to product, public and employer's liability together with evidence of financial viability;
- the contractor must have sufficient resources in terms of trained/qualified staff and the necessary equipment to carry out proper pest control services. It is a legal requirement for all of the contractor's staff to be trained to a competent level;
- clear reporting procedures and accountability must be established;
- the methods and materials used for pest control treatment have to be approved under The Control of Pesticides Regulations, 1986 and it is illegal to use as a pesticide any substance not so approved. Contractors should provide the relevant Safety Data Sheets and demonstrate knowledge of COSHH. Some contractors now provide a pest risk assessment as part of the contract, which may assist clients taking

advantage of the due-diligence defence in food safety legislation;
- the ability of the company to provide a complete service, including preventive measures such as proofing and the installation, maintenance and cleaning of electronic flying insect killing equipment. Reports of inspections should include advice on good housekeeping, storage and any preventive work required; and
- the company should be a member of the British Pest Control Association (BPCA) and employ staff that hold the Association's certificate of proficiency or its equivalent.

Liaison with the contractor

The contractor should provide the client's management with details of all preparations used, supply a written report on each visit and make any necessary recommendations with regard to proofing, waste disposal, stock control, housekeeping, cleaning or access. Action points should then be agreed and follow-up visits made to ensure the remedies are carried out. Whichever firm is chosen, to ensure successful control, it will need the full cooperation of the client and the contractor should be called in immediately when evidence of a pest is discovered. Dates of visits should be recorded. Additionally:
- close contact and regular liaison should be made with the contractor;
- written notes on any unsatisfactory housekeeping should be made and immediate action taken to remedy defects;
- the position of bait boxes should be noted; ideally maps should be produced;
- regular inspections of bait boxes should take place to look for droppings and dead bodies, which should be removed immediately;
- if necessary, fresh bait should be laid at the end of a week, by which time weekly inspections may well suffice; and
- when no further evidence of infestation appears, intensive baiting can be discontinued and proofing carried out. Permanent baits should be maintained.

Development of an in-house pest control competency

An alternative to contracted pest control is to develop an in-house pest control expertise. This option is used by a number of larger food manufacturers and processors.

The perceived advantages of developing such a strategy are that:
- the responsibility for the pest control work remains directly within the line and functional management responsibility of the unit;
- management has greater control over response times and flexibility of response; and
- staff used are more familiar not only with the structure of the facility, but also with manufacturing processes used and the work practices employed. They will also know and be known by both managers and workers on site, increasing opportunities for cooperation between departments.

Ownership of a pest problem is very clear, for example, if pest control responsibility lies within the hygiene unit, which is also responsible for work relating to hygiene and proofing, there is no doubt about who is responsible for pest control-related problems.

It is essential, however, that all the criteria used for employing a contractor, in terms of skills, competency, reporting procedures and safety procedures, are applied equally to an in-house scheme so that standards are maintained. In-house pest control must not be seen as a cheap alternative.

HYGIENE for MANAGEMENT *Pest control*

The role of management in pest control

The role of supervisors and managers will depend on the size of the operation and the pest control strategy employed. However, responsibilities may include:

- ensuring the provision of proactive pest control management to assist a due-diligence defence, if required. This will include effective systems for checking deliveries of raw materials and laundry;
- arranging for the instruction of staff, especially cleaners to recognize pests or signs of pests;
- routinely inspecting vulnerable areas for pests, signs of pests and poor maintenance or proofing;
- arranging for defects and poor housekeeping to be remedied and for the contractor to be requested to deal with any infestation;
- ensuring that any contaminated food is destroyed and any food contact surfaces are cleaned and disinfected before use if they may have been in contact with rodents;
- ensuring that the contractor does not expose food to risk of contamination during treatment, for example, spraying near open food;
- walking the site with the contractor, probably after the treatment, and acting on his recommendations. Ensuring that all visits and actions are recorded and the correct number of visits are undertaken;
- ensuring that any food-contact surfaces that may have been contaminated are cleaned and disinfected after treatment;
- being aware of the position of bait boxes and traps and arranging for them to be checked regularly; and
- knowing how to deal with bait boxes and dead bodies.

RODENTS AND RODENT CONTROL

The three rodents that may infest food premises in this country are:
- *Rattus norvegicus* (Common Rat, Norway Rat or Brown Rat);
- *Rattus rattus* (Ship rat or Black rat); and
- *Mus domestics/musculus* (House Mouse).

The provision of food stores gives rodents the ideal conditions for rapid multiplication; food, shelter and no predators or competition. Furthermore, the transportation of food in containers has meant that rodents can easily be brought into premises, if suitable precautions are not taken during unloading and emptying.

Surveys

Certain members of staff should be specifically trained to identify evidence of rats or mice. These people should carry out regular inspections of the premises, both internally and externally, to look for signs such as:

- droppings, if very recent they are shiny and soft;
- footprints in dust;
- gnawing marks and damage, for example, holes in sacks;
- smear marks from the fur of rodents where their bodies are in regular, close contact with surfaces, for example, horizontal pipes adjacent to light-coloured walls;
- holes and nesting sites;

HYGIENE for MANAGEMENT *Pest control*

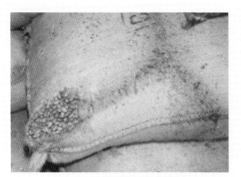

Mouse tracks on sacks of imported peanuts.
(Courtesy of Vic George.)

Rodent damage caused by gnawing.

Rodent damage and smears.

- rat runs in undergrowth; and
- the animals themselves, either dead or alive.

The reporting of signs should not be restricted to specific people, all staff should be aware of the problems of rodents and must observe their responsibility to notify their supervisor, if they believe them to be present.

The House Mouse

The cosmopolitan *Mus domesticus* is normally found inhabiting buildings, where it finds harbourage, warmth, food and nesting materials. Unlike the Norway Rat, the House Mouse is an inquisitive animal, investigating new objects in its environment rather than avoiding them, it does, however, rapidly pass on to other objects. For this reason the mouse tends to feed little and often, possibly feeding at some 20 points in a night. Effective control is therefore best achieved by placing many, small bait points or traps. The House Mouse is also a very good climber, spending much of its time off the ground if it can. This three dimensional activity should be reflected in the distribution of control points such as baits.

House Mouse. Norway Rat.

The success of the House Mouse to coexist with man can be attributed to its ability to live in a wide range of habitats, to its immense reproductive capacity and its omnivorous feeding habits (a pair of mice can, given ideal conditions, produce up to 2,000 young within a year).

The Norway or Brown Rat

Rattus norvegicus is the predominant rat in the UK and is essentially an outdoor species, sometimes moving indoors to feed. It is rarely found living only indoors. When it is found indoors, it will usually be entering through, under, over or round doors, holes in walls, through the drains (the usual route when rats are found in or on a roof), or burrowing under foundations and entering through cavity walls.

Rats pose a serious public health threat as they often carry pathogens. The Norway Rat is omnivorous but has a preference for cereals. Theoretically a pair of rats can produce hundreds of offspring within a year; fortunately many fail to achieve maturity. However, once a pair of rats becomes established, failure to implement immediate eradication measures can soon result in a major infestation.

Norway Rats prefer stable environments and are neophobic, (they will avoid new objects placed in their environment, such as baits and bait boxes, until they get used to them). Rat control therefore requires patience.

The Black Rat

Rattus rattus is also omnivorous but has a preference for fruit and vegetables. It is rarely found in mainland Great Britain, although some infestations can still be found in port areas. It is an excellent climber and is often found in the upper storeys of buildings and roof spaces. The Black Rat is more difficult to control because of its food preference and its ability to climb, making access for control difficult. In many parts of the world, particularly in the tropics, it remains the dominant commensal rodent species.

Black Rat.

Barbecued mice – caused by gnawing a live electric cable.

A COMPARISON OF COMMON RODENTS

	Norway Rat	Black Rat	House Mouse
Body/weight	Thick set 100-500g	Slender 100-300g	Up to 20g
Snout	Blunt	Pointed	Slightly pointed
Ears	Small	Large	Large
Tail	Shorter than head plus body	Longer than head plus body	Much longer than head plus body
Colour	Variable: brown on back, grey belly	Variable: black or brown back, dark/white belly	Grey back, light grey belly
Feet	Large	Large	Small

Survey and monitoring

The key to all good rodent control is an effective survey, to know when an infestation is present and to identify the extent of the infestation. There are obvious signs of rodent infestation and those involved with pest control should be actively seeking out these signs on a regular basis. However, all staff, especially cleaners, should be aware of rodent signs and what to do if they see them.

Monitoring, together with the survey data, is a more planned and conscious attempt to identify activity. It uses a range of specially developed baits and traps that are used in high-risk areas particularly and are visited and inspected on a regular basis. These include:

- bait points, using both toxic and non-toxic baits;
- activity indicators on bait boxes;
- tracking dust, including fluorescent dust in some rare situations;
- break back traps; and
- live capture traps.

The information gained from effective surveying and monitoring should detect infestation as soon as it develops and will assist in monitoring the progress of the control operations.

Control and treatment

Can be divided into chemical and physical.

Chemical control
Chemical control involves the use of toxic rodenticides. Rodenticides are chemicals used for killing rodents and may be divided into three groups:
- chronic rodenticides;
- acute rodenticides; and
- sub-acute rodenticides.

Chronic rodenticides
These chemicals are all anticoagulants and are designed to be ingested by rodents in small doses over several days, to ensure that a lethal dose is absorbed. Even the more toxic anticoagulants work as multi-dose (chronic) poisons and are not usually lethal after the first feed that a rodent takes. This is mainly because the behavioural characteristics of both Norway Rats and House Mice ensure that first feeds are small. Correct and thorough use of anticoagulants can provide 100% mortality.

The anticoagulants work by preventing blood coagulation or at least by prolonging the time that coagulation takes, so that the animal haemorrhages, mostly internally, and dies. Whilst this sounds unpleasant, the symptoms are far less dramatic than the acute rodenticides and this, together with the relatively slow-acting nature of the anticoagulants, means that the rodents do not develop poison or bait shyness as they do with the acute rodenticides if sub-lethal doses are consumed.

THE ANTICOAGULANTS AVAILABLE IN THE UK CAN BE DIVIDED INTO TWO GROUPS

First generation	Second generation
Warfarin (S)	Difenacoum (S)
Chlorophacinone (S)	Bromadiolone (S)
Diphacinone (S)	Brodifacoum (P)
Coumatetralyl (S)	Flocoumafen (P)

The second generation anticoagulants are the more recent anticoagulants and are usually more toxic to rodents and to non-target species than the first generation anticoagulants, although this is not always the case. Most anticoagulants should be applied using what is known as saturation or surplus baiting (S), essentially making sure that the rodents always have bait on which to feed. Two may be applied using pulsed baiting (P) because they are more toxic than the rest. This means that less bait is required and a break of a few days in feeding is not so critical to success. However, weekly visits to check on all baits are still recommended. Both the anticoagulants that are recommended for pulsed baiting (brodifacoum and flocoumafen) must, however, only be used "indoors" in the UK, thus restricting their use.

Resistance
Physiological resistance to the chronic rodenticides is a problem in some areas. This is recognized in the field by continued feeding of anticoagulant baits over many weeks, without the expected mortality. In house mice resistance to the first generation anticoagulants is now so widespread that there is little point in using them. There is also resistance to both difenacoum and bromadiolone, but the extent and distribution is not known.

In Norway Rats, resistance to the first generation anticoagulants is mainly in central southern England, where there is also resistance to both difenacoum and bromadiolone.

Similar resistance is found elsewhere but the problem is less intense. To date, resistance in Norway Rats has only been found on farms and holdings, none in the UK in urban areas.

Acute rodenticides

There are only two acute rodenticides currently registered for use in the UK. Zinc phosphide (used for Norway Rat control) and alphachloralose (works through hypothermia and only used for House Mouse control). Both work fast and cause relatively painful symptoms. Unless a lethal dose is taken very rapidly, the animal stops feeding, eventually recovers and avoids the bait and poison if it encounters them again(bait shy).

Zinc phosphide should only be used following prolonged prebaiting, alphachloralose should be used following a thorough removal of alternative foods and then both intensively and extensively in the infested area. Neither is likely to give 100% control, an average mortality of about 70% to 80% is more likely.

Sub-acute rodenticides

Only one rodenticide belongs to this group, calciferol (vitamin D2/D3). This rodenticide causes death through hypercalcaemia and kidney failure. It is very effective against house mice, but is not effective against Norway rats, unless prebaiting is used.

Rodenticide formulations

The rodenticides may be used in a number of different formulations. By far the most frequently used are edible baits. There are a range of bait formulations, the most palatable of which are the loose cereal baits. Canary seed is particularly palatable to house mice. Pellets and wax block formulations are also available. Wax blocks usually have holes in them so that they can be tied down.

Contact rodenticides are also available. Both contact dust and contact gels are used, although the dust formulation is more commonly available. Both can be placed in the rodent's normal runs and the rodent picks them up on their fur and ingests the rodenticide whilst grooming. These contact formulations should not be used where food or food preparation surfaces may become contaminated.

A variation on this theme is the development of a "wick" formulation. Wicks are impregnated with a waxy formulation of brodifacoum, and positioned in the middle of a long tube with entrances at both ends. The mice push past the wick and get the wax on their fur, which is then ingested when grooming.

Liquid rodenticide formulations may also be used. These are placed so that the rodents are able to drink from them and thus ingest the rodenticide. They can be productive in very dry situations but great care must be taken in their use.

Physical control

There are a number of physical control techniques available for rodent control. By far the most widely used are traps.

Traps

Traps may be used if there is a particular risk of contaminating food, to remove a very small infestation or to catch a few survivors of a treatment. They are unlikely to lead to the control of larger infestations. The advantage of the trap is that it prevents rodents dying in inaccessible places and causing offensive odours or Blowfly problems. Traps should be placed

The use of rodenticide wicks within a tube to control House Mice. (Courtesy of Sorex)

The use of "activity indicators" incorporated into the entrance of a mouse bait box to detect mouse activity - not active. (Courtesy of Sorex)

Live capture rodent traps.

on runs, at the entrance of harbourage or at right angles to walls, with the treadle nearest the wall. They should be examined daily so that if necessary they can be reset or dead rodents removed. The policy of some companies only allows the use of "live capture traps".

Rodent sticky boards

These may be used to eliminate the occasional survivor of a treatment. A piece of hardboard is coated with 3mm of a very sticky substance, which holds any rodent that comes into contact with it. Boards should be fastened down and baiting around the edge of the boards increases their effectiveness. The BPCA's Code of Practice for operational procedures concerning the humane use of boards, should be adhered to.

Safety precautions

The position of all bait points should be numbered. All bait boxes should be given a corresponding number and be dated to show when the bait was last inspected or replaced. A plan of the premises, showing the location of all bait points, should be kept by the senior person responsible. It is important to keep accurate details of visits by contractors, and a pest control book should be kept on the premises under the control of management. The following information should be recorded:

- the results of the initial survey;
- the work carried out as a result of the survey;
- the degree of infestation found and the type of pests;
- details of each treatment carried out and the rodenticide used;
- the recommendations made by the technician on each visit and the action taken;
- a record of any special or emergency visits made by the contractor; and
- all reported sightings by staff of pests on or around the premises.

Baits must not be positioned where they could expose food to risk of contamination and all boxes must be labelled "poison". Most rodenticide baits now contain bitrex, a human taste deterrent, to reduce the chances of ingestion by humans. Suitable protective clothing and waterproof gloves must always be worn when handling rodenticides, rodents or traps. It is essential that the label recommendations be followed when using rodenticides.

All persons involved with pest control must have regard to the Health and Safety at Work, etc. Act, 1974 and the Control of Substances Hazardous to Health Regulations, 1999.

INSECTS AND INSECT CONTROL

Any insect in food premises is a pest. However, apart from those insects that wander in, there are many that will cause extensive problems if they become established. Insect pests can attack and destroy large quantities of food, which become contaminated with their bodies, webbing and excreta. In addition, several insects are capable of transmitting pathogens, including food poisoning organisms. No food is safe from insect attack, although beans, cereals, flour, dried fruits and some dried meats are amongst the most susceptible to infestation.

Common insect pests of food premises include:
- flies;
- wasps;
- stored product insects, including moths;
- psocids and mites;
- silverfish;
- cockroaches; and
- ants.

Webbing from moths indicating a major infestation.

FLIES AND FLYING INSECTS

The order Diptera contains approximately 78,000 species. Those of particular importance to the food industry are: *Musca domestica* (Common Housefly), *Fannia canicularis* (Lesser Housefly), *Calliphora spp.* (Bluebottle), *Lucilia spp.* (Greenbottle) and *Drosophila spp.* (Fruit Fly).

Flies contaminate food in four ways:
- to feed, they regurgitate enzymes and partly-digested food from the previous meal;
- they continually defecate;
- they carry bacteria on the hairs on their body and legs; and
- pupal cases, eggs and dead bodies end up in our food.

The danger to health from flies must not be underestimated. Many pathogens have been found on and in flies and their droppings. Furthermore, flies have been allegedly involved in the transmission of *E. coli* O157, *Shigella spp.* and food poisoning organisms. The close proximity of sources of pathogens, such as faecal material or raw meat/poultry, the number of flies landing on high-risk food and the temperature abuse of contaminated food will significantly increase the risk of illness.

The life cycle of flies depends on the breeding site temperature and the amount of available food. Typical breeding sites are accumulations of waste organic matter, such as refuse, and refuse tips. The female Housefly deposits around 600 eggs during her life span. From egg to adult is normally less than two weeks in warm weather. Maggots grow to around 9mm. Blowflies usually breed on decaying matter of animal origin, especially meat.

Fruit Flies generally occur in bakeries, fruit-canning factories and beer cellars, as fermenting organic materials attracts them. Food debris swept under inaccessible machinery bases and kept wet by "spray" from regular wet cleaning or condensation provides ideal conditions for larvae. Uncleaned pipes connecting beer or fruit juice dispenser drip trays to drains may also cause problems. Some species lay their eggs in unwashed milk bottles, when the pupal cases may be found cemented to the inside of the bottles.

Drain Flies may cause similar problems because of a failure to clean up or remove organic residues. Condensation and a failure to clean underneath uninsulated ice bins may result in a wet, slime fungus, which produces good conditions for larvae.

Housefly and pupae.

The presence of these flies should generally be regarded as bad management. They are more of a nuisance than a health hazard, control is usually achieved by removal of the breeding material and keeping potential breeding sites clean and dry. In the case of an infestation, suspect areas such as drains may be checked by using an inverted jar when closing the food premises and inspecting the jar for the presence of flies on arrival the following morning. Wet accumulations of debris should be inspected for the presence of larvae and pupae, using a magnifying glass if necessary. Glue board type UV fly killers may also provide an early warning of the presence of small flies.

Wasps

Wasps may contaminate food by transferring bacteria from their legs and are a nuisance during the late summer in such premises as bakeries and fruit factories. Proofing is essential. Nests should be located and destroyed, but wasps fly considerable distances. External perimeter attractants, which drown wasps, may be successful.

Flying insect control

Wherever possible, emphasis should be placed on environmental and physical control methods to reduce the risk of food contamination. The areas around food premises should be kept clean and tidy and all possible breeding sites should be removed. Drainage gullies, effluent treatment plants and waste disposal areas can all cause problems if neglected. Within kitchens, the high-risk areas tend to be the base of lift shafts, behind deep freezers, fridges and cookers. Air curtains are occasionally used to keep out flying insects.

All refuse containers should be kept clean and in good repair. Lids should always be tight fitting. If skips are used, completely enclosed, compacting types are preferable. Waste-food containers, such as syrup tins, which attract insects, should be washed thoroughly before storage outside. All refuse areas should have well-drained hard-standings, which are kept clean. Polythene sacks are recommended for internal use.

Proofing

Proofing is usually designed to prevent access by highly mobile, adult stages of the insect. Food safety legislation requires that windows that can be opened in rooms where food is prepared, treated or processed, must, where necessary, be fitted with insect-proof screens, which can be easily removed for cleaning.

Windows and other openings used to provide ventilation must, where necessary, be fitted with cleansable fly screens. Roof access at apexes and eaves should also be screened. Doors should be kept closed or provided with cleansable screens or clear, heavy-duty plastic strips. Self-closing doors and double door air locks are useful.

Electronic flying insect killers

Those insects gaining entry to food areas should be destroyed using suitably sited electronic fly killers (EFKs). Flies are attracted by an ultra-violet light and then electrocuted on charged grids. Performance is proportional to total light wattage of each unit. Units are most effective in subdued light, away from windows and fluorescent lights. They should not be positioned over food or food equipment or in draughts, as dead flies may be blown out. Neither should they be positioned in such a way that they might attract insects from outside, opposite doors or windows for instance.

Some units use **insecticide** and many now incorporate a system to prevent insect particles falling out, for example, the use of UV light-transmitting polymer film. Glue-board EFKs are becoming the most popular as, provided the board is changed when necessary, they reduce the risk of dead flies dropping into open food.

Catch-trays should be emptied frequently and the units regularly serviced. Tubes should be replaced every six months, at the beginning of the fly season, or as recommended by the manufacturers. Due to risk of explosions, they should not be sited where there are high concentrations of flour or sugar dust.

An important function of EFKs is that they may be used as a means of monitoring the flying insect activity in their vicinity. To do this the contents of the catch trays should be identified on a regular basis and action taken as necessary.

A badly sited electronic fly killer.

Sticky flypapers

These are very useful to supplement EFKs in areas to which the public are not admitted. Useful protection can be given to storage areas, refuse areas and bakeries. They should not be positioned near heat sources and should be changed as often as necessary.

Chemical control of flying insects

Insecticides are chemical substances, that kill insects. They should only be used as a back-up to physical control methods. Only insecticides that are cleared for use in areas where food is prepared, stored or processed should be used where food may become contaminated. The insecticide labels will identify where this is possible. Before treatment, food and equipment should be removed, or protected, to prevent risk of contamination from chemicals or insects. After treatment, all food-contact surfaces should be cleaned and disinfected and dead insect bodies should be removed.

The use of residual insecticides in food rooms is not recommended because of the danger of dead insects dropping into food. However, treatment of corridors, or vestibules in non-food areas, with residual insecticides may be useful. Synthetic pyrethroids are the most common of the residual insecticides used in these areas. Where there is a severe problem of flying insects that requires immediate control, ultra low volume (ULV) application of insecticide may be used. ULV mists are designed to control flying or exposed insects, which collide with the very small droplets produced. ULV formulations do not produce significant insecticide residuality.

It is essential that COSHH assessments are undertaken prior to the use of any pesticides and that label recommendations are followed at all times.

STORED PRODUCT INSECTS (SPIs)

This is a large group of insects, that attack foodstuffs in storage, transport and manufacture. It includes beetles, weevils and moths. Cereals, flour, beans, dried products and nuts may all be attacked.

These insects have no direct health significance, although they do cause considerable economic loss. Infestations are often difficult to detect due to the small size of the insects and because the SPIs often live within the food or commodity itself. They can also remain hidden in the building structure, concealed in the crevices and inaccessible, dark corners. Signs to watch for include tiny moving adult insects, unusual debris or speckling, fine strands of webbing and tunnels in some foods. If evidence of infestation is discovered, a specialist pest control contractor must immediately be asked to carry out treatment.

Infestations may be introduced into the food premises with the raw materials or possibly arise due to breeding in materials that are poorly rotated. Inaccessible, poorly cleaned structures also provide useful breeding sites. Panelling of walls increases problems significantly. Infested pallets may be a source of infestation. It is essential that all raw materials be thoroughly inspected when unloaded at the premises. Containers or sacks showing evidence of damage or holes should be regarded with suspicion. If adequate inspection is not possible when unloading, the new materials should be segregated from existing stocks. All raw materials, which are held for considerable periods, should be inspected at least weekly. This includes paper and other stationery stocks. All spillages and residues must be removed as soon as possible.

Insect control

To successfully control SPIs it is necessary to know:
- the species involved;
- the source of the infestation; and
- the size of the infestation and its location.

To identify this information, it is essential to introduce an effective monitoring strategy. This may be based upon a range of techniques including sticky traps and live capture traps, incorporating species-specific pheromone lures. Alternatively, specially designed probe or pitfall traps may be used.

Should an infestation occur, despite environmental control, insecticides will have to be used. Trained specialists using approved insecticides should undertake the treatment. The formulation and insecticide used will depend on the site of infestation and the type of insect. Infestations of raw materials should be dealt with immediately. In some cases treatment may be successfully employed, for example, fumigation. However, quite often the food will need to be destroyed.

Fumigation is the application of a toxic chemical in the form of a gas, vapour or volatile liquid in a closed container or to a food stack under gas-proof sheets. The required concentration of gas must be maintained for a specific period. For some pests, such as insects in cereals, fumigation is the only satisfactory method of treatment.

Psocids or Booklice

Psocids are small (1–2mm), cream, light brown or dark brown insects, which are being reported more frequently in food premises and are resulting in increasing numbers of food complaints. They are omnivorous and commonly infest flour, grain, nuts, chocolate, fish and meat products. They also feed on moulds and yeasts and infestations are often associated with packaging materials and pallets. Their presence usually indicates conditions of high humidity. They are mainly of nuisance value but they may be responsible for disseminating spoilage bacteria.

Control involves high standards of hygiene, which prevent mould development. Adequate ventilation and dry conditions are particularly important. Residual insecticides may be applied to pallets, walls or other surfaces, although fumigation of product may also be necessary.

Fumigation may be required to control infestation of stored products.

Mites

Mites are similar to insects but adults have eight legs. They appear to the naked eye as coarse dust and can only be detected by their movement and through the use of specially designed traps incorporating attractive baits. A variety of food may be infested including flour, cheese and smoked meats. They are particularly likely to be found in damp, poorly ventilated stores where they taint food. They may cause dermatitis and allergies.

Mites require high humidity in order to survive and are best controlled by drying and high temperatures, although some insecticide applications are available for treating infested food (acaricides).

Silverfish and firebrats

These essentially nocturnal insects are either bright or dull silver and are the shape of miniature carrots about 10mm long. Identification is made easier because of the two large antennae on the head and the bristles on the abdomen. They move about very quickly and, although they are frequently found in food premises, they are of little health significance. However, as with all other insects, infestations cannot be tolerated.

Good housekeeping and denial of harbourage are essential for successful control. Crumbs and spillages must be cleared up and particular attention paid to waste disposal. Sources of moisture, including condensation, must be removed. Hundreds of Silverfish can live under

HYGIENE for MANAGEMENT *Pest control*

Stored product pests
a) *Stegobium paniceum* (Biscuit Beetle). b) *Sitophilus granarius* (Grain Weevil).
c) *Liposcelis bostrychophilis* (Booklice). d) *Ptinus tectus* (Australian Spider Beetle).

defective surfaces, especially badly fixed tiles, and in cracks and crevices. Infestations thrive on tea and coffee waste trapped behind sink units.

Live insects may be killed with aerosol sprays for specific crawling insects, provided that the usual precautions are taken. Residual insecticides may need to be applied by a pest control operator where other control measures fail.

a) Larva, webbing and damage by *Ephestia elutella*.
b) *Plodia interpunctella* (Indian Meal Moth).
c) *Tineola bisselliella* (Clothes Moth).
d) *Hofmannophila pseudospretella* (Brown House Moth).
e) *Ephestia kühniella* (Mill Moth).

COCKROACHES

Only two species of cockroach are widely distributed in this country:
- the German Cockroach (*Blattella germanica*); and
- the Oriental Cockroach (*Blatta orientalis*).

Occasional problems are caused by the larger American Cockroach (*Periplaneta americana*), for example, in hotels, dock areas, zoos and schools, where they have escaped from laboratories. Some other cockroach species may also occur and it is essential if effective control is to be achieved, that correct identification is undertaken. Incorrect identification is likely to result in inappropriate and incorrect use of control techniques.

Cockroach habits

Cockroaches are gregarious, omnivorous, nocturnal insects and give off an unpleasant characteristic odour. During the day, they hide in cracks, pipe ducts, electric motors, behind skirtings and in stores. Their presence is normally detected by faecal pellets or their smell. The periods of maximum activity are just before dawn and just after dusk. Neither species fly, although the German cockroach may glide for short distances.

Cockroaches are capable of carrying various pathogens, which may lead to outbreaks of disease, although there is little evidence that they are a common vector for spreading infection. Over 40 pathogenic organisms have been isolated from cockroaches collected from a variety of premises, either in the faeces or contaminating their legs or antennae, including:
- *Staphylococcus aureus*;
- *Salmonella* Typhimurium; and
- *Salmonella* Typhi.

For this reason and the fact that their faecal pellets, moult debris and dead bodies contaminate food, cockroaches in food premises must be destroyed.

The German Cockroach (Steam Fly)

Probably now the most commonly encountered of all the cockroach species in the UK. The adult is 10 to 15mm long and is yellowish-brown. It is capable of climbing smooth vertical surfaces such as painted walls. German Cockroaches prefer warm, moist conditions and are commonly found in kitchens, pantries, restaurants and especially ships' galleys. Particular problems have been encountered over recent decades in blocks of flats, where the warm conditions and relatively easy access between accommodation units have led to their rapid spread and difficulties with control. At 30°C, the life span is around three to 12 months and the female produces approximately 250 to 300 eggs. The eggs hatch to produce nymphs, which are miniature versions of adult cockroaches. The nymphs pass through a series of moults, when they shed their rigid exoskeleton, to become adults. The number and size of nymphs, egg cases, discarded exoskeletons and adults are useful to assess the size and age of the infestation.

The Oriental Cockroach

The adults grow to about 24mm long and are shiny dark brown to black. They can climb rough vertical surfaces, such as brickwork, and often congregate around water sources. Oriental Cockroaches are frequently found in cellars, kitchens, bakeries and drains. Long-term survival and breeding externally, for example, in refuse tips and gardens, is possible. For this

Oriental Cockroach, with protruding ootheca.

reason, it may be necessary to incorporate external control operations when treating a known infestation within a building or within ducting. Thorough survey and monitoring will help identify the extent of the infestation.

The eggs are laid in an **ootheca** (egg case) and take around two months to hatch at 25°C, but this period is extended in cooler conditions. It takes between six and 12 months to reach adult stage in heated buildings with good food supplies. The female can produce over 150 young in its lifetime.

Monitoring

Any effective cockroach control programme must incorporate an effective monitoring strategy. Such monitoring must identify the species involved, the size and distribution of the infestation and the progress of the control operations.

Sticky traps are the most effective method of monitoring even minor infestations. Traps may incorporate a hormone or other attractive tablet. The traps are placed throughout the areas where infestation is suspected, and beyond. It is essential that the full scope of the infested area, both vertically and horizontally, be identified. These small adhesive cockroach detectors are very effective in identifying even minor infestations and the species of cockroach involved.

Torch light surveys at night or the application of an insecticide with a flushing action, such as pyrethrum, during the day should also form a part of the survey and monitoring programme.

Cockroach control

Prevention is of prime importance when considering cockroach control in food premises. Cockroaches may be brought into the building with food containers, raw materials or laundry. It is imperative to ensure that commodities entering the food premises are not a constant source of supply of cockroaches.

Sound building structure is important as it precludes the entry of a large number of cockroaches through openings and helps eliminate harbourage. Crevices may be sealed with putty, and service pipes or conduits passing through walls should be cemented in position. Ill-fitting panels that are not removed, are often likely to harbour cockroaches. Strict hygiene should discourage infestations. Food should be stored in containers with close-fitting lids and spillage should be removed promptly.

Areas inaccessible for cleaning should be eliminated. Particular attention should be paid to drains and refuse areas.

There are number of insecticide formulations available for the control of crawling insects such as cockroaches. It is essential to select the one most suitable, both for the species involved and the circumstances of the infestation. Formulations include wettable powders, dusts, emulsion concentrates and space sprays. It is essential to use these in such a way that the life cycle is broken and that repeat applications target susceptible stages of the life cycle.

Because of problems of natural repellency to spray insecticides and the move to environmentally friendly controls, more emphasis is being placed on the use of edible baits for cockroach control, these contain insecticides such as boric acid, fipronil and hydramethylnon.

It is advisable to sketch a plan of the rooms, including details of the plant layout, to achieve effective control. Established infestations are often difficult to eliminate and reinfestation from isolated pockets commonly occurs. Sound planning and the "block control" approach will be necessary for larger scale infestations.

Cockroach traps

Several types of cockroach trap have been developed with varying degrees of success. They may be useful to monitor the extent of infestations and to determine whether any cockroaches have survived a treatment with insecticide.

ANTS

The Garden Ant (*Lasius niger*)

The black Garden Ant nests outside but often becomes a persistent pest of food premises, as it forages for food. It is mainly of nuisance value, although contaminated food must be discarded. Successful control depends on the destruction of the nest. This may be achieved using residual sprays, dusting powder or insecticidal baits.

Pharaoh's Ant (*Monomorium pharaonis*)

Pharaoh's Ants are pinkish and approximately 2mm in length. They are often found in seemingly impenetrable food containers. Infestations are usually restricted to permanently heated buildings, especially hospitals, bakeries, hotels, residential properties and kitchens. All kinds of food may be attacked, although there is a preference for sweet and high-protein food. Physical transmission of pathogens to food is possible as they may visit drains, excreta and soiled dressings.

Effective control requires the destruction of the ants' nests, rather than simply a proportion of the working ants. However, nests are very difficult to detect and destroy. Control may eventually be achieved using baits containing boric acid or hydomethylnon as an insecticide or methoprene as a growth regulator. Baits are more effective because the ants take the bait containing the active ingredient back to the nest. Residual sprays, powders and lacquer formulations are not usually effective as they only affect the ants that wander on to the treated surface.

BIRD PESTS AND THEIR CONTROL

Birds that commonly gain access to food premises are sparrows and feral pigeons, although other species, such as starlings, occasionally attain pest status. Sometimes, more unusual species, such as seagulls, may cause problems, but this is usually associated with their roosting in outside areas where they may cause fouling and contamination. Warehouses and large food factories are prime targets, although bakeries and supermarkets may also be affected.

Reasons for control:
- to prevent the contamination of food or equipment by droppings, feathers, regurgitated pellets and nesting materials;
- to prevent the transmission of food poisoning organisms such as salmonella and campylobacter;
- to remove sources of insect and mite infestation provided by nests, excreta and the birds themselves;
- to prevent blockages of gutters, which may result in flooding and expensive maintenance;
- to prevent defacement of buildings; bird droppings encourage the growth of microflora and fungi, which attack stone and brickwork;
- to prevent roosting on fire escapes and similar structures, which may result in a safety hazard for human occupants; and
- to prevent damage to food packaging.

Proofing and exclusion

Control of birds is best considered during the planning stage of new buildings. Food premises should be designed to prevent ingress of birds, as it may be extremely costly to proof against entry once the building is completed. Ledges and perches should be eliminated.

Good housekeeping

There is clear evidence that the size of any localized bird population, particularly feral pigeons, is related directly to the amount of food available. Remove the food source and the bird activity will be significantly reduced.

As with other forms of pest control, prevention is better than cure. Good housekeeping is essential. All food spillages, which attract birds, must be removed as soon as possible. Waste receptacles should be provided with tight-fitting lids and not overfilled. Waste areas must be kept in a clean condition.

Proofing

It is not always possible to solve a bird access problem with hygiene improvements alone. They usually have to be used in conjunction with proofing and exclusion techniques. All openings, whether large or small, should be proofed to prevent the entry of birds. The roof apex, open eaves, louvres and ventilation and other openings should be protected with 15mm galvanized chicken wire or preferably knotted polyethylene netting.

There are a range of techniques available for fixing these nets, but whichever technique is used, it is essential that the correct mesh sizes are used to protect against the species causing the problem.

THE FOLLOWING MESH SIZES ARE RECOMMENDED

seagulls	100mm (4") mesh net
feral pigeons	50mm (2") mesh net
starlings	28mm (1 1/8") mesh net
house sparrows	19mm (3/4") mesh net

It is essential that the nets be installed correctly, leaving no opportunity for further bird access. Netting probably provides the best means of preventing access to areas.

Plastic strips
Doors should be self-closing and heavy-duty. Overlapping plastic strips should be fitted if factory doors are constantly left open or are used for forklift trucks.

Point systems
These involve the use of strips of plastic into which upward pointing, thin steel wires have been embedded. The strips are secured, usually by adhesive, on to the surfaces being protected. The wires interfere with the bird's ability to use the ledge or surface and they move elsewhere! It is essential that such systems be correctly installed if they are to be successful. The technique is not usually effective against smaller birds and is most effective against feral pigeons. Special point systems for use against gulls are also available.

Birdwire systems
This involves the erection of stainless steel wire, tensioned with a spring, supported on a variety of posts and brackets. Once again, the wire suspended across and over the surfaces used by the birds causes them to move elsewhere. The technique is recommended only for use against feral pigeons and gulls, if special equipment is used.

It is important when using any deterrent device, such as spike or birdwire systems, to bear in mind the degree to which the birds want to be at any particular point. Sometimes the birds are not particularly worried if they use an area or not, at other times, usually because they feed from the point or nest at a point, they have a strong urge to return. It is necessary to use the more intense and effective systems when the birds have a strong urge to return.

Repellent gel
Thick inert gels have been used, with limited success, to prevent birds perching. Birds feel insecure on contact with the gel and look for alternative sites.

Scaring
There is a range of visual and acoustic scarers on the market. These include both live predatory birds and artificial predatory birds, distress calls that copy the species-specific distress call that some birds make and simply loud scary noises!

These vary in terms of their effectiveness. It is essential when using any of them to ensure that their application procedure is both varied and reinforced, otherwise the birds rapidly acclimatize to their presence and they lose any effectiveness that they may have.

Scaring is not usually an effective method of solving serious bird activity problems.

Maintenance

All buildings should be well maintained and any holes that remain after alterations should be filled as soon as possible. Broken windows must be repaired immediately. Regular checks should be made in roof spaces, as roof linings and girders often provide ideal nesting or roosting sites; nests must be removed and a search made to locate points of entry.

Culling of birds

Clearly one way of reducing the number of birds in an area or within a building is to kill them, as we do with other hygiene pests. With birds this approach is unlikely to be effective in anything but the short term. Birds can be killed, but if the food supply remains the same, the population will rapidly return to the original size, as a result of breeding or immigration from elsewhere.

It is also essential to ensure that when culling birds, all legal constraints are both fully understood and complied with. In the UK all birds are protected, although "authorized" persons in certain circumstances may take some. "Authorized" in this context means that those killing the birds have the approval of the owner/occupier of the land on which the work is being undertaken.

Prevention of the spread of disease is an approved purpose for which feral pigeons, house sparrows and starlings may be taken. It is essential that anybody becoming involved with the culling or taking of birds, familiarizes himself/herself with all aspects of the relevant legislation under the Wildlife and Countryside Act 1981.

Traps

Traps may be used against both house sparrows and feral pigeons. The birds are attracted into correctly positioned traps using food. This often results in birds being caught that were not part of the original problem. In addition the traps should be regularly visited, ideally twice a day. There is frequently strong adverse public reaction to this method of control.

Shooting

The use of suitable air rifles may be quite successful, especially if control is carried out at night, when birds are roosting. Foods must not be exposed to risk of metal pellet contamination during shooting. Reinvasion is likely unless proofing is carried out. The benefit of shooting over other removal techniques is that the birds actually causing the problem can be targeted. It is essential that only those who are trained and competent to use an air rifle be used on such work.

Narcotizing

The use of poisons against birds is prohibited by legislation. However alphachloralose is available for use under licence from government departments in special circumstances. Alphachloralose is a stupefying substance, which may be used successfully in a bait base attractive to birds. Those that succumb may be disposed of by humane methods and protected species can be released. Chances of success are increased if all alternative sources of food, such as spillages, are removed. It is important to ensure that all baits are collected after treatment.

It is essential that correct procedures be applied. The efficiency of narcotics against birds is very low.

Mist netting used internally to control sparrows.

Mist netting

The use of mist netting is rather like fishing for birds. The nets are very fine and are not detectable by birds. They are fitted over doorways and other flight paths of birds. It is a humane technique; protected species can be released and pest birds disposed of. Once again, only licensed operators are permitted to use mist nets. Nets must be removed when the operator leaves the premises.

The control of bird problems requires a high degree of expertise and it is essential that legal constraints be recognized at all times.

Most of the photographs used to illustrate this chapter were provided courtesy of Anticimex and Rentokil Initial plc.

16 Control & monitoring of food standards & operations

Control and monitoring are integral parts of all food safety management systems and essential functions of management, which complete the circle of responsibility. Cost-effective operations that produce **safe food** will only be achieved if managers:
- set the requisite standards/objectives;
- provide the resources and establish systems and controls, including documentation, to achieve the standards;
- communicate the standards required to staff;
- train and motivate staff to ensure their competence to produce safe **food**, especially in relation to the control of **hazards**;
- provide effective supervision; and
- monitor, analyse, compare actual standards with those required and, if necessary, take corrective action and improve performance to facilitate achievement of objectives.

Monitoring of all food operations is essential to:
- confirm that expected standards/controls are achieved;
- ensure the production of safe, **wholesome food** of good quality and **shelf life**;
- ensure **compliance** with specific legislation and to facilitate the use of the due-diligence defence;
- identify problems, for example, **sources** of **contamination**;
- satisfy customers and enforcement officers and to minimize complaints;
- facilitate modification of procedures;
- encourage commitment and improve motivation of staff; and
- assist in the development of a food safety culture.

Commitment

Notwithstanding the legal and moral responsibilities of managers, satisfactory standards of **hygiene** will not be achieved without the commitment of all involved in the **food business**. Commitment is required from the owners, the board of directors, the managing director, the **food handlers** and all levels of management. If this commitment is not forthcoming, then inadequate resources will be made available to provide the basic framework essential to achieve the necessary standards. Resources are required to plan, design and construct food premises so that they can operate hygienically, to provide satisfactory facilities and to secure the employment and training of suitable staff. Commitment of owners will only be achieved when they believe that poor hygiene, food complaints, **food poisoning**, prosecutions and adverse publicity will threaten the profitability, or even the existence, of the business.

The commitment of management is essential to motivate and effectively supervise staff, and develop a food safety culture. Managers must always lead by example and provide staff with suitable incentives to encourage the maintenance of standards.

Food safety policies

A planned approach is a prerequisite of hygienic operations. The standards required and the way they can be achieved should be incorporated into a food safety policy in the same way that safety matters are included in the safety policy required by the Health and Safety at Work etc. Act, 1974. This policy document would be extremely useful to support a **due-diligence** defence. It can be used as an effective way of communicating the requisite standards to staff and for determining their training requirements. To remain effective the document must be reviewed regularly.

A responsibility flow chart showing management structure and individual responsibilities with regard to hygiene should be included. The document should be brought to the attention of all staff and written in a way that demonstrates company commitment to producing **safe food**. All aspects of hygiene should be covered by the policy, which should include:

- a commitment to produce safe food;
- a commitment to observe all relevant legal requirements, industry guides to good hygiene practice and government codes of practice;
- a commitment to implement a food safety management system based on the principles of HACCP;
- staff training and the implementation of a planned **food hygiene** training programme that results in the competency of all staff (training records should be maintained);
- procedures to ensure that all food and water suppliers are satisfactory/approved. Suppliers should provide a copy of their food safety policy and customer references. Procedures for removing unacceptable suppliers from the approved list;
- a commitment to provide the necessary premises, equipment, facilities and maintenance to achieve high standards of hygiene, including **personal hygiene**;
- satisfactory temperature control and monitoring systems for food ingredients and products during storage, preparation/processing, distribution and display. Procedures should be identified for safe alternatives if equipment is defective;
- systems to ensure satisfactory **cleaning** and, where necessary, disinfecting of the premises, equipment and facilities (cleaning schedules will be required);
- adequate **pest** control measures including proofing, the use of specialist contractors and maintaining records;
- procedures and systems for health screening and the reporting of staff illness, dealing with visitors, contractors, enforcement officers, food poisoning incidents, customer complaints, delivery of raw materials, traceability and product recall, hazard warnings and waste management;
- effective quality assurance/control systems, including **stock rotation**, foreign body control, **organoleptic** assessment, sampling, food labelling and in-house **audits**; and
- a commitment to provide the resources for and training of managers to ensure the implementation, updating and enforcement of the policy throughout the business.

HAZARD ANALYSIS CRITICAL CONTROL POINT (HACCP)

Food product safety is ultimately a management responsibility and is regarded as an absolute requirement by the customer. If a manager is to do more than hope that the hygiene standards he/she has set are being achieved consistently and that every item of food sold is safe, it is necessary to establish a food safety management system.

A food safety management system includes the policies, procedures, practices, controls and documentation that ensure the food sold by a food business is safe to eat and free from contaminants.

It is possible to define a safe food product by setting acceptable limits for relevant pathogenic micro-organisms. It is then theoretically possible to control the product safety within these limits, whatever the quality of ingredients or process, by not releasing the food for sale until it has passed microbiological and/or toxicological analysis. Control by such end-product testing, however, is not physically possible if the gap between production and consumption is short, as in conventional catering, and little information is provided as to why a product has failed to meet a standard. Consequently, a preventive approach, involving control of the ingredients, process and processing environment, is always preferable.

HACCP is a food safety management system designed to control hazards at points critical to food safety. HACCP systems prioritise controls so that resources can be concentrated at fewer points in the process. It was developed in the early 1960s by Pillsbury, NASA (National Aeronautics Space Administration) and US Army Natick Laboratories to apply a nil-defects programme for the manufacture of food for astronauts. As you will appreciate, safe food was essential - the consequences of astronauts suffering from diarrhoea and vomiting in a space suit with no gravity would be horrendous.

Prior to the introduction of HACCP, most hygienic food businesses were aware of potential food safety problems and implemented appropriate controls. However, resources were often spread very thinly over many controls. Very little attempt, especially in catering and retailing, was made to determine which controls were the most important to reduce the risk of food poisoning. Furthermore, it was rare to find food safety management systems which analysed food preparation to determine which controls early in the process were not essential because of controls later in the process which could be relied on to remove the problem before the food was consumed. In manufacture, control relied heavily on end-product testing. Internal and external inspections/audits were also used to determine the safety of operations. However, many inspections concentrated on walls, floors and ceilings, and bad practices, but not necessarily on those points which were critical to food safety.

Terminology associated with HACCP
- *Acceptable level* – the presence of a hazard at a level that is unlikely to cause an unacceptable health risk.
- *Control measures* – actions or activities required to prevent or eliminate a food safety hazard or reduce it to an acceptable level.
- *Control point* – a step in the process where control may be applied, but a loss of control would not result in an unacceptable health risk.
- *Corrective action* – the action to be taken when results of monitoring at a CCP indicate loss of control, i.e. a critical limit is breached.
- *Criteria* – specified characteristics of a physical (for example, time or temperature), chemical (for example, pH) or biological (for example, sensory) nature.
- *Critical control point (CCP)* – a step in the process where control can be applied and is essential to prevent or eliminate a food safety hazard or reduce it to an acceptable level.
- *Critical limit* – a monitored criterion which separates the acceptable from the unacceptable.
- *Decision tree* – a sequence of questions applied to each process step with a potential hazard to identify those which are critical to food safety.

- *Deviation* – failure to meet a critical limit.
- *Flow diagram/chart* – a systematic representation of the sequence of steps or operations involved with a particular food item or process, usually from receipt of raw ingredients to consumer.
- *Food safety management system* – the policies, procedures, controls and documentation that ensure the food sold by a food business is safe to eat and free from contaminants.
- *HACCP (hazard analysis critical control point)* – a food safety management system which identifies, evaluates and controls hazards which are significant for food safety.
- *Hazard* – a biological, chemical or physical agent in, or condition of, food with the potential to cause harm (an adverse health effect) to the consumer. (N.B. most biological hazards are microbiological.)
- **Hazard analysis (Codex Alimentarius)** – the process of collecting and evaluating information on hazards and conditions leading to their presence to decide which are significant for food safety and should therefore be addressed in the HACCP plan.
- *Monitoring* – the planned observations and measurements of control parameters to confirm that the process is under control and that critical limits are not exceeded.
- *Prerequisite programmes* – the good hygiene practices a business must have in place before implementing HACCP, to enable the HACCP plan to concentrate on the most significant hazards.
- *Severity* – the magnitude of the hazard or the seriousness of the possible consequence.
- *Target level* – control criterion that is more stringent than the critical limit, and which can be used to reduce the risk of a deviation.
- *Tolerance* – the specified degree of latitude for a control measure, which, if exceeded, requires immediate corrective action.
- *Validation* – obtaining evidence that elements of the HACCP plan are effective, especially the critical control points and critical limits.
- *Verification* – the application of methods, procedures and tests, and other evaluations, in addition to the monitoring, to determine compliance with the HACCP plan. (Includes prerequisite programmes.)

Advantages of HACCP
- proactive remedial action can be taken during processing/production, i.e. before serious problems occur;
- control parameters are relatively easy to monitor, for example, time, temperature, pH, texture, appearance;
- cost-effective because it targets resources to the control of critical points and reduces the need for end-product testing. Wastage, complaints and recalls should be reduced.
- the operation is controlled on the premises;
- reduces business risk and demonstrates management commitment;
- all staff can be involved with product safety;
- safety introduced during product development;
- reduced product loss and avoidance of expensive reprocessing;
- complementary to quality management systems, for example, ISO 9000;
- useful in demonstrating due diligence;
- provides greater confidence in product safety and management;

- compliance with legal requirements* and the requirements of third party auditors;
- internationally recognized (facilitates international trade);
- generates a food safety culture; and
- brand protection.

Reasons for failure of HACCP
- general hygiene standards (prerequisite programmes) poor;
- lack of management commitment;
- lack of knowledge/training and understanding, especially science-based knowledge;
- lack of resources, especially time for implementation, monitoring and verification;
- too complex, not understood by staff involved in implementation;
- too much paperwork or poor inaccurate records;
- not science-based;
- the imposition of an external "off-the-shelf" consultancy-based package when there has been little contribution from relevant staff within the organization;
- the HACCP manual is left in the manager's office. HACCP should be brought into food rooms and CCPs clearly identified; and
- too many control points wrongly identified as critical control points.

Prerequisite programmes for HACCP

Prior to the implementation of HACCP, it is essential that the food business is operating in accordance with good hygiene practice or, in the case of factories, good manufacturing practice. Management commitment, adequate resources and suitable facilities must be provided. It is usually considered that to successfully implement HACCP and reduce the number of critical control points to a manageable level, the following prerequisite programmes will be required:
- approved suppliers;
- premises and equipment well designed, constructed and maintained. Contingency plans in case of breakdown. Equipment calibrated. Where practicable there should be a unidirectional product flow;
- water and ice used in food production must be potable;
- staff and managers must be trained commensurate with their work activities i.e. they must be competent. They should have high standards of **personal hygiene**, especially in relation to handwashing. A health and exclusion policy should exist to screen new employees and ensure that food handlers with diarrhoea and/or vomiting do not handle food until they are symptom-free for at least 48 hours **and, when they return to work, they can be relied on to thoroughly wash their hands after using the toilet;**
- effective, planned cleaning and **disinfection** and the use of cleaning schedules for monitoring purposes;
- **integrated pest management**;
- stock rotation;
- thorough washing and disinfecting of all ready-to-eat fruit and salad vegetables;
- effective waste management; and
- labelling, traceability and recall procedures.

*Legislation relating to HACCP is dealt with in Chapter 17.

The seven HACCP principles (as defined by Codex Alimentarius)

HACCP is a food safety management system based on the following seven principles.

1. Conduct a **hazard analysis**. Prepare a flow diagram, identify the hazards and specify the control measures.
2. Determine the **critical control points (CCPs)**.
3. Establish critical limits.
4. Establish a monitoring system for each CCP.
5. Establish corrective actions to be taken when a CCP is breached.
6. Establish verification procedures to confirm that HACCP is working effectively.
7. Establish documentation and records concerning all procedures appropriate to these principles and their application.

The implementation of HACCP

1. Assemble and train the HACCP team

The HACCP team must be proportionate to the size, risk and complexity of the business. In a large business the team will be multi-disciplinary and will include people who are aware of the hazards, risks and controls associated with the products forming part of the study. Team expertise will be required for all aspects of the process including user target groups and abuse potential. Knowledge of microbiology, hazards, controls, monitoring, corrective action, the technology used in the process including the equipment, the product characteristics (a_w, pH, composition and use of preservatives), packaging, distribution and HACCP will be required. The role of each team member and the team leader should be clearly defined. In small businesses, one person may be the sole team member, although external consultants may also assist.

Team responsibilities

Team members will need to undertake research and obtain as much information as possible in relation to the products and processes. Short-term members may be invited to join the team because of their specific expertise. The team will need to communicate effectively with the rest of the food handlers and may need additional training. They will need to draw up flow diagrams relating to the processes and check they accurately reflect what happens in practice. Records relating to monitoring should be carefully considered and verification and documentation will be required.

Team members may obtain information from a number of sources, including websites, seminars, The Food Standards Agency (web-based), consultants, laboratories, food poisoning data, enforcement officers, codes of practice and industry guides. Television, radio, books, trade associations and national statistics relating to foodborne illness may also be useful.

Define the terms of reference and scope of the HACCP study

The starting point is to decide which operations or processes, which products and which hazards are to be included within the HACCP study.

It is sometimes preferable to concentrate on one type of hazard, for example, microbiological. It is also important to decide the starting point, for example, the purchase of foods from a cash and carry and the finish point, for example, the consumption of the food by the consumer.

Grouping of products

Although HACCP is usually applied to each specific product in manufacturing, this is rarely the case in catering or retailing. Because of the large numbers of varied products and frequent

menu changes, products prepared in a similar way or subject to the same process may be dealt with as a group. Typical types of catering processes, which can be used as a basis for producing flow diagrams, include:
- high-risk food prepared and served cold;
- perishable raw food, cooked and served or held hot then served;
- high-risk food reheated and served or held hot and served;
- frozen high-risk food, served cold or reheated and served or held hot and served;
- low-risk food, cooked and served hot or held hot and served or cooled and served cold;
- frozen raw food, cooked and served or held hot and served or cooled and served cold;
- perishable raw food, cooked, cooled and served cold or reheated and served or held hot and served.

2. Describe the products or processes

Products will need to be described in relation to their composition, the potential hazards and risks associated with them, their suitability for bacterial multiplication, methods of processing, cooking, storage and distribution. The intended shelf life of the product, packaging and labelling instructions, presence of allergens such as nuts, and legal requirements will need to be considered.

The abuse potential from customers is also important. Will food be taken home immediately? Will it be left in the car boot on a hot day? Will the food be reheated and will it be eaten immediately after cooking? Is it likely to be reheated again?

3. Identify intended use

The likely customers should be considered. They will include vulnerable (sensitive) groups such as babies, elderly people, pregnant women, ill people, those who suffer allergic reactions and those with immune deficiency such as drug abusers. Institutional feeding will need to be considered.

4. Construct a flow diagram

A flow diagram is a systematic representation of the sequence of steps or operations involved with a particular food item or process, usually from purchase of raw ingredients to the consumer.

5. Validate the flow diagram

Most flow diagrams will be written in the office and it is important to ensure that the flow diagram accurately represents what happens in practice. Over a period of time it is possible that some steps in the process have been omitted or substituted with other procedures.

In catering in particular, different chefs may produce a dish in a variety of ways, which may be different from the flow diagram. This should not occur and the flow diagram may need to be modified.

Once it has been confirmed that the flow diagram is accurate you are ready to start your hazard analysis.

GENERIC FLOW DIAGRAM

PURCHASE → DELIVERY → STORE (CHILLED) → PREPARE → COOK → COOL → SERVE

6. Conduct a hazard analysis (Principle 1)
Hazard analysis involves:
- identifying the hazards that may affect the process;
- identifying the steps at which the hazards are likely to occur;
- deciding which hazards are significant i.e. their elimination or reduction to acceptable levels is essential to the production of safe food; and
- determining the measures necessary to control the hazards.

Hazards
Food safety hazards are biological, chemical or physical contaminants with the potential to cause harm to the person who consumes the contaminated food. The most common biological hazards are microbiological.

Biological hazards include bacteria or their toxins, viruses, moulds and parasites that may cause foodborne illness. They involve:
- the contamination of ready-to-eat food by sufficient numbers of pathogens to cause illness;
- the multiplication of micro-organisms; and
- the survival of micro-organisms, for example, as a result of undercooking.

Chemical hazards
Chemicals rarely cause food poisoning; however, major outbreaks occasionally arise. Chemical hazards may be delivered with the raw materials or may be introduced during processing. Chemical hazards include pesticides, herbicides, fungicides, fertilizers, heavy metals, antibiotics, hormone residues, allergens, natural toxins and industrial chemicals. Chemical hazards introduced during processing include fumes, cleaning chemicals, lubricants, refrigerants, allergens and excessive additives. Chemicals may migrate from plasticisers, inks or adhesives from packaging.

Common allergens include peanuts, tree nuts, eggs, milk, fish, shellfish, cereals containing gluten, soya, sesame seeds, mustard and celery/celeriac.

Physical hazards
Physical hazards occasionally make food unsafe to consume. Problems include choking, cuts, penetration, burning and broken teeth. People may require surgery to remove the hazards. Physical hazards may be brought into the food business with the raw materials or be introduced by food handlers. Sources of physical contaminants include the building, equipment, notice boards, packaging, pests, maintenance operatives, visitors and cleaning activities. An example of a serious physical hazard in 2001 involved a dome-shaped jelly sweet with a hard fruit centre, which was linked to the deaths of at least 16 children, in several countries, through choking.

A comprehensive hazard analysis for any food product should include a consideration of likely events in distribution, storage and retailing as well as the potential for consumer abuse. The risks associated with ingredients can be estimated in terms of the following three general characteristics:
- the ingredient can usually be assumed to be a potential source of contamination by toxins or pathogenic micro-organisms;
- the ingredient has not been subject to a controlled process to destroy pathogens; and

- the **pH**, a_w and nutritional content of the ingredient make it favourable to the multiplication of **pathogens**.

Once the detailed **hazard analysis** has been completed, it may be possible to eliminate some of these **hazards** by making changes to the ingredients, process or processing environment. This is easier to do if a hazard analysis of proposals is undertaken at the planning stage of a new product. Actions such as the creation of physical barriers to prevent **cross-contamination** are extremely beneficial because alternative attempts to control the hazard will probably involve a complex set of activities which will be expensive in time and resources throughout future production.

In catering, hazard analysis is usually applied generically or to specific types of catering processes. It is rarely necessary or feasible to apply **HACCP** to each individual menu item.

Control measures

Control measures are those actions taken to prevent, eliminate or reduce hazards to an acceptable level. In the case of **micro-organisms**, controls to prevent multiplication can be applied to temperature, time, pH (measure of acidity or alkalinity), a_w (water activity - the amount of moisture available for micro-organisms), the size, shape or weight, for example, of a joint of meat and the quantity and type of **additives** used, especially preservatives.

Most factories will have glass and wood policies in place to reduce the risk of contamination from these substances. They may also have metal detectors fitted to remove any products that may be contaminated with pieces of metal. Chemical control measures involve storing chemicals separately from food, not storing chemicals in unmarked or food containers, not storing food in chemical containers, rinsing following chemical cleaning and following strict rules when using **pesticides** or **cleaning**.

7. Determine the critical control points (CCPs) (Principle 2)

Critical control points in the process are those steps where control measures must be used to prevent, eliminate or reduce a hazard to an acceptable level. A failure to apply the necessary control measures will result in **food poisoning**, injury or harm. By contrast, loss of control at a control point will not result in an unacceptable health risk.

Careful consideration of each step in the process and having regard to control measures applied later in the process, will enable critical control points to be established. This determination can be assisted using the **Codex Alimentarius** decision tree, although a simplified decision tree is much more likely to be used in catering or retailing.

For example, control measures should be applied to the preparation of raw poultry to prevent contamination by, and/or multiplication of, salmonella. However, because the poultry will be cooked, and proper cooking will destroy any salmonella present, the preparation of raw poultry is a control point. By contrast, the preparation of cooked poultry is a critical control point as contamination by, and/or multiplication of, salmonella is likely to result in food poisoning. Unless the poultry is thoroughly reheated before serving, there is no further step in the process after the preparation of cooked poultry that would destroy salmonella. Good hygiene practice, high standards of **personal hygiene** and effective cleaning and **disinfection** will apply at control points and critical control points. However, the time food is at ambient temperature and the temperature of food will be strictly monitored at critical control points. Additional training and effective supervision are also important for staff involved with steps which are critical control points.

The number of critical control points should be as few as possible, so that resources can be

SIMPLIFIED DECISION TREE

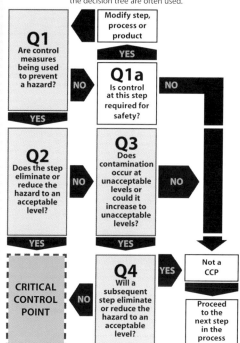

Critical control point(s)
A step in the process where a control can be applied and is essential to prevent or eliminate a food safety hazard or reduce it to an acceptable level.

concentrated at these points. One way of checking the selection of CCPs in manufacturing and retailing, especially for physical and chemical hazards, is to consider whether or not a product that had been distributed and was later discovered to have failed a critical limit at a CCP, would have been recalled. If it is decided not to recall the product then perhaps it should not have been a CCP.

8. Establish critical limits for each CCP (Principle 3)

Critical limits are the values of monitored actions at critical control points which separate the acceptable from the unacceptable. Critical limits must be measurable and the results obtained on site, preferably instantaneously, for example, temperature, time, pH and physical parameters such as weight and size of food. Bacteriological sampling may take several days to obtain the number of specific bacteria that are present and therefore this can not be used as a critical limit.

Critical limits can be determined empirically from the basis of experimentation. However, they can also be derived from theory and published information and guidelines.

Temperature limits may apply to a position within the food but may also, for practical reasons, need to apply to the air or heat transfer medium surrounding the food. In the latter case it is still the temperature of the food which is significant. Where such indirect temperature limits are used, it is essential to understand the relationship between these and the actual temperature of the food.

Target levels and tolerances

If a critical limit is breached a significant amount of food may need to be destroyed. It is preferable to set target limits which may enable a potential breach of a critical limit to be detected and remedied before the food becomes unfit. The difference between a critical limit and a target level is known as a tolerance.

For example, a critical limit for refrigerated storage of high-risk food could be 8ºC for 4 hours. If this critical limit was exceeded all the food in the refrigerator would have to be destroyed. It is therefore sensible to set a target figure of 5ºC. If the target limit is breached, for example, if the food is at 6ºC this would allow action to be taken before the 8ºC was breached and avoid the need to throw the food away. Adjustment of the thermostat may be the only action necessary to bring the process back under control.

9. Monitoring of control measures at each CCP (Principle 4)

Monitoring or checking of control measures at critical control points is essential to confirm that the process is under control and critical limits are not exceeded. Whichever type of observation or measurement is used for monitoring it must permit rapid detection and correction.

Monitoring methods include the monitoring of time/temperature, pH, physical dimensions, organoleptic assessments (smell, touch and appearance), auditing, visual inspections and checking records and competency. Microbiological analysis is more appropriate for verification.

Monitoring systems should state: WHAT the critical limits and target levels are; HOW the monitoring should be undertaken; WHERE the monitoring should be undertaken; WHEN the monitoring should be undertaken; and WHO is responsible for monitoring. The frequency of monitoring must be cost-effective and sufficient to ensure that the hazard is controlled.

Monitoring can be automatic or manual and continuous or at set frequencies. Arguably, continuous automatic monitoring, for example, the use of temperature data-loggers fitted with alarms, may be more cost-effective and the records provided are likely to be more accurate than manual monitoring with a probe thermometer that is hand-recorded. Busy periods of work or forgetfulness often result in the bad practice of records being completed at the end of the day or even the end of the week. Monitoring of data-logger records by supervisors will be necessary at daily or weekly intervals.

In manufacturing it is important to be able to relate the results of monitoring to discrete batches of food. This will reduce waste and assist in recalls in the event of a failure of critical limits.

Managers are often somewhat distant from production, and the monitoring of CCPs is carried out by operatives who may be the same people responsible for the activities that are designed to limit the hazards. It is therefore essential that understanding and commitment to the system are fostered in all levels of staff within the establishment.

10. Establish corrective actions (Principle 5)

Corrective action is the action to be taken when a critical limit is breached. Corrective action usually involves two distinct actions: firstly, dealing with any affected product and secondly, bringing the critical control point and the process back under control. Procedures for corrective action should specify the action to be taken, the person responsible for taking action and who should be notified. Management usually take the decision on whether production/sales should be stopped and when they can be restarted. A clear chain of command is required to avoid delays and ensure the correct action is taken.

There are several ways in which the product outside the critical limit can be treated, for example, the product may be quarantined and subject to microbiological analysis. The process may be continued, for example, the cooking time can be extended, the product may be destroyed or the shelf life may be reduced, for example, eat within 12 hours instead of 3 days. Similar actions might be taken following testing of food held in quarantine, including using the product for a different purpose, releasing the product or destroying it.

If a product has been released and it is discovered that a critical limit has been breached, for example, cooked meat contaminated with glass or poisonous chemicals, it will need to be recalled. All products must therefore be clearly labelled and traceable.

11. Establish verification procedures (Principle 6)

Verification involves the use of methods, procedures and tests, in addition to those used in monitoring, to determine compliance with the **HACCP** plan. Part of verification is validation i.e. obtaining evidence that elements of the HACCP plan are effective, especially the critical control points and critical limits. For example, what evidence has been obtained to prove that 75ºC is a satisfactory temperature for cooking raw meat and destroying any **pathogens** that are likely to be present? In practice, small catering and retail businesses are likely to use generic critical limits which are generally accepted by the industry, the scientific community and the enforcement agencies.

However, manufacturers must carry out detailed tests to ensure that, for example, the time and temperature of processing will be sufficient to reduce the numbers of specific pathogens that may be present to a safe level. The time and temperature may vary considerably between factories producing the same product, depending on composition, type of packaging, type of processing equipment and cooling systems employed.

Verification will usually involve **auditing** against the HACCP plan to ensure the correct implementation and ensuring that the flow diagram remains valid, **hazards** are being controlled, monitoring is satisfactory and, where necessary, appropriate corrective action has been or will be taken. The scientific data relating to hazards and **risks** may be re-examined along with monitoring records. Random bacteriological sampling, end-product testing and analysing complaints for types and trends are also verification techniques. Verification frequency should ensure confidence in the system to provide safe food and may require external expertise.

The HACCP plan should be reviewed periodically to ensure that it remains effective. New epidemiological information or the availability of new technology or scientific information may trigger reviews.

Reviews may be required if:
- raw ingredients are changed, for example, the use of fresh chicken instead of frozen;
- the recipe changes, for example, salt is removed;
- equipment changes, for example, a blast chiller is introduced;
- packaging or distribution changes, for example, refrigerated transport is utilized; and
- a justified complaint is received or illness occurs.

Reviews are useful to identify new products that have not been added to the HACCP plans. They should also be used to identify weaknesses in the system and to eliminate unnecessary or ineffective controls.

12. Establish documentation and record-keeping (Principle 7)

The amount and type of paperwork required to support HACCP systems varies considerably depending on the type and size of food business and the risks involved with the processes. Generally, documentation and record-keeping must be proportionate to the particular size and type of business. Documentation is useful to demonstrate that food safety is being managed and, provided records are completed accurately at the appropriate time, they are useful to support a **due-diligence** defence if this is required in court.

Records are very useful for investigating food complaints and alleged **food poisoning** claims. Managers use them when auditing and enforcement officers/external auditors will wish to examine them.

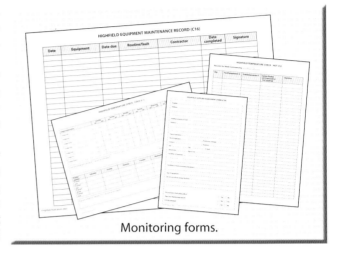

Monitoring forms.

Documentation includes the HACCP plan and details of the prerequisite programmes.

Records include:
- CCP monitoring activities;
- deviations and corrective actions;
- audit reports;
- customer complaints;
- calibration of instruments; and
- prerequisite programme monitoring activities.

All monitoring records should be signed and dated by the food handler undertaking the monitoring and countersigned by the supervisor.

The HACCP plan

The HACCP plan is a written document that describes all of the procedures involved in implementing the seven HACCP principles. The HACCP plan may include:
- the details and responsibilities of the HACCP team;
- the scope and terms of reference;
- a description of the process, product and intended use;
- flow diagrams clearly identifying the CCPs;
- consumers;
- all information relating to the hazard analysis and CCP determination;
- the critical limits and the evidence used to establish these limits;
- the target levels, taking into account the accuracy of monitoring and processing equipment;
- monitoring and record-keeping procedures and determination of frequency of monitoring;
- the action to take when target levels are exceeded;
- the corrective action to take when a critical limit is exceeded, i.e. a deviation occurs. This will include testing, destruction, recall and traceability;
- a record of the reasons for critical decisions and modifications to the plan; and
- validation, verification and review procedures.

A HACCP control chart can be used to record a significant amount of the information within the HACCP plan.

HACCP CONTROL CHART

Date .... Approved by.................................... Page .... of ....

Process step	Hazards (Sources & causes)	Controls	CCP	Critical limit	Monitoring (Frequency & person responsible)	Corrective action (Person responsible)

Flexibility and HACCP

Guidance from the EU and the Food Standards Agency (FSA) make it clear that Article 5 does not require all food businesses to implement a Codex based HACCP system. In practice, the Codex approach has not proved successful for small catering or retail businesses. In certain food businesses it is not possible to identify critical control points and good hygiene practices may be sufficient to comply with the law, for example, a newsagency selling sweets and small business only handling low-risk foods.

In small catering and retail businesses, food safety management systems which ensure that safe food is produced may:
- avoid using HACCP jargon;
- use generic controls, e.g. cook to 75°C;
- use pre-validated procedures (e.g. by the FSA or the food industry);
- combine good hygiene practice and HACCP;
- use practical craft skills, sensory observation and supervision instead of numerical critical limits;
- use minimal documentation; and
- use exception recording, i.e. writing down important changes, e.g. to equipment or menu items or when something goes wrong, e.g. food complaint/poisoning.

The Food Standards Agency have developed the following food safety management systems, all of which use the above approach to take advantage of the flexibility allowed: Safer food, better business for small catering and retail businesses, CookSafe (Scottish model) and Safe Catering (Northern Ireland model).

A science-based and validated approach to implementing HACCP in catering, developed by Dr Peter Snyder, is based on his assessment of the twelve most important controls to prevent hazards.

1. Effective handwashing (1,000,000 to 1 reduction of **pathogens**) on entering the kitchen and when the hands become contaminated.
2. Effective **cleaning** and **disinfection** of food and hand contact surfaces (100,000 to 1 reduction of pathogens).
3. Procedures for preventing physical hazards.
4. Procedures for preventing chemical hazards, including allergens.

HYGIENE for MANAGEMENT Control & monitoring of food standards & operations

AN EXAMPLE OF A HACCP CONTROL CHART SHOWING STEPS IN THE OPERATION CRITICAL TO FOOD SAFETY

STEP	Hazards	Controls & critical limits/target levels	Monitoring+	Corrective action*
PURCHASE	Contamination with harmful bacteria, toxic chemicals or foreign bodies.	Use only reputable suppliers. Select least hazardous ingredients. Product specification. Purchase branded products.	Ensure supplier has effective HACCP system and good hygiene practices. Obtain HACCP documentation. Inspect supplier's premises. Utilise customer references.	Revisit/audit supplier's premises. Review product specification. Warn supplier. Change supplier.
DELIVERY AND UNLOADING	Contamination with harmful bacteria, toxic chemicals or foreign bodies.	Food covered/protected or in suitable containers. Deboxing area. Use only approved suppliers. Date codes/labels. Specific delivery requirements. (NOT with toxic chemicals.)	Visual and sensory checks. Condition of packaging/food. Condition of vehicle/driver. Supplier on the approved supplier list. Food is within date code. Food is delivered in accordance with specification.	Change supplier. Review/reissue product specification/instructions. Refuse delivery. Return stock. Inform chef or manager. Review systems/training.
	Multiplication of food poisoning bacteria.	Chilled food below 5°C. (Critical limit 8°C.) Frozen food below −18°C. Move to storage within 15 minutes of unloading.	Temperature of food using calibrated probe thermometer. Time to move to storage.	
CHILLED STORAGE	Contamination with harmful bacteria, toxic chemicals or foreign bodies.	Food covered/protected. Segregation of raw and high-risk food. Use disinfected temperature probe. Food not stored in open cans. Load food correctly.	Visual checks. Condition of food. Condition of packaging.	Inform chef or manager. Discard unfit or out-of-date food. Review systems/training.
	Multiplication of food poisoning bacteria.	Chilled food below 5°C. (Critical limit >8°C for 2 hours.) Don't overload refrigerator. Good stock rotation procedures. Date codes/labels.	Air temperature. In-between pack temperature. Temperature of food. Time above 8°C. Door seals. Food is within date code.	Adjust thermostat. Reorganise refrigerator in order to allow good air circulation. Call out refrigerator engineer. Repair/replace fridge. Implement contingency plan.
FROZEN STORAGE	Multiplication of food poisoning bacteria.	Food frozen below −18°C.	Air temperature. In-between pack temperature. Temperature of food.	Inform chef or manager. Discard unfit or out-of-date food. Review systems/training. Adjust thermostat. Call out refrigerator engineer. Replace freezer. Implement contingency plan.
DRY STORAGE	Contamination with harmful bacteria, toxic chemicals or foreign bodies.	Food covered/protected or in suitable containers. Care in handling.	Condition of packaging. Condition of food. Condition of canned goods. Housekeeping.	Inform chef or manager. Discard unsatisfactory food. Review systems/training. Building maintenance to remove dampness. Store out of direct sunlight.
	Multiplication of food poisoning bacteria.	Maintain in a cool, dry condition. Food within date code. Good stock rotation.	Visual checks.	

Personal hygiene and training, **pest control** and **cleaning** and **disinfection** are dealt with as prerequisite programmes.

HYGIENE for MANAGEMENT Control & monitoring of food standards & operations

STEP	Hazards	Controls & critical limits/ target levels	Monitoring+	Corrective action*
PREPARATION	Contamination with harmful bacteria, toxic chemicals or foreign bodies.	Separation of raw and high-risk food. Use separate equipment. Colour coding. Organisation/workflow. Disposable cloths/paper roll. Minimise handling of ready-to-eat food where practical. Thoroughly wash all ready-to-eat produce and garnishes. Don't top up sauces.	Visual checks.	Inform chef or manager. Discard contaminated food. Review systems/training.
		Exclude staff with food poisoning symptoms.	Staff sickness record and exclusions.	
	Multiplication of food poisoning bacteria.	Maximum time at room temperature 30 minutes.(Critical limit 2 hours)	Temperature of food. Time at room temperature.	Discard high-risk food left at room temperature for >2 hours.
		Pre-cool salad sandwich ingredients.		
		Minimise quantities prepared. Portion control.		Reduce quantities prepared.
THAWING	Multiplication of food poisoning bacteria.	High-risk food should be thawed in a refrigerator. Use thawed food within 24 hours. (Label)	Temperature of food.	Inform chef or manager. Discard food which has thawed and then been left at room temperature for >2 hours.
	Contamination with harmful bacteria, toxic chemicals or foreign bodies.	High-risk and raw food should not be thawed in the same area. Raw food should not be thawed in areas used for cooling cooked food.		Discard contaminated food. Review systems/training.
	Survival of food poisoning bacteria (during cooking).	Raw frozen food must be completely thawed, especially poultry. (Critical limit - absence of ice.)	Temperature/visual and physical checks.	Continue thawing. Allow longer thawing time in future.
COOKING/REHEATING	Survival of harmful bacteria/spores and/or toxins.	Cook thoroughly to 78ºC. Critical limit 75ºC (or equivalent time and temperature.) Boil liquids. Don't reheat more than once.	Core temperature of food. Visual checks.	Inform chef or manager. Extend cooking time until core temperature of 78ºC is reached. Allow longer cooking time/higher oven temperature in future. Carry out cooking trials.
		Contingency plans. Cook/reheat just before eating.		Discard simmering high-risk liquids that have not been stirred for more than 2 hours. If less than 2 hours, stir liquid and bring to boil.
		Follow microwave procedures. Follow instruction on packaging. Stir liquids.		
		Ensure frozen meat and poultry is completely thawed prior to cooking.		Call out engineer. Implement contingency plan. Review systems/training.
	Contamination with toxic chemicals or foreign bodies.	Avoid copper and aluminium for acid food. Protect from contamination.		Repair/replace unsatisfactory equipment and pans.

HYGIENE for MANAGEMENT *Control & monitoring of food standards & operations*

STEP	Hazards	Controls & critical limits/target levels	Monitoring+	Corrective action*
COOLING	Contamination with harmful bacteria, toxic chemicals or foreign bodies.	Separation of raw and high-risk food. Protect from contamination.	Visual checks.	Inform chef or manager. Discard contaminated food. Review systems/training. Discard food if not cooled and refrigerated within critical time. Call out engineer. Repair/replace blast chiller.
	Multiplication of food poisoning bacteria, formation of toxins and/or germination of spores.	Rapid cooling (blast chiller) and refrigeration. Minimise weight/thickness of joints, e.g. 2 kgs. Cool liquids in clean, shallow trays (maximum depth 25mm.) Blast chiller (90 minutes.) 63°C to 20°C in <2 hours. (Critical limit in >3 hours.) 20°C to 7°C in <4 hours. (Critical limit in >5hours.)	Time to cool. Temperature of food.	
HOT HOLDING	Contamination with harmful bacteria, toxic chemicals or foreign bodies.	Protect from contamination. Sneeze guards etc. if hot holding for service. Use disinfected temperature probe.	Visual checks.	Inform chef or manager. Discard contaminated food. Review systems/training.
	Multiplication of food poisoning bacteria, formation of toxins and/or germination of spores.	Maintain at the correct temperature (minimum 63°C). (Critical limit <63°C for 2 hours.) Minimise quantities. Preheat hot-holding equipment. Contingency plans.	Time and temperature of food.	Adjust thermostat. Discard any food that has been maintained at below 63°C for more than 2 hours. Call out engineer. Repair/replace equipment. Implement contingency plan.
COLD DISPLAY	Contamination with harmful bacteria, toxic chemicals or foreign bodies.	Food covered/protected, e.g. sneeze screens. Segregation of raw and high-risk food. Use disinfected temperature probe. Minimise handling, use long-handled tongs/serving utensils.	Visual checks.	Inform chef or manager. Discard contaminated food. Review systems/training.
	Multiplication of food poisoning bacteria.	Food stored below 5°C. (Critical limit >8°C for 4 hours.) Good stock rotation.	Air temperature. In-between pack temperature. Temperature of food. Time above 8°C. Food is within date code.	Adjust thermostat. Discard any food >8°C for 4 hours. Call out engineer. Implement contingency plan.
SERVICE	Contamination with harmful bacteria, toxic chemicals or foreign bodies.	Protect from contamination. Minimise handling. Separate staff handling raw and high-risk food. Colour coding. Handwashing between handling soiled crockery and/or cutlery.	Visual checks.	Inform chef or manager. Discard contaminated food. Review systems/training.
	Multiplication of food poisoning bacteria.	Minimise time high-risk food is at room temperature. Serve within 15 minutes of cooking, hot holding, chill storage or preparation.		

+ Frequency of checks and person responsible should be included. Monitoring records will usually be needed at most steps.
* Person responsible for corrective action should be included.

N.B. Effective supervision, instruction and competency training are controls at each step. High standards of personal hygiene, integrated pest management, effective cleaning and disinfection and maintenance are prerequisites to the implementation of HACCP and are not included within the HACCP control chart. However, they are controls that apply at most steps. Chemical and physical hazards are mainly dealt with through the prerequisite programmes. Allergenic hazards will require specific controls.

5. Double wash of ready-to-eat fruit and vegetables (100 to 1 reduction of pathogens).
6. Refrigeration of all raw perishable food below 13°C.
7. Cooking to achieve a salmonella kill of 100,000 to 1 (71°C for 5 seconds)
8. Hot holding above 54°C.
9. Cooling from 54°C to 7°C within 15 hours (<1 log multiplication of C. *perfringens*).
10. Refrigeration of high-risk food based on the multiplication of listeria: 2°C 19 days; 5°C 7 days; 7°C 4 days.
11. Cold salad sandwiches precooled to <10°C (prevent the multiplication/toxin production of *S. aureus*).
12. Leftovers - never add fresh to old (microbiological and allergen contaminants).

Reasons for modifying HACCP applied to catering

- no multidisciplinary team, small numbers of employees;
- little technical/microbiological and scientific expertise;
- little knowledge of ingredient characteristics, for example, a_w and pH;
- no linear product flow and little consistency;
- large numbers of different suppliers;
- lack of auditing skills (required for verification and validation);
- major potential for cross-contamination;
- large numbers of constantly changing menu items;
- language, literacy problems and high staff turnover; and
- catering based on flair and imagination not science, laboratory testing and documentation.

The effective inspection of food premises

An inspection of a food business will only be effective if the person undertaking the inspection has a clear understanding of the reason for the inspection, and he/she has the relevant technical knowledge, skills and experience. It is essential to plan and prepare adequately and allow sufficient time to achieve the objectives of the inspection.

After the inspection, the data collected should be analysed to determine the appropriate action to be taken to rectify the defects. A comprehensive report of the inspection should always be written and problems must be followed up.

The purpose of the inspection will dictate the timing of the inspection, the time spent inspecting, whether or not to give advance warning of the inspection and the equipment needed during the inspection.

As with any type of inspection, proficiency in food safety inspections of food premises will only be achieved through the experience of inspecting a large number of similar types of premises, to enable the inspector to develop a mental picture of satisfactory premises. A comparison between this "ideal model" and the food premises being inspected enables data to be obtained, findings considered, results analyzed and conclusions drawn. This will then enable recommendations to be made regarding corrective action.

Auditing versus inspection

The terms "auditing" and "inspection" are often used interchangeably, although auditing is more fashionable and perceived to be of higher status. In practice, audits are essentially concerned with comparing what is actually being done with a specific standard. They may be

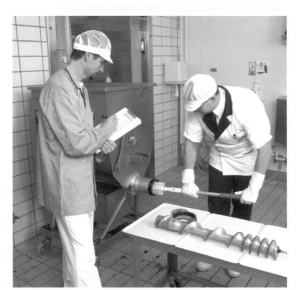
Auditing the cleaning process.
(Courtesy of Vikan Ltd).

complex and require a considerable amount of skill; however, they are often undertaken by people with limited food safety experience using a checklist.

A food safety inspection may be considered as the systematic gathering and recording of data from observations, examinations and discussions with food handlers and managers. It also involves the analysis and interpretation of data collected and the preparation of an understandable and actionable report.

Purpose of the inspection process

The main purpose of a food safety inspection should be to ensure the food business is capable of producing safe food, and to identify and prevent foreseeable incidences of food poisoning or injury. This will involve assessing the food safety knowledge and commitment of the manager, the effectiveness of the food safety management system, including HACCP, and the competency of the food handlers to produce safe food.

It may not be possible to achieve the objective during a single visit as the whole operation, from delivery to service, will need to be observed. It is usually appropriate, therefore, to consider the inspection as part of a process as opposed to a single visit. Several visits will be required at varying times to ensure everything is seen.

Knowledge of the inspector

If the primary focus of the inspection is the prevention of food poisoning, then the inspector must be fully aware of the causes of food poisoning. This will include the proximate causes of food poisoning, as described by Roberts, and the management failures that all too frequently result in food poisoning, as described by North (see chapter 1). It is essential to keep up to date with the latest information on sources, organisms, food **vehicles** and causes of food poisoning.

The inspector will also require knowledge of the type of operation, including relevant technology, for example, cook-chill, as well as an understanding of HACCP and its application to the business. A significant amount of research may sometimes be required to obtain sufficient knowledge to ensure an effective inspection can be undertaken.

Equipment required during the inspection
Equipment required may include a carrying case, protective clothing (coat, hairnet, hard hat, ear plugs and gloves), a calibrated digital thermocouple thermometer and probes, a pH meter, disinfectant wipes, a torch and spare batteries, a magnifying glass, a camera, a small mirror, pyrethrum spray, cockroach traps, tweezers, sterile sample bags or jars, a scraper, a screwdriver, a penknife, a clipboard, notebook or dictaphone, a mobile telephone and blue waterproof dressings. It is also useful to have the last completed report.

Skills of the inspector
The inspector should be able to recognize hazards and assess risks. He/she should have the ability to investigate, analyze and resolve problems. He/she will need to make appropriate and consistent judgement, proportionate to the risk. He/she must be competent to use all equipment and, in particular, to take accurate temperatures and other measurements. All inspectors must have good communication skills, including listening, questioning, interviewing and advising, as well as accurate recording of observations and effective report writing.

Planning and preparation
Planning of inspections is an important but often overlooked part of inspections. The scope of the inspection must be determined. Which hazards will be considered - physical, chemical and microbiological? Is it intended to inspect all of the business or just part of the business? Enquiries into the two major food

A Comark UKAS accredited laboratory, used to calibrate and certify thermometers. (Courtesy of Comark Ltd)

poisoning outbreaks, John Barr and Stanley Royd Hospital, both concluded that a cursory, superficial inspection was a waste of time and could give a misleading impression.

Having decided on the type of inspection, the records of previous inspections should be examined, relevant codes of practice, industry guides and legislation should be checked and inspectors should ensure they have the knowledge and skills to undertake a competent inspection. From all this information and the experience of the inspector visiting similar premises, he/she will be able to build a picture of what standards he/she expects from the type of operation. Any shortfall between this theoretical model and the actual conditions found during the inspection will form the basis of the report and the necessary corrective action.

Modelling
Getting the specific model correct is essential. Further information to complete the specific model can be obtained at the pre-meeting with the manager, prior to the physical inspection. For example, different standards will be expected in a raw vegetable area compared with a

cooked chicken area. If it is a catering unit, do they also undertake outside functions? The standards at the beginning of the production/preparation will vary from standards at the end. The facilities may be satisfactory for a small butcher's shop but what if there is a major wholesale supply element from the same operation? Several well-known outbreaks of food poisoning have arisen because the inspector failed to appreciate the full extent of the business and therefore inspected using the wrong model.

Timing of inspections

Over a period of time the inspection should cover all hours of operation including early morning starts, evenings, weekends and shift work. It is common to observe different procedures during the day, compared with an evening operation. It is important to be present at peak periods of activity to ensure there is a sufficient numbers of staff and adequate temperature control facilities, which minimizes the time high-risk food is in the danger zone.

The frequency of inspections should be related to the risks, the level of complaints and the standard achieved on previous inspections.

Pre-meeting

Prior to undertaking the inspection it is very useful to have a pre-meeting with the manager/proprietor. The manager can be advised of the purpose of the inspection and will be able to provide full details of the business activities, including the customers supplied, for example, wholesale, the elderly, children or hospitals. The manager should be asked about the food safety management systems in place and the documentation available, for example, the HACCP plan. The inspector should assess the commitment of the manager to food safety and his/her understanding of the hazards and risks associated with the business and the controls implemented. A cursory examination of paperwork and records at this stage should suffice, as the effective assessment of food safety can only take place where food is actually being handled and prepared, not in the manager's office.

The inspection

Ideally, an inspector would like to observe the food handlers carrying out their work activities as they normally would. Unfortunately, the very presence of an inspector, who is there to assess, among other things, whether or not practices are satisfactory, will change the behaviour of those being observed. A good inspector tries to be as unobtrusive as possible, keeps out of the way and is courteous and friendly. The "royal tour" with a high profile entourage is the worst possible type of inspection, which will achieve very little.

The inspector should lead by example, always ensuring he/she wears the appropriate protective clothing, washes his/her hands properly on entering each new food room, introduces himself/herself to the manager of each area and always ensures that the best hygiene practices are followed.

The inspection should be methodical. A typical routine will be to start at a defined point within a room, for example, the wash-hand basin, and progressively examine all relevant items around the perimeter of the room until arriving back at the wash-hand basin. Then follow the same ordered routine and examine all central fittings, installations and equipment.

The inspection is hard work, and must be thorough. Look under work surfaces and equipment, get steps to look at high shelves and the tops of cupboards, pull open drawers, check ducting and if you need to get on your knees or even lie on your back to view something, so be it. Remember; if you cannot see it, you cannot inspect it. It may not be

HYGIENE for MANAGEMENT Control & monitoring of food standards & operations

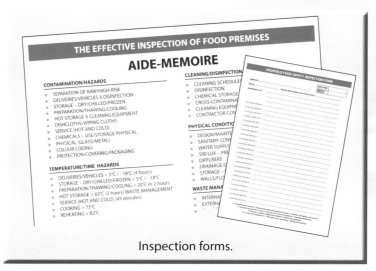

Inspection forms.

enough to look at a piece of equipment; instead, it may need to be moved or even dismantled. Cupboards may need to be partially or wholly emptied and surfaces may need to be scraped or tapped to assess soundness. In this respect full use should be made of the senses: sight, smell, hearing, and touch.

An *aide-memoire* can be useful but a detailed inspection form may be detrimental, unless extremely well designed.

Observations

Observation during inspection can be direct, for example, when looking at surfaces and equipment or indirect, which is the best method for watching people. Food handlers can be watched discreetly, the inspector appearing to be doing something else.

Semi-covert observation could involve, for example, the inspector going into the premises, as a cleaner or joining the maintenance team. Covert observation would involve using CCTV. This can be an excellent training medium provided the staff, and unions, are aware of the cameras and the purpose for which they will be used. The main benefit of CCTV is that, after a short while, food handlers will behave completely normally.

The HACCP approach to inspection

One way of ensuring a thorough and orderly inspection is to use the HACCP approach. Start with deliveries and follow the passage of food through storage, preparation, cooking or processing, preparation, service or distribution. At each stage contamination, multiplication and survival hazards should be considered, together with controls, monitoring, especially at CCPs, and corrective action. Standard of suppliers, design of premises and equipment, workflow, water supply, personal hygiene, competency of food handlers and supervisors, pest control, cleaning and disinfection and waste management should all be assessed.

Documentation and records will need to be examined but should not be relied on in isolation. Many cases of inaccurate and falsified records have been reported by experienced inspectors. Discuss systems and procedures with relevant staff, ask for a demonstration, for example, how do they clean the equipment? Can they show you how they take centre temperatures of cooked food? What is the company policy if staff are suffering from symptoms of food poisoning?

Recording of data

Extensive note-taking throughout the inspection or immediately thereafter is essential. Descriptions should be quantified, for example, heavy, medium or light soiling, the depth or area, whether soiling is fresh or long standing, the colour and odour of soil, etc. Both good points and bad practices should be noted. Sketches and plans can be very useful. Areas of **cross-contamination** can be highlighted by crossover of, for example, raw food workflows drawn in red and high-risk food workflows drawn in blue. As far as practicable, data should be verified. Conclusions are more suspect if based on uncorroborated data.

Examples of key questions to answer
- Does the design and layout of the premises minimize the risk of contamination of high-risk food?
- Are the staff competent as far as food safety is concerned?
- Are the procedures and practices effective in preventing the contamination of ready-to-eat food or the multiplication of bacteria?
- Is the equipment provided capable of preventing the multiplication or destruction of food poisoning bacteria?
- Do the management have the knowledge and commitment to ensure the sale of safe food free of contaminants?

Analysis and interpretation

The simple findings during the inspection should be analysed before corrective actions are taken.

For example, if a soap dispenser is empty, the written instruction in a report three days later to "fill the soap dispenser" is almost irrelevant. It should be filled immediately and will probably have been refilled several times before the report arrives. The inspector needs to consider why the soap dispenser was empty. Has it happened before? How many soap dispensers are empty? If the soap dispensers are regularly found empty, in a high-risk area, this paints a picture of a management that have little regard for food safety and indicates that there are no effective controls and monitoring in place.

All items of a similar nature can be grouped together in an effort to determine common causation. For example, several work surfaces in different rooms were not properly cleaned. This may be because of an ineffective cleaning team or inadequate time allowed for cleaning by the food handlers themselves.

A single fault may have multiple causation, for example, a dirty slicer may be because of the **culture** of the organization, lack of resources, lack of training, inadequate time allowed for cleaning or poor supervision. Always look for the real problem behind the symptom in an effort to achieve a permanent solution. Why were staff wearing dirty protective clothing? Was it because they didn't know the replacement procedure, or they had to launder it themselves, or because adequate replacement clothing was unavailable?

Remember things are not always what they seem. Allow for intelligent compromises and do not jump to conclusions. Do not be offended or annoyed if someone challenges your view and asks you to justify your reasons for requesting a particular course of action. If a small business has a sickness problem and two out of the four staff are unexpectedly ill, the manager may carry out a risk assessment and prioritise their work to deal with temperature control and cross-contamination. There may be a pile of dirty crockery and the floor may need to be cleaned. However, an inspector who orders the manager to clean up immediately, by removing staff from high-risk tasks, will be giving a totally inappropriate instruction.

Post-inspection discussions

Discussions with management following an inspection are essential. Summarize your preliminary findings, discuss failure to meet standards, and discuss unidentified hazards, risks and any other deficiencies in the food safety management system. Invite a response; do not assume; obtain further information. Ensure the accuracy of your records. Recommendations should be practical. Agree the corrective action required and the timescale, for example, immediate, one week, one month or three months. Keep a record of the agreement.

The report

The report is a powerful tool of communication, which is intended to achieve change. It is much more than a list of faults. Readability and clarity are essential. A summary can be helpful and jargon should be avoided. Compliment good practices and achievements. Apply rigorous science and avoid ritualistic, impractical solutions. Distinguish recommendation and requirements and prioritise the work required. Offer cost-effective solutions that work and try to assist rather than condemn. A report that makes life easier is much more likely to be favourably received. Although consistency of outcome is important, advice may change as science, experience and understanding progress.

Before sending the report, read it through again. How would you react if you received it? Will it promote change? Can it be implemented within the required timescales and, if so, what will be the outcome?

Suppliers and raw materials

The quality and safety of the final product depends, among other things, on the standard of the raw materials. Consequently, food should be obtained from reputable sources. Hygiene officers may be employed to inspect the premises of main suppliers to ensure that they are capable of providing a satisfactory product, which consistently meets the agreed specification. When it is impracticable to inspect the suppliers' premises, managers have to use alternative assessment techniques to determine acceptable suppliers. Requesting a copy of the food safety policy, the HACCP plan, the prerequisite programmes and the monitoring records would be an excellent start. References from other customers, absence of complaints and bacteriological sampling of results provide useful indicators. The appearance of delivery staff, their adherence to basic hygiene rules and the condition of the delivery vehicle may be considered as a likely reflection of the hygiene standards of the supplier. Carriage of cooked and raw meats in the same container, high-risk food delivered at unsafe temperatures or deliveries placed in unsuitable areas are an indication of a poor supplier.

Specifications should be set for raw materials and may include bacteriological standards, flavour, colour, temperature and permissible levels of contamination. For example, an occasional pod in a delivery of peas may be acceptable, a rat dropping would not. Certain foodstuffs carry official certificates of inspection, for example, all red meat carcases intended for human consumption must bear an official inspector's stamp. Imported food must be accompanied by documents certifying fitness.

Raw materials should be checked on arrival before unloading and regular samples should be obtained for bacteriological and chemical assessment. Dry herbs and spices are a frequent source of undesirable and pathogenic micro-organisms and must not be overlooked. The time of delivery, the weight and condition of the load and the type and condition of packaging should all be noted. The presence of decomposition, mould, unusual odours, dampness, undesirable levels/types of micro-organisms, dirt, insects, parasites, foreign bodies, pesticides, veterinary drugs, blown or leaking containers, and severely damaged packaging will usually require the rejection of the delivery (as would unacceptable temperatures of refrigerated food stuffs), unless normal, hygienic sorting and/or processing would remove or reduce the contaminants to an acceptable level.

Organoleptic assessment of food

The appearance, smell, texture, taste and other physical characteristics of food are valuable for obtaining a rapid assessment of food standards (quality, taint and spoilage). However, food contaminated by pathogenic bacteria may appear in all respects fit to eat and suspect food should not be tasted. Specific indicators include:

- *Smell* – good food should smell fresh, pleasant and natural. Unusual, stale, musty or rancid smells should invite suspicions. Chemical smells may indicate chemical contamination. Ammonia smells in some fish are an early sign of decomposition.
- *Taste* – unusual bitterness, sweetness, a soapy taste or any untypical flavour may indicate unfitness.
- *Appearance* – food should be visibly free from signs of spoilage, fungal growth, slime, darkening or other change in colour, untypical wetness and mechanical damage. Absence of foreign objects and dirt in finished goods, including pests, pest debris and parasites, is important. Meat, poultry and fish should be free from signs of disease or other pathological conditions. In frozen food, excessive ice can be an indicator of mishandling, as can large ice crystals within the texture; loose foods such as peas should not weld together. A final judgement of frozen food can only be made after defrosting.
- *Sound* – many packed foods, especially canned goods, emit a characteristic sound on being tapped or shaken. Any such food or pack emitting an untypical sound is suspect.
- *Texture* – unusual softness, hardness, brittleness or change in texture may be indicative of unfitness. Meat, fish and certain other products such as cheese should display a springy texture. Light pressure from a finger that causes an indentation to remain can be significant.

In all situations, the key to a sound judgement is to use all the relevant senses and to make a decision on the basis of all the available evidence. In cases of doubt, advice should be obtained from local authority environmental health officers (EHOs) or consultants.

Bacteriological monitoring

Bacteriological monitoring is concerned with more than end-product testing and can be used to:

- build up a profile of product quality;
- indicate trends in product quality;
- ascertain whether handling techniques are satisfactory;
- indicate product safety and the absence of specific organisms (pathogens);
- determine effectiveness of cleaning and disinfection;
- determine effectiveness of processing; and
- confirm that legal standards or customers' specifications are being met.

The number and types of bacteria in a finished food product depend on:
- the microbiological quality of the raw materials;
- the standards of hygiene implemented during storage, manufacture or preparation;
- the type of food, preservatives used and the processes to which it is subjected;
- the type and quality of packaging; and
- the time/temperatures observed during handling, distribution, storage and service.

All large organizations that handle high-risk foods should have access to facilities that enable the bacteriological assessment of food and food equipment. Whether or not in-house laboratory facilities are maintained is mainly an economic judgement. A number of consultant laboratories provide an *ad hoc* or contract service and a certification scheme run by the National Physical Laboratory at Greenwich will direct enquirers to the appropriate laboratory.

Bacteriological monitoring must not rely solely on the examination of the final product. By the time the defect is discovered, a large amount of unsatisfactory food may have been produced and sold. This is usually the case with regard to high-risk products with short shelf life.

To provide information as early as possible, a bacteriological sampling programme based on the type and amount of food produced should be designed and this should include raw materials, efficacy of cleaning and disinfection, water supplies, food containers and samples of the product at all appropriate stages of production, distribution and retail sale. The maximum number of bacteria permissible at each stage of production should be established. Departures from this standard normally indicate a breakdown in hygiene or unsatisfactory processing. Unsatisfactory trends may allow remedial action to be taken before a product becomes unsaleable.

Bacteriological monitoring provides targets to achieve and can lead to a gradual improvement in hygiene. It can also be used to demonstrate the safety and quality of a product to customers and demonstrate the effectiveness of controls. However, there are some disadvantages.

Disadvantages associated with bacteriological sampling
- It is usually retrospective and cannot be used to verify product safety where there is a short time between production and consumption, for example, in conventional catering or for products with short shelf lives.
- It is relatively expensive.
- Considerable expertise may be needed to interpret results and relate them to product age.
- The non-uniform distribution of bacteria in foods and the effect of different laboratory techniques and sampling methods significantly affect the results.
- The operation is being controlled by a laboratory technician, who may be remote from the food production.
- Only a limited number of samples can be taken.
- Not all hazards are identified.
- Only a small section of the workforce assumes responsibility for product safety.

The development of rapid assay techniques is reducing the time to obtain results and may overcome some of the above disadvantages. One of the few rapid bacteriological sampling tests available is the resazurin test for milk, which enables dairies to accept or reject raw milk supplies from the farm within ten minutes of testing.

When selecting laboratories, preference should be given to those that are National Measurement Accreditation Service (NAMAS) accredited, to provide assurances that work undertaken will be accurate and reliable.

RAPID METHODS FOR MICROBIOLOGICAL TESTING*

The term "rapid method" in **microbiology** has numerous connotations. To microbiologists, a rapid test is one that has a significant time-saving over more traditional methods. However, a rapid test for one manager (producing a long shelf life product) may be unacceptable to another. The selection of a particular rapid method therefore depends upon the individual circumstances of a particular company/product.

Rapid versus traditional methodology

Assessment of the microbiological quality of food has traditionally been determined by the use of standard **plate counts**. This has often incorporated the use of specialist media to isolate a particular organism, for example, salmonella or listeria, followed by a confirmatory test. These methods involve a growth stage and hence can take several days between analysis of the sample and obtaining the result.

Rapid methods are usually based on a few technologies such as those of DNA detection, immunoassay and biochemical determination, although other novel methods do exist. However, for the vast majority, enrichment (broth) culture is required to allow organisms to be detected, which can take up to 24-48 hours. It should, of course, be stressed that for routine purposes, culture is still the most sensitive method with the lowest limit of detection being one **colony** forming unit (CFU). One has to realize that detection of an organism at a high level of sensitivity is significantly different from the traditional numbers game where microbiological specifications are designed to reflect the number of permissible viable organisms within a food. This leads to the main problem with rapid methods – drawing inappropriate conclusions from the results obtained.

Even with significant advances in technology, however, a truly rapid method (taking only a few hours), requiring minimal resources and with the ability to detect low levels of specific pathogens, does not really exist at the time of writing this book.

Advantages and disadvantages of rapid methods

Rapid methods:
- provide real time results, leading to rapid clearance of raw materials and products. This provides savings in warehouse/cold storage space due to shorter release time;
- are in some cases very easy to use, for example, adenosine triphosphate (ATP) - photometry can be carried out by non-specialist staff after a short training session. However, it should be stressed that in the usual way of using ATP-photometry, the test is non-specific and will not discriminate between pathogens and other microbial flora;
- will not solve hygiene/spoilage problems; they will only allow them to be detected over a shorter time span – they will **not** make the problem easier to find;
- have to be commissioned and this may be relatively slow as protocols will have to be

*Originally written by Dr Paul Beers (University of Humberside) and edited by Dr Adrian Eley (University of Sheffield)

designed and optimized to fit in with the production process. This will ensure that the information obtained is relevant to the process being monitored; and
- require the education of line managers, and supervisors since a test result will probably not be expressed as CFU/ml or gram and there is often confusion in relating metabolic measurements to the real life situation.

Rapid methods available

Impedance method

This method refers to the measurement of impedance (the degree of resistance to an electric current) within a culture medium. Micro-organisms will metabolize substrates as they grow, decreasing the impedance of the medium. It is possible to directly relate the changes in impedance to cell numbers. The bactometer is a commercial instrument that uses this technique. Numerous specific methods have been developed for the detection of coliforms, including the detection in meat samples. Individual organizations can develop their own protocols in consultation with the companies producing the instrumentation.

DNA methods

Different DNA methods include the use of a DNA-probe, designed to detect a single organism or a group of organisms. The future of rapid methods lies in the DNA-based methods, as they can be used to detect and amplify bacteria, for example, the polymerase chain reaction (PCR). Although PCR is a rapid technique, insufficient sensitivity for detection of pathogens in foods has necessitated enrichment culture for up to 24 hours. Additional problems have been difficulties over specificity of the target, which may also be found in non-pathogenic organisms. Improved sensitivity may be achieved by TaqMan (Tm), which is a newer real-time PCR method, giving results in only a few hours. This technology combines DNA amplification and detection of the products in a single tube, which reduces the contamination risk (false-positives). It is also less time-consuming than gel-based analysis and can supply a quantitative and qualitative result.

Adenosine triphosphate measurement

This is the growth area in rapid methods. However, it has to be remembered that this method is really a test of hygiene and not for specific pathogens. The technique relies upon the presence of ATP in bacterial cells. All bacteria contain ATP, as it is the primary energy store of all living organisms. Its measurement usually involves the use of the firefly luciferin-luciferase system, which emits light that is measured by the use of a luminometer. There are manufacturers of dedicated instrumentation that enable this method to be used on the factory floor. The main disadvantage is that food also contains ATP, making it difficult to use this technique to estimate cell numbers. It is used widely in the industry for the monitoring of cleaning efficiency, as microbial cells and/or food contamination will give a high reading.

Swabbing a knife blade for ATP testing
(Courtesy of Biotrace.)

Culture media

The addition of identifiable substrates (i.e. substrate that is used by a particular group of micro-organisms leading to an observable change within the media) to existing culture media is another expanding area. These methods may shorten the time to make a positive identification and have the advantage, for food industry personnel, of expressing the results in the same format as traditional methods. One such method is the addition of 4-methylumbelliferyl-D-glucuronide (MUG) to culture media. Most *Escherichia coli* convert this non-fluorescent substrate into a fluorescent product, causing the media to fluoresce under a simple UV lamp. More recently, there has been a growing interest in the use of chromogenic media, for example, 5-bromo-4 chloro-3 indoxyl-B-D-glucuronide (BCIG) to detect *E. coli* O157. They can, however, seem rather costly although they may save time and expense in confirming suspect colonies.

Immuno methods

There are many types of immunological assay available, which can be divided into those based on **enzyme** linked immunosorbant assay (ELISA) with or without immunomagnetic separation, and latex agglutination methods. The ELISA method involves the use of antibodies to detect a particular organism or group of organisms that are linked to an enzyme system, which changes colour if the organism is present. A wide variety of organisms can be detected using ELISA, although the method can take 24-48 hours for a result, is not particularly sensitive and is labour-intensive. Immunomagnetic separation increases sensitivity as the **antigen** is concentrated, using paramagnetic beads coated with an **antibody** to the target organism or toxin and then applying a powerful magnet. In latex agglutination, the antibody is attached to an inert **carrier** particle, such as latex, which enhances visible clumping of the antigen. This is a rapid method of identifying the target organism but because the inherent sensitivity is poor, an enrichment procedure is often required. Alternatively, this method known as reversed passive latex agglutination (RPLA), can be used to detect toxins such as *Bacillus cereus* diarrhoeal enterotoxin.

Biochemical methods

Another common way of rapidly identifying isolated organisms is by the use of biochemical profiles, achieved through addition of the test organism to a range of different substrates, often in a kit form. The biochemical profile of an organism will denote the "best fit" identification. Several automated systems are available, allowing for testing of a large number of organisms.

Selection of a rapid method

Further information can be obtained from food microbiology textbooks, reviews of current methodology and more importantly from trade literature. If managers intend to introduce rapid microbiological testing, the following points should be considered.

Firstly, identify if current methods are adequate for the company's needs. If they are, then a change is probably unnecessary. If existing methods are inadequate then:
- review the current information available on rapid techniques and identify any that may be of benefit;
- collect as much information as possible on reliability, ease of use and sensitivity (including false positive and negative rate). However, one needs care in interpreting data. Information provided depends on the gold-standard or reference method,

which can only be reliably compared if all data was generated in the same trial;
- contact the manufacturers and ask for a demonstration;
- set up a trial of the method to run in parallel with existing tests; and
- ascertain if the results of the trials lead to an improvement in the information obtained

If this is the case the test may be adopted, otherwise continue to use existing methods or select an alternative rapid method.

Food MicroModel

This is not a rapid method in itself but is a computer-based system for predicting the growth, death and survival of foodborne pathogens. It can be used to help determine the microbiological safety of food and reduce the number of challenge tests when developing new products.

Bacteriological food sampling by enforcement officers

The Food Safety Act, 1990 formalized the administrative procedures for the taking and examination of food samples. Food liaison groups, involving EHOs from adjacent authorities, Public Health Laboratory food examiners, trading standards officers, and sometimes, public analysts and have introduced coordinated sampling programmes within their areas. These groups also cooperate with the Food Standards Agency and the Local Authorities Coordinators of Regulatory Services (LACORS) to obtain food samples to comply with the requirements of the EU with regard to national sampling programmes.

EU legislation is increasing the emphasis on bacteriological standards as a way of improving food safety. Standards exist for many foods including egg products, milk and milk products and shellfish. However, EU policy dictates that inspection and sampling are primarily undertaken at the production premises in the country of origin and not at frontiers between member states (EU Directives 89/397 and 89/662).

Samples of food may be taken for bacteriological examination to determine the acceptability or safety of a batch of food intended for human consumption, for example, a consignment of imported food or a quantity of food suspected of being contaminated by pathogens. As bacteria are unevenly distributed throughout food, a statistical sampling plan should be used to ensure that an adequate number of random samples are taken, on which to base a decision regarding the whole consignment.

However, most routine bacteriological sampling by enforcement officers is carried out at food shops, factories and restaurants, etc., to support observations regarding the standard of hygiene within the premises. When sampling from retail outlets, it is preferable to take samples of several different high-risk foods from one or two premises, as opposed to taking samples of the same type of high-risk food from many premises.

Unacceptable numbers of bacteria in the sample usually reflect poor handling techniques, under-processing, post-process contamination or incorrect storage. Due to the uneven distribution of bacteria, unsatisfactory results should not usually be used to support an opinion about any food remaining on the premises. However, the presence of pathogens or a suspicion of under-processing would require the detention and further sampling of any of the original food that had not been used or sold.

If only one or two samples out of many taken from a premises are unacceptable, this may be due to an unsatisfactory delivery of the specific food. If several samples from a single premises are unsatisfactory, for example, because of the presence of *Escherichia coli*, and all of the food originated from different suppliers, this would indicate that it was more likely a handling or storage problem at the retail outlet or restaurant.

The routine use of statistical sampling plans requiring many samples of the same food to be taken from one premises is not practicable or necessary to determine whether a particular food sample has been handled hygienically. However, a sampling programme should be designed to ensure the appropriate frequency of sampling of specific foods from particular premises. The programme may be based on the risk of food hygiene associated with particular foods, the history of premises with regard to food poisoning and food complaints and the nature of the food industry within a particular area.

When procuring a food sample, the officer obtains as much information as possible concerning its handling. This would include method of distribution, date of delivery, date of production, temperature of storage, type and condition of packaging, and any significant information regarding hygiene associated with the product, for example, risks of cross-contamination, handling techniques, personal hygiene of operatives and hygiene of premises and equipment.

The food is usually placed in an insulated container, to avoid the multiplication of bacteria prior to testing, and taken to the public health laboratory. The bacteriologist examines the food to establish the total number of bacteria present per gram of food and whether any indicator or pathogenic organisms are present. On completion of incubation, typically 48 hours at 37ºC, the number of bacteria are counted and a report forwarded to the environmental health department. As none of the methods used by laboratories can be guaranteed completely accurate, tolerances must be allowed. Furthermore, if comparisons are to be made or standards established, it is important that the same methods and procedures are used by all laboratories involved in analysis, as well as by all enforcement officers.

Interpretation of results
Bacteriological counts

The total number of living bacteria in a food sample is usually expressed as the **total viable count** (TVC) and is used as a microbiological indicator of food quality. The temperature of incubation used varies depending on the type of bacteria likely to be present if the food has been mishandled. A temperature of 37ºC is usually employed by the PHLS, as they are interested in the presence of mesophilic bacteria and the opportunity for pathogen growth. A TVC in excess of 10^4g in most high-risk foods indicates either gross contamination or old product, together with storage at incorrect temperatures. High counts at 20ºC on chilled foods indicate inadequate, or lack of, refrigeration.

Indicator organisms

Indicator organisms are organisms that are usually present in very large numbers in environments inhabited by pathogens. The usual indicator organisms used by food microbiologist belong to the Enterobacteriaceae family, many of which live in the intestine of man and animals, and include coliform organisms, faecal coliforms and *Escherichia coli*.

Faecal streptococci

Faecal streptococci include *Streptococcus faecalis, Streptococcus faecium, Streptococcus bovis* and *Streptococcus equinus* – this group is serologically defined as Lancefield group D. They are normally present in mammalian faeces but are more resistant to drying, high temperatures and disinfectants than Enterobacteriaceae. Their presence usually indicates poor plant hygiene or a failure of a specific heat process to destroy them; post-process contamination may or may not have occurred.

THE ENTEROBACTERIACEAE FAMILY

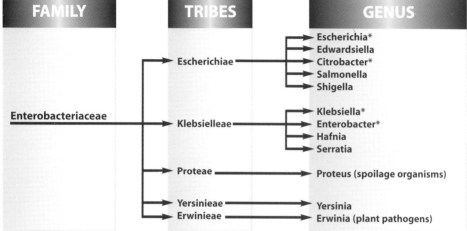

*Lactose fermenting coliforms (also some species of *Salmonella* and *Shigella*).

Coliforms/enterobacteriaceae

Coliforms are those members of the enterobacteriaceae that **ferment** lactose and produce gas. They are sensitive to heat and their presence in food usually indicates post-process contamination or poor hygiene. The presence of coliforms in water, or food, may be a result of recent faecal contamination but as some coliforms exist in soil and on plants, this is not always the case. However, the presence of *E. coli* does indicate recent human or animal faecal contamination and also the likelihood of pathogenic contamination. Faecal coliforms are those coliforms that can be successfully incubated at the higher temperature of 44 to 45.5°C and include *E. coli*.

Pathogenic contamination

Pathogens should not be present in hygienically handled and effectively processed high-risk food, even though some microbiologists believe that levels of 10^3 *Staphylococcus aureus*, *Clostridium perfringens* or *Bacillus cereus* per gram are insignificant. However, notwithstanding the fact that such low numbers are extremely unlikely to cause food poisoning, such food is potentially hazardous and has not been handled hygienically.

In high-risk food presence of:
- *Staphylococcus aureus* usually indicates a breakdown in personal hygiene or the handling of food;
- *Clostridium perfringens* usually indicates prolonged warm storage after cooking; and
- *Salmonella* indicates cross-contamination, under-cooking/inadequate thawing or faecal contamination from a carrier.

Microbiological guidelines

The Health Protection Agency has published microbiological guidelines for **ready-to-eat foods** sampled at point of sale. The microbiological quality of the food is based on the **aerobic** colony count (ACC), indicator organisms and common foodborne pathogens, and ranges from satisfactory to potentially hazardous. Foods are divided into five categories; however, the guidelines relating to pathogens are the same for categories one to five. Category one foods have stricter guidelines for aerobic plate counts than category two foods, which have stricter guidelines than category three etc. Although the guidelines are not a legal requirement, they may be referred to in the event of enforcement action.

GUIDELINES FOR THE MICROBIOLOGICAL QUALITY OF CATEGORY 1 READY-TO-EAT FOODS

Beefburgers/kebabs, sausage rolls, pickled fish, scotch eggs, bhajis, pasties, meat pies, pork pies, mayonnaise, mousse

Criterion	Microbiological quality (CFU per gram unless stated)			
	Satisfactory	Acceptable	Unsatisfactory	Unacceptable/ Potentially hazardous*
Aerobic colony count† 30°C:48h∓2h	$<10^3$	$10^3 - <10^4$	$\geq 10^4$	N/A
Indicator organisms‡				
Enterobacteriaceae§	<100	$100 - <10^4$	$\geq 10^4$	N/A
E. coli (total)	<20	20 – <100	≥ 100	N/A
Listeria spp. (total)	<20	20 – <100	≥ 100	N/A
Pathogens				
Salmonella spp.	absent in 25g			present in 25g
Campylobacter spp.	absent in 25g			present in 25g
E. coli O157 & other VTEC	absent in 25g			present in 25g
V. cholerae	absent in 25g			present in 25g
V. parahaemolyticus¶	<20	20 – <100	$100 - <10^3$	$\geq 10^3$
L. monocytogenes	<20**	20 – <100	N/A	≥ 100
S. aureus	<20	20 – <100	$100 - <10^4$	$\geq 10^4$
C. perfringens	<20	20 – <100	$100 - <10^4$	$\geq 10^4$
B. cereus and other pathogenic *Bacillus spp.*#	$<10^3$	$10^3 - <10^4$	$10^4 - <10^5$	$\geq 10^5$

[*] Prosecution based solely on high colony counts and/or indicator organisms in the absence of other criteria of unacceptability is unlikely to be successful.
[†] Guidelines for aerobic colony counts may not apply to certain fermented foods – for example, salami, soft cheese, and unpasteurized yogurt. These foods fall into category 5. Acceptability is based on appearance, smell, texture, and the levels or absence of indicator organisms or pathogens.
[‡] On occasions some strains may be pathogenic.
[§] Not applicable to fresh fruit, vegetables and salad vegetables.
[¶] Relevant to seafood only.

| # | If the Bacillus counts exceed 10^4 CFU/g, the organism should be identified.
| ** | Not detected in 25g for certain long shelf life products under refrigeration.
| N/A | Not applicable.

Enterobacteriaceae tests have replaced coliform tests and will include non-lactose fermenting organisms such as salmonellas. However, the criteria listed for enterobacteriaceae do not apply to fresh fruits, vegetables or sandwiches containing salad vegetables, as they often carry high levels of the organisms as part of their normal flora.

GRADES OF MICROBIOLOGICAL QUALITY

The terms used to express the microbiological quality of ready-to-eat foods are:
- satisfactory ~ test results indicating good microbiological quality;
- acceptable ~ an index reflecting a borderline limit of microbiological quality;
- unsatisfactory ~ test results indicating that further sampling may be necessary and that authorized officers may wish to undertake a further inspection of the premises concerned, to determine whether or not hygiene practices for food production or handling are adequate;
- unacceptable/potentially hazardous ~ test results indicating that urgent attention is needed to locate the source of the problem; a detailed risk assessment is recommended. Such results may also form a basis for prosecution by environmental health departments, especially if they occur in more than one sample. Food examiners will wish to draw on their own experience and expertise in determining the advice and comments they wish to give and they will be required to do this if invited to give an expert opinion during legal proceedings.

If the food fails to achieve a reasonable bacteriological quality, then an investigation is usually carried out to ascertain the reasons for this and advice will be given on how to prevent further unsatisfactory results by ensuring that:
- the product is acceptable when delivered to the premises;
- stock rotation is effective;
- handling techniques are hygienic;
- cleaning and disinfection is satisfactory;
- proper temperature control is exercised at all stages of manufacture, distribution and sale;
- packaging is satisfactory; and
- staff training is effectively carried out.

Repeat samples should be taken until a sequence of satisfactory results is obtained.

Official control of food premises

In the UK, central government has given the responsibility for protecting public health and ensuring food businesses comply with food hygiene legislation to local authorities. Authorized officers with a wide range of qualifications, experience and expertise are necessary to enable authorities to carry out the significant range of food hygiene and food safety controls that now exist. The most common local authority officer involved in food hygiene control is an environmental health practitioner/officer. Authorities may also appoint technical officers with specialist food qualifications.

The functions of authorized officers in the field of food hygiene and food safety include:
- ensuring product safety and fitness for consumption;
- reducing possible sources of contamination entering the food environment;
- monitoring conditions and hygienic operations within the food environment;
- ensuring compliance with relevant legislation;
- establishing the competency and attitude of management and effectiveness of control procedures; and
- offering professional guidance, particularly when legislation is changing.

Authorized officers undertake the above functions:
- during routine visits to and inspections of food premises;
- whilst investigating food poisoning outbreaks and incidents;
- whilst investigating food complaints;
- by lecturing on hygiene courses and seminars and giving related talks;
- by using the media, for example, press releases, committee reports and hazard warnings;
- whilst dealing with planning and licence applications;
- by developing partnerships with businesses' decision-making bases in the local authority's area (often referred to as the "Home Authority Principle"); and
- by developing local business forums for the exchange of information and the provision of advice.

Hygiene inspections of food premises by authorized officers

Inspections should be carried out at all stages of production, processing and distribution to ensure compliance with food law. The main purpose of a hygiene inspection is to ensure food is being handled and produced hygienically and to identify forseeable incidences of food poisoning or injury resulting from consumption of the food.

The main objectives of primary food hygiene inspections are the:
- determination of the scope of business activities and the relevant food safety legislation, which applies to the operations taking place at the premises;
- thorough and systematic gathering and recording of information, from observations of practices, procedures and processes, including procedures based on HACCP principles, and discussion with food handlers, contractors, food business operators and managers;
- identification of potential hazards and associated risks to public health;
- assessment of the effectiveness of process controls to achieve safe food;
- assessment of the HACCP based food safety management system operated by the business;
- identification of actual or potential breaches of food law and, if appropriate, the gathering and preserving of evidence;
- consideration of appropriate enforcement action, (proportionate to risk), to secure compliance with food safety legal requirements;
- provision of advice and information to food business proprietors and food handlers, in accordance with industry guides and codes of practice; and

- determination of the need to collect samples of food or materials and articles in contact with food for analysis and/or examination.

Secondary inspections may involve visits: for sampling; to investigate food complaints, to discuss food safety management systems; for training; and to check work carried out (revisits).

LACORS has issued advice on food hygiene inspections, which reflects the main purposes of an inspection. LACORS guidance emphasizes the need for local authorities to have an appropriate management system, to ensure enforcement officers undertake inspections in a consistent manner. The various stages of a food hygiene inspection are specifically addressed, including planning for the inspection, preliminary interviews, the inspection of premises, the post-inspection interview and post-inspection administration.

Before carrying out a food hygiene inspection, authorized officers will take account of a number of issues. These will include:
- reviewing the premises' previous history, including information on its operations and systems, previous complaints and responses to previous inspection outcomes;
- timing of inspection – generally unannounced, although advance notice may occasionally be appropriate to ensure relevant persons are present;
- equipment availability, for example, calibrated temperature recording equipment;
- appropriate protective clothing; and
- assessing the need for additional expertise, for example, food examiners.

Before commencing an inspection, the officer will explain the purpose of the inspection and what it will entail to the manager. Inspections will include:
- a preliminary assessment of the food hazards associated with the business; and
- determining whether the business has a satisfactory system for assessing food hazards and controlling risks at those points that are critical to food safety.

The approach to inspection will depend significantly on whether such a system exists. LACORS guidance on food hygiene inspections therefore emphasizes the central role of a preliminary interview. Discussion at an early stage of the inspection about the hazards associated with the business, and any system in place for assessing those hazards and controlling the risks, allows an officer to properly plan the subsequent detailed inspection of the business operations and premises. The officer should examine any food safety policy, the HACCP plan and other relevant documentation.

Businesses may have various types of hazard analysis systems in place. Only a minority will have a formal documented HACCP system; most are more likely to have implemented some less formal system. Different approaches to auditing such systems are necessary. LACORS guidance on risk assessment provides detailed advice on assessing different types of systems. It emphasizes the flexibility officers must have in recognizing that food businesses may use a variety of methods to identify and assess relevant food hazards and control risks. It also recognizes that with this new requirement, many businesses will need advice on how to develop systems and officers are seen as central to enabling proprietors to develop relevant systems.

In addition to considering any systems in place, an inspection will include a visual and physical examination of the premises and its operations. Officers will have particular regard to the food hazards and the control and monitoring procedures in place at the critical points.

LACORS guidance on risk assessment includes advice on the broad stages of a risk-based inspection and the judgements officers must take. The guidance emphasizes going beyond simply identifying hazards. It recognizes that action taken to control hazards, without any consideration of the risk, can result in unnecessary controls being imposed on food businesses.

Food hygiene inspections should include:
- a review of the information held on record by the food authority in relation to the food business;
- a preliminary discussion with the duty manager/proprietor, which should include:
    - an explanation by the officer of the purpose of the inspection;
    - identification of all the food-related activities undertaken by the business, for example, the areas of the premises used for the preparation/production/storage of foodstuffs, the processes used, and the staff involved;
    - identification of the customer base of the business;
    - identification of any food safety management systems that may be in use;
    - an assessment by the officer of the hazards posed by the business's activities;
    - an assessment of the manager's/proprietor's understanding of the hazards posed by the business and the application of appropriate controls;
    - an examination of any documented food safety management system/hazard analysis;
    - an assessment of the provision of supervision and instruction and/or training of staff;
- a discussion with any staff responsible for monitoring and corrective action at CCPs to confirm that control is effective;
- a physical examination of the premises to assess if all the critical controls have been identified, whether those controls are in place and to assess compliance with the relevant legislation; and
- an assessment whether to take microbiological or chemical samples.

The physical examination of the premises will usually consider the following areas:
- temperature controls during storage, cooking, processing, cooling, reheating, thawing, preparation and distribution. Physical checks will be made and records examined, including action taken in the event of breakdown, for example, the failure of refrigeration;
- the absence of cross-contamination and the use of good handling techniques. Work flows will be examined to ensure separation of high-risk food from raw food and waste. The use of staff and equipment will be considered and particular attention paid to thawing and cooling. The protection/covering and packaging of food will be included;
- cleansing and disinfection. The physical and bacteriological cleanliness of premises and equipment, evidence of a planned cleaning programme/cleaning schedules;
- personal hygiene and training. The hygiene awareness of managers and staff, the standard of supervision, procedures for appointments, medicals and exclusions. Procedures for replenishment of soap, towels and replacement of soiled protective clothing. The availability of training records;

- the delivery and handling of raw materials, including the use of specifications, checking food on arrival, dealing with non-food items such as packaging, to avoid food contamination and reject procedures;
- pest control including absence of pests, proofing and control. Records of visits, treatment and recommendations will be required;
- complaints. An examination of records and action taken on receipt of complaints.
- recall systems and product traceability will be checked. Foreign body detection equipment such as metal detectors, and systems will be examined;
- waste and refuse. Procedures for handling and removal of waste and for dealing with detained or unfit food;
- visitors. Procedures for dealing with visitors, including enforcement officers, engineers and contractors;
- food storage and stock rotation. Systems and documentation, including staff awareness;
- finished product. Handling, wrapping, storage and distribution;
- management control systems, including the hygiene policy and the application of hazard analysis and control. Monitoring of procedures at critical points will be examined. Contracts, for example, pest control, cleaning, equipment maintenance, waste disposal and catering. Quality assurance/quality control systems including microbiological monitoring, records and action taken in event of adverse results. Hygiene audit records should be available; and
- structure, equipment and facilities including design, lighting, ventilation, drainage, water supply, staffrooms, first aid, storage areas, facilities for washing hands, equipment and food, and external buildings and yards.

It will be clear from the above that comprehensive inspections of food premises by authorized officers may take several hours or even days, depending on the size of premises, the type of operation and the standards of hygiene. Furthermore, inspections may take place at any time of the day or night depending on the hours of operation.

Some form of post-inspection interview should occur at a closing meeting, to enable the officer to discuss any significant findings including any contraventions of food hygiene laws with the manager/proprietor. At all times, officers should clearly distinguish those matters that are contraventions from recommendations of good practice. The closing meeting should include:
- a discussion regarding any hazards that have been identified by the officer and have not been covered by the business's systems;
- a discussion regarding failures to implement or monitor any critical controls that have been identified by the business;
- a discussion regarding any contravention of the relevant legislation;
- any recommendations of best practice the business may wish to consider; and
- a discussion regarding the timescale for any remedial work needed and any follow-up action the officer intends to take.

Authorized officers should report back in writing after every relevant inspection. The reports should include the following information:
- name and address of premises;
- person seen/interviewed;
- type of premises;

- date and time of inspection;
- specific legislation under which inspection was conducted;
- areas inspected (whole or part of premises (specify areas));
- records/documents examined (and outcome);
- details of any samples procured including description and batch number;
- summary of matters discussed at closing meeting; and
- summary of action to be taken by the authority, for example, standards satisfactory, a letter or improvement notice.

It is quite common for such reports to be provided at the end of the inspection, in handwritten form. The summary of action should confirm whether any further measures will follow, for example, an advisory letter or service of a notice.

Action taken as a result of an inspection

During an inspection, an officer may identify contraventions of food hygiene legislation and/or poor or unsafe **food handling** practices. Several options exist to remedy the contravention and detailed guidance is found in the Food Law Code of Practice as to the most appropriate action. In addition, LACORS has published guidance on food enforcement policies. The guidance stresses the range of options available to local authorities and provides detailed criteria that may be applicable to them. The options available and potential outcomes for all food hygiene inspections include:

- verbal advice/warnings or informal written advice/warnings where the officer is confident the work will be carried out;
- a hygiene improvement notice, for contraventions of food hygiene legislation; allowing not less than 14 days to comply;
- the detention or seizure of unsafe food, where food does not comply with the food safety requirements;
- a hygiene emergency prohibition notice, where there is an imminent risk of injury to health, requiring closure of the premises or prohibition of processes or use of equipment;
- a formal caution where an offence exists but it is not considered in the public interest to prosecute through the courts; and
- prosecution, where it is considered in the public interest.

Frequency of inspections

Effective inspection programmes recognize that the frequency of inspection will vary according to the type of food business, the nature of the food, the degree of handling and the size of the business. Essentially, those premises posing potentially a higher risk should be inspected more frequently than those premises with a lower risk. The Food Law Code of Practice details an inspection rating scheme for assessing premises for this purpose.

Premises are rated in the following areas:
- potential hazards and significance of risk associated with the premises:
    - type of food and method of handling, including risks from *E. coli* O157, other VTEC, *Clostridium botulinum*, salmonella or *Bacillus cereus*;
    - method of processing, including cook-chill, small-scale production of cooked meat or milk and dairy products, purification of shellfish, thermal processing and aseptic packaging of **low-acid foods**, vacuum packing including sous vide but excluding raw and unprocessed meat and dried foods; and

- consumers at risk;
- level of current compliance with regard to:
  - food hygiene/safety, including practices, procedures and temperature controls; and
  - structure, including cleanliness, layout, lighting, ventilation and facilities;
- confidence in management/control systems including:
  - previous history;
  - management attitude;
  - the technical knowledge available on hygiene and food safety matters, including HACCP; and
  - documented procedures and HACCP-based food safety management systems.

The officer determines the score for each of the categories, in line with the code's detailed guidance. The total score is used to determine the minimum frequency of inspection and the date of the next inspection. Businesses processing high-risk food, are at risk from *E. coli* O157, other VTEC or *Clostridium botulinum*, or supplying vulnerable consumers, are likely to be inspected at least every six months. Establishments scoring less than 31 points are deemed to be low-risk and may be subject to an alternative enforcement strategy instead of routine inspections, for example, hygiene questionnaires. However, inspections must be carried out in the event of consumer complaints, application for registration and significant changes in activities. It would be beneficial if the scoring system was simplified to improve consistency.

Provided premises comply with hygiene legislation, it should be noted that the interval between inspection will be greater if the officer has confidence in the management and the control systems. This enables local authorities to target their resources at those businesses that fail to make efforts to comply and in which the authorities have little or no confidence.

The investigation of food complaints

Food safety legislation contains a number of offences that may be relevant to food complaints. There is a general offence regarding selling foods that do not comply with food safety requirements, which includes food that is either injurious to health, unfit for human consumption or so contaminated (whether by extraneous matter or otherwise) that it would not be reasonable to expect it to be used for human consumption. The act also contains an offence of selling food not of the nature, substance or quality demanded by the purchaser. These offences cover a whole range of food complaints, including food that would be harmful if it were consumed, food contaminated by foreign bodies, contamination by mould and food in dirty containers (a particular problem with reusable bottles).

Local authorities investigate food complaints as part of the general enforcement service to the public. The purpose of investigating complaints includes:
- resolving problems that pose a risk to public health;
- the provision of information to the food industry in order to raise and maintain standards;
- the prevention of future complaints; and
- fulfilling the duty of enforcement contained in the act.

To ensure a consistent approach to the investigation of food complaints, LACORS has published specific guidance on dealing with and investigating those made to local authority environmental health departments. A structured and systematic approach is necessary in investigations because what may appear to be a simple isolated problem could have significant, possibly national or international, implications.

The first stage of any investigation will be receipt of the complaint, when information will be obtained from the complainant. Various details are necessary to assist the future investigation. At this early stage, a decision must be made confirming that it is the responsibility of the authority that has received the complaint to investigate.

The relevant local authority officer, usually an environmental health practitioner/officer, will then inspect the food in question and arrange, if necessary, to further interview the complainant. The complainant will usually be asked to provide a written statement, detailing various matters including information on when, where and by whom the food was purchased, what happened to the food after purchase and when, where and by whom the problem with the food was discovered. Details of the vendor and, if relevant, the manufacturer and importer will be noted. Some form of proof of purchase is desirable to supplement the statement of the complainant, for example, a till receipt or distinctive packaging.

At this early stage, the officer will consider whether the complaint may relate to national/widespread problems or malicious contamination. The Food Law Code of Practice provides guidance on further action to be taken in the event of a food incident or hazard. The officer will also assess whether the complaint is associated with illness and, therefore, whether it is necessary to initiate a food poisoning investigation.

The retailer, caterer or manufacturer where the food was purchased or produced will then be visited. Where the retailer is within the officer's local authority area but the manufacturer is outside that authority's area then, after visiting the local retailer, the officer will make contact with the local authority environmental health department responsible for the manufacturer (often referred to as the home or originating authority) to obtain further information. Food legislation empowers **authorized officers** such as environmental health practitioners/officer to enter business premises outside their area to identify evidence of contraventions, which may encompass, amongst other things, the investigation of food complaints. However, authorities generally rely on the reports from home or originating authorities for detailed information on such premises as this is seen as a more cost-effective option.

At this stage, further information obtained may suggest that a national or widespread problem exists. In most cases, however, the officer will proceed with a detailed investigation and inspection of all relevant premises. The investigation will seek to confirm if an offence has been committed, the alleged person(s)/enterprise responsible and the precautions taken to prevent such a complaint.

Food complaint investigations will, on occasions, necessitate formal interviews having regard to all the relevant law on criminal evidence and, in particular, the Police and Criminal Evidence Act, 1984 and its associated codes of practice. This will occur when the officer considers that an offence exists and the outcome of the investigation is likely to result in a recommendation to prosecute. In addition to interviews with persons, officers may formally request confirmation, in writing, of details from manufacturers and other relevant enterprises relating to the complaint and, in particular, precautions taken to prevent such complaints.

The usual outcome of food complaint investigations is that advisory or warning letters are sent to the business responsible. However, if an officer believes a prosecution may be

appropriate, the following factors should be considered:
- the sufficiency of the evidence including:
    - the reliability and credibility of witnesses and their willingness to cooperate;
    - the alleged persons(s) have been identified;
    - any explanation offered by the suspect;
    - the likely success of a defence, especially a due-diligence defence; and
    - that it is in the public interest.
- the seriousness of the offence;
- the likelihood of a nominal penalty;
- the offence was committed as a result of a genuine mistake or misunderstanding (balance against seriousness); and
- whether other action such as issuing a formal caution or service of a hygiene improvement notice would be more appropriate.

The reference in the Food Hygiene (England) (Scotland) (Wales) (NI) Regulations 2006 and the Food Safety Act, 1990 to the defence of "**due diligence**" with regard to food-related offences is worthy of specific note. The legislation creates a number of offences known as "strict liability". In such a case, it does not matter that the person accused did not intend to break the law. The mere fact that there is clear evidence that the statute has been contravened is sufficient for a conviction. This regime of strict liability was perceived as causing injustice if a person was held to have committed an offence for which he/she had no responsibility at all, or because of an accident or some cause completely beyond his/her control. In order to create a balance of fairness, a defence was included which has become known as the "due-diligence defence".

The legislation specifically states that it is a defence to prove that all reasonable precautions were taken and all due diligence exercised to avoid the offence. Through legal precedent, various principles have been confirmed as necessary if a defence is to succeed. Some positive steps will always be required. Taking reasonable precautions involves the setting up of a system of control, having regard to the nature of the risks involved. Due diligence involves securing the proper operation of that system. Where there is a reasonable precaution then it should be taken. Nevertheless, what is reasonable will depend on particular circumstances including the size of the business. All relevant aspects of the operations of the business must be included in any control system and the operation of the system must be kept under review and amended as necessary.

Food complaint investigations may, therefore, involve considerable time and effort on the part of the investigating authority. In addition to determining whether there is admissible evidence to provide a realistic prospect of conviction for any food complaint offence, authorized officers will also consider whether it is in the public interest to prosecute. It is likely that only those food businesses that have failed to fulfil their legal and moral responsibilities will find themselves in court for food complaint offences.

The role of the consultant environmental health practitioner (EHP) - formerly EHO

Some of the largest food manufacturers, retailers and caterers employ EHPs or hygiene technologists to help them formulate and manage their own hygiene policies.

However, many companies use the services of hygiene consultants. There are a number of experienced and self-employed EHPs providing consultancy services and some companies offer a national service.

Consultants provide an independent and objective assessment of a food operation and can help companies formulate policies to improve the management of their hygiene standards. Consultants often fulfil a troubleshooting role but increasingly their advice is sought on a proactive basis, as the food industry recognizes the benefits that accrue from a proper investment in hygiene. They may be utilized to assist with the implementation of ISO 9000 and HACCP and to approve potential suppliers. The considerable demand for hygiene training created by food safety legislation has resulted in many consultants specializing in this area of work.

The introduction of compulsory competitive tendering has resulted in some local authorities using consultants in areas not subjected to competition, including food hygiene inspections, and this trend may increase.

The role of the expert witness *by Dr Slim Dinsdale*

Evidence given by expert witnesses on the technical issues of a case, in both criminal and civil litigation, can play an important role in assisting the court to arrive at a fair decision regarding guilt or liability.

The expert's role begins with instructions to produce a report from a solicitor on behalf of the defence or the prosecution. A bundle of documents is provided, consisting of the charges against the food business, witness statements, and various exhibits such as microbiological results from food samples, public analyst's reports, inspection reports, etc. These must be critically examined and any errors, unsubstantiated assumptions, and lack of continuity noted. Nothing should be taken for granted and every avenue explored to identify the strengths and weaknesses of the case. Unused material held by the prosecution should always be requested, as it may contain information that could assist the defendant's case, and a thorough examination of all the material will show where further information and/or experimental work is needed. For example, assumptions may have been made that food that has been cooked may have been cooled too slowly. Simple experiments using recording thermometers, used with the food produced and stored as stated, can provide convincing information. Genetic profiling of bacteria isolated from suspect food, and stool samples of patients, can also help to establish whether or not a link exists between the food and the food poisoning.

The skills of expert witnesses can only be fully used when they are part of a team, dedicated to maximizing the chances of success. From a defence viewpoint, the inter-relationships of lawyer, the food company, and the expert witness are crucial and are outlined in the illustration below.

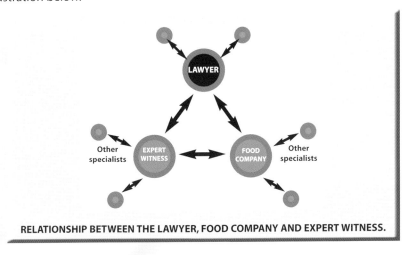

RELATIONSHIP BETWEEN THE LAWYER, FOOD COMPANY AND EXPERT WITNESS.

With the lawyer occupying the key position, it is essential that only appropriately qualified lawyers are instructed to deal with the case, and this is amply illustrated in the Oakwood Deli case (R-v-Kashioulis, 2000) where the specialist food law firm was Birmingham-based Margetts and Ritchie.

The Oakwood Deli case

The Crown alleged that chopped liver sold by the Oakwood Deli, in North London, was at the centre of an outbreak of *Salmonella* Enteritidis phage type 4 (SET4) food poisoning involving more than 150 people. The Crown's case was based upon microbiological and epidemiological evidence, both of which had been "tainted" to such an extent as to make them unreliable sources of evidence. In particular, the sampling protocol for food adopted by the local authority bringing the charges used a spoon, which was disinfected with sanitizing wipes, instead of being sterilized. This was a clear breach of Code of Practice No. 7, "Sampling for Microbiological Analysis and Examination", and any results from such foods should have automatically been regarded as unreliable. In an attempt to recover the situation and demonstrate that the sampling method was adequate, the prosecution commissioned the PHLS to carry out an experiment to determine the validity of the disinfection protocol.

A minced meat sample was spiked with SET4 and sampled with the original spoon; subsequent examination of the mince showed that SET4 was detected. The spoon was then wiped with sanitizing wipes and used to sample liver pâté known to be free from salmonella. The subsequent examination of the pâté did not identify any SET4. The conclusion was that the sanitizing wipes had removed any SET4 that was on the spoon and therefore the method was acceptable for taking formal food samples.

There was a fundamental flaw in this experiment, in that it was assumed that the spoon was contaminated with SET4 after sampling the 'spiked' mince, but this was not demonstrated since the spoon itself was not sampled. Further, the SET4 in the spiked minced meat sample was diluted greater that 100,000:1 with the indigenous bacterial flora. This would have further reduced the chances of any SET4 getting on to the spoon during the sampling of the spiked minced meat. This mistake was one of a number that fatally weakened the prosecution's case, and ended with an acquittal on all the charges.

Giving oral evidence

Oral evidence enables complex issues, such as those involving micro-organisms and microbiological results from food, to be explained to the court in an understandable way. Therefore these issues can be seen in the light of the whole food process, enabling a clear picture to be painted in support of any argument for or against the allegations. Oral evidence given under oath may carry as much weight in the court as written statements or records.

Conclusion

In conclusion, there are three pieces of advice that the expert witness should heed. First, be wary of accepting instructions from lawyers who appear to know little about the issues involved, and who seem reluctant to spend the time necessary to construct a good case. Second, carry out a thorough investigation, assume nothing, and adopt some lateral thinking when approaching brick walls. Third, and most importantly, be independent, produce a fair report and do not hesitate to state that there could be difficulties supporting a particular aspect of the case. Doing this will make the expert a much more credible witness and greater weight will be given to the evidence when the time comes for the court to make a decision.

Organizations involved in food control

The food control service operated by EHPs relies from time to time on advice and assistance from other organizations including LACORS (Local Authorities Coordinators of Regulatory Services), the Health Protection Agency, the public analyst, the Food Standards Agency, the Communicable Disease Surveillance Centre, research establishments, and, with regard to animals, veterinary investigation centres. In addition, a considerable amount of work on the microbiology of food is undertaken by the Food Hygiene Laboratory at the Central Public Health Laboratory at Colindale.

LACORS and the coordination of local authorities

LACORS (www.lacors.gov.uk) is a local government central body charged with the task of coordinating local authority trading standards and food enforcement activity. Its terms of reference were extended in 1992 to encompass local authority food hygiene and food safety functions. This reflected the need for local authorities to ensure delivery of an effective enforcement service that was both coordinated and consistent.

LACORS is funded by local government monies and is answerable to a group of local authority elected members, appointed through the local authority associations, which make up the LACORS Management Committee. It has a small administrative and professional secretariat but is reliant on the hard work, experience and expertise of a large group of local authority advisers, who volunteer, through the associations, their time and efforts to assist LACORS in its work. Advisers include directors, chief officers, experienced managers and practitioners, who are organized into various technical panels and working groups to address specific enforcement issues requiring coordination. By involving local authority officers, who are responsible for the policy and detail of local authority enforcement work, LACORS ensures its strategies and guidance are relevant and acceptable.

LACORS has identified three key coordinating strategies, which have been applied to food hygiene and food safety. Firstly, through LACORS advisers, it emphasizes the need for common-sense, pragmatic interpretations and application of food hygiene laws and a consistent approach to inspections, investigations and outcomes. LACORS identifies and promotes existing good enforcement practice, encouraging local authorities to ensure enforcement policies are based on protecting the public.

The second strategy involves encouraging the development of local authority food liaison groups. These groupings of local authorities enable EHPs to raise enforcement difficulties and potential inconsistencies and resolve them through peer group discussion. Where difficulties cannot be resolved, or the issues have national implications, the groups can seek the view of LACORS and its advisers.

The third strategy is known as the Home Authority Principle. This involves local authorities taking particular responsibility for those food businesses whose decision-making bases are located in their areas (usually the head office). The role of a home authority is to assess a food business's own system for ensuring that food hygiene and food safety issues, including legal compliance, are properly addressed. The home authority also provides preventive advice, particularly when legislation changes. It acts as a focal point for liaison with other authorities, who inspect the business's decentralized branches or units situated in their area. However, the final decision to pursue legal action in the event of a contravention always remains with the authority where the offence occurred.

The strategies of LACORS have been recognized by central government as an essential part of the local authority food control system.

The public analyst

Food authorities appoint a public analyst to provide, among other things, a service of chemical analysis and advice. Work is also undertaken for private companies. However, public analysts must not engage in any food business that is carried on in the area to which their appointments apply. Food and water are analyzed using extremely sensitive and sophisticated equipment to confirm the presence, or absence, of specific contaminants, pesticides, additives, trace metals or taints. Tests can be carried out to ascertain whether an item has undergone a particular process - for example, the phosphatase test is used to determine whether an insect has been subjected to heat treatment or cooking. The composition of food and the amount of water present can also be ascertained.

17 Food safety legislation

The law is a complex subject and most acts and regulations affecting the **food** industry are difficult to interpret. However, ignorance of the law is no defence in the event of a prosecution and all managers must make a special effort to become conversant with legislation that affects their business.

This chapter is intended only as a general outline of the most important legislation relating to **food** and **food hygiene**. There are differences in the way that food law has been implemented in England, Wales, Scotland and Northern Ireland. However, since this law originates from the same European regulations and directives, such differences are minor in nature.

It is not intended to provide a concise legal document and as legislation is constantly changing, regard must always be had to any new legislation that may have been introduced since this book was written. Should information or advice be required regarding the interpretation of a particular section, or the current legislation applicable, the local environmental health practitioner/officer or a solicitor should be consulted. UK legislation consists of:

Acts of Parliament

These are statutes passed by Parliament that can only be modified by parliamentary procedure. Acts are normally concerned with principles of legislation and must pass through the House of Commons and the House of Lords, before receiving royal assent.

Regulations and orders

Regulations and orders are delegated legislation made by the appropriate Minister, who is empowered to do so under a specific act. For example, The Food Labelling Regulations, 1996 were made under the Food Safety Act, 1990. Regulations normally deal with specific premises or commodities in much greater detail than acts.

Local acts or bylaws

These are made or adopted by local authorities and are legally binding only within the area of the particular authority. Bylaws must be formally approved by a specified minister before they can take effect.

The European Union (EU)

When the UK joined the European Union, most food safety legislation originated from directives which required member states to provide regulations to achieve the objectives of the specific directive. It now appears that the EU prefers to issue regulations which are binding on all member countries without the need for each country to introduce their own legislation.

Copies of legislation can be purchased from Her Majesty's Stationery Office. Regulations are known as statutory instruments (SI) and are numbered to enable rapid identification when ordering.

Acts and regulations applicable to the food industry are concerned with:
- preventing the production or sale of injurious, unsafe or unfit food;
- preventing the **contamination** of food and food equipment;
- the **hygiene** of food premises, equipment and personnel (including training);

- temperature control and the prevention of contamination;
- the implementation of food safety management systems based on the principles of HACCP;
- the provision of sanitary accommodation, water supplies and washing facilities;
- the control of **food poisoning**;
- the importation of food;
- the composition and labelling of food; and
- the registration and licensing of food premises and vehicles.

LEGAL TERMS

Court of Appeal (Criminal Division)

This division deals with appeals from the crown court against conviction or sentence. Usually three or five judges sit to hear an appeal.

Crown court

This court is responsible for serious criminal trials of indictable offences and also hears appeals from magistrates' courts.

County court

The county court is responsible for dealing with civil offences, for example, claims for damages from persons who have suffered food poisoning.

Indictable offences

Serious crimes usually tried by a judge and jury in the crown court. The accused is given a full written statement setting out each alleged offence. This document is known as an indictment and each offence is called a count.

Information

A statement of the alleged contravention(s), which is sent to the justices' clerk by the prosecuting authority. On receipt of the information the clerk issues the summons. This is the usual method of instituting criminal proceedings before a justice of the peace.

Judge

An officer of the Crown, appointed on the advice of the Lord Chancellor, who sits in court to administer justice according to the law. Most judges are barristers and will have had several years of experience before appointment; however, some are solicitors.

Justices of the peace (JP)

Persons from all walks of life, appointed by the Crown to preside over magistrates' courts in a particular area. They act judicially in the case of summary offences and initiate proceedings and commit the accused for trial in the case of indictable offences. They receive expenses only and must attend training courses prior to appointment.

Magistrates' court

A court that deals with summary offences in a particular geographic area. Usually presided over by three justices of the peace or a single stipendiary magistrate. Each bench of magistrates is assisted by a legally qualified justices' clerk.

At the court hearing the accused has to plead guilty or not guilty. If he pleads guilty the magistrates will then sentence him, taking into account previous convictions. If he pleads not guilty the prosecutor will call witnesses to prove his guilt. The solicitor acting for the accused will cross-examine the prosecution witnesses, the accused will give his own evidence and may produce his own witnesses to prove his innocence.

The prosecution has to prove the guilt of the accused beyond all reasonable doubt. After the closing speech by the defence solicitor, the magistrates will state whether or not they find the case proved. If the accused is found guilty, he will then be sentenced.

Statutory instrument
Subordinate legislation, such as regulations and orders, which the appropriate minister is empowered to lay before Parliament.

Stipendiary magistrate
A salaried magistrate who is a legally qualified barrister or solicitor with at least seven years' experience.

Summons
A document signed by a magistrate and issued by the justices' clerk on receipt of the information, requiring the attendance of the accused in court on a particular date.

THE LAW GOVERNING THE SALE OF FOOD

Regulation (EC) No 178/2002 general principles of food law and procedures in matters of food safety

Definitions
"Food" includes drink, chewing gum and any substance, including water, intentionally incorporated into the food during its manufacture, preparation or treatment.

"Food business" means any undertaking, whether for profit or not and whether public or private, carrying out any of the activities related to any stage of production, processing and distribution of food.

"Food business operator" means the natural or legal persons responsible for ensuring that the requirements of food law are met within the food business under their control.

"Retail" means the handling and/or processing of food and its storage at the point of sale or delivery to the final consumer, and includes distribution terminals, catering operations, factory canteens, institutional catering, restaurants and other similar food service operations, shops, supermarket distribution centres and wholesale outlets.

"Primary production" means the production, rearing or growing of primary products including harvesting, milking and farmed animal production prior to slaughter. It also includes hunting and fishing and the harvesting of wild products.

"Final consumer" means the ultimate consumer of a foodstuff who will not use the food as part of any food business operation or activity.

"Placing on the market" means the holding of food for the purpose of sale, including offering for sale or any other form of transfer, whether free of charge or not, and the sale, distribution and other forms of transfer themselves.

The general food law relates to all stages of the production, processing and distribution of food. Food law must achieve a high level of protection of human health and life and should be based on risk analysis (risk assessment, risk management and risk communication).

Article 7 (the precautionary principle)
In cases where the possibility of harmful effects on health have been identified but scientific uncertainty persists, proportionate risk management measures may be adopted, to protect health, pending further scientific information for a more comprehensive risk assessment (a scientifically based process consisting of hazard identification, hazard characterisation, exposure assessment and risk characterisation).
Food law should involve open and transparent public consultation.

Article 14
Food shall not be placed on the market if it is unsafe i.e. if it is injurious to health or unfit for human consumption. The long-term and cumulative toxic effects and the health sensitivities of the intended consumers must be taken into account.
If part of a batch is found to be unsafe the whole batch shall be considered to be unsafe unless a detailed assessment provides evidence to the contrary.

Article 16
Labelling, advertising and presentation of food must not mislead consumers.

Article 18
Food business operatives must have systems in place to enable the suppliers and business customers to be traced by the **competent authority**.

Article 19
Food business operators must withdraw unsafe food from the market and, if necessary, recall unsafe food already supplied to consumers. The competent authority must be notified.

This regulation established the European Food Safety Authority including its mission, tasks and organisation. It also established a rapid alert system to facilitate crisis management and for dealing with emergencies throughout the EU.

The General Food Regulations 2004 (No.3279)
These regulations align the Food Safety Act, 1990, with the general principles and requirements of Regulation (EC) No. 178/2002.
They confirm the competent authorities as the Food Standards Agency, the Port Health Authority or the Food Authority. The competent authorities are provided with new enforcement powers in respect of Articles 14, 16, 18 and 19 of Regulation (EC) No. 178/2002. Penalties for contraventions are also confirmed.
Regulation 4a. It is an offence to place food on the market if it is unsafe (injurious to health or unfit for human consumption) i.e. in contravention of Article 14 of Regulation (EC) No. 178/2002.

Regulation (EC) No 852/2004 on the hygiene of foodstuffs
Article 1
This regulation repeals Directive 93/43/EEC which was enforced by the Food Safety (General Food Hygiene) Regulations 1995. It applies directly to all members states and lays down general hygiene rules for food business operators based on the following principles:

(a) primary responsibility for food safety rests with the food business operator;
(b) food safety must be ensured throughout the food chain, starting with primary production;
(c) the cold chain must be maintained for food that cannot be stored safely at ambient temperatures;
(d) procedures should be based on good hygiene practices and HACCP principles;
(e) guides to good practice are invaluable to aid compliance with hygiene rules and the application of HACCP principles;
(f) microbiological criteria and temperature control based on a scientific risk assessment;
(g) imported food must be of the same hygiene standard as Community food.

The regulation does not apply to:
(a) primary production for private domestic use;
(b) domestic preparation, handling or storage for domestic consumption;
(c) the direct supply, by the producer, of small quantities of primary products to the final consumer or to local retail establishments supplying the final consumer.

N.B. Recital 9 of Regulation 852/2004 states that community rules should only apply to undertakings, the concept of which implies a certain continuity of activities and a certain degree of organisation. The scope, therefore, excludes activities such as the occasional preparation of food by individuals or groups for gatherings or for sale at charitable events or where, for example, hairdressers offer refreshments to their customers.

Article 2
Definitions
"Food hygiene" means the measures and conditions necessary to control hazards and to ensure fitness for human consumption of a foodstuff taking into account its final use.
"Establishment" means any unit of a food business.
"**Contamination**" means the presence or introduction of a hazard.
"Wrapping" means the placing of a foodstuff in a wrapper or container in direct contact with the foodstuff concerned, and the wrapper or container itself.
"Packaging" means the placing of one or more wrapped foodstuffs in a second container, and the latter container itself.
"Processing" means any action that substantially alters the initial product, including heating, smoking, curing, maturing, drying, marinating, extraction, extrusion or a combination of those processes. (Excludes mincing, chilling and freezing).

Article 3 (General obligation) Food business operators shall ensure that all stages of production, processing and distribution of food under their control satisfy the relevant hygiene requirements.

Article 5 (HACCP) Food business operators must implement a food safety management system based on the following HACCP principles:
(a) identifying hazards;
(b) identifying the **critical control points** (CCPs) at the steps at which control is essential to prevent or eliminate a hazard or to reduce it to acceptable levels;
(c) establishing critical limits at CCPs which separate acceptability from unacceptability;
(d) implementing effective monitoring procedures at CCPs;

(e) establishing corrective actions when a CCP is out of control;
(f) establishing verification procedures;
(g) establishing documents and records commensurate with the nature and size of the food business. (Documents must be kept up-to-date and retained for an appropriate period.); and
(h) the system must be reviewed if the product, process or any step is modified.

This article does not relate to primary production and transport related to primary production.

Article 6 Food business operators must register with the **competent authority** (usually the local authority).
 Significant changes in activities must be reported to the competent authority.

Article 8 National guides to good practice shall be developed by food business sectors in consultation with competent authorities and consumer groups. Guides should have regard to relevant codes of practice of the Codex Alimentarius.

Article 9 relates to the production of Community guides to good hygiene practice or the application of HACCP principles.

Annex II (General hygiene requirements)
Chapter I (General requirements for food premises)
(1) Food premises are to be kept clean and maintained in good repair and condition.
(2) The design, layout, construction, siting and size of food premises are to:
 (a) permit adequate maintenance cleaning and/or disinfection, minimize airborne **contamination**, and provide adequate working space to allow for the hygienic performance of all operations;
 (b) be such as to protect against the accumulation of dirt, contact with toxic materials, the shedding of particles into food and the formation of condensation or undesirable mould on surfaces;
 (c) permit good food hygiene practices, including protection against contamination and, in particular, pest control; and
 (d) where necessary provide suitable temperature controlled handling and storage conditions of sufficient capacity for maintaining foodstuffs at appropriate temperatures to be monitored and, where necessary, recorded.
(3) An adequate number of flush lavatories are to be available. Lavatories are not to open directly into rooms in which food is handled.
(4) An adequate number of washbasins are to be available, suitably located and designated for cleaning hands. Washbasins are to be provided with hot and cold running water, materials for cleaning hands and for hygienic drying. Where necessary, the facilities for washing food are to be separate from the handwashing facility.
(5) There is to be suitable and sufficient means of natural or mechanical ventilation. Mechanical air flow from a contaminated area to a clean area is to be avoided. Ventilation systems are to be so constructed as to enable filters and other parts requiring cleaning or replacement to be readily accessible.
(6) Sanitary conveniences are to have adequate natural or mechanical ventilation.

(7) Food premises are to have adequate natural and/or artificial lighting.
(8) Drainage facilities are to be adequate for the purpose intended. They are to be designed and constructed to avoid the risk of **contamination**.
(9) Where necessary, adequate changing facilities for personnel are to be provided.
(10) Cleaning agents and disinfectants are not to be stored in areas where food is handled.

Chapter II (Specific requirements in rooms where foodstuffs are prepared, treated or processed)
   (1) The design and layout of the rooms are to permit good food hygiene practices, including protection against contamination between and during operations. In particular:
      (a) floor surfaces are to be maintained in a sound condition and be easy to clean and, where necessary, to disinfect. This will require the use of impervious, washable and non-toxic materials, unless the food business operators can satisfy the competent authority that other materials used are appropriate. Where appropriate, floors are to allow adequate surface drainage;
      (b) wall surfaces are to maintained in a sound condition and be easy to clean and, where necessary to, disinfect. This will require the use of impervious, washable and non-toxic materials and require a smooth surface up to a height appropriate for the operations, unless food business operators can satisfy the competent authority that other materials used are appropriate;
      (c) ceilings and overhead fixtures are to be designed, constructed and finished so as to prevent the accumulation of dirt and to reduce condensation, the growth of undesirable moulds and the shedding of particles;
      (d) windows and other openings are to be constructed to prevent the accumulation of dirt. Those which can be opened to the outside environment are, where necessary, to be fitted with insect-proof screens which can be easily removed for cleaning. Where open windows would result in contamination, windows are to remain closed and fixed during production;
      (e) doors are to be easy to clean and, where necessary, to disinfect. This will require the use of smooth and non-absorbent surfaces, unless the food business operators can satisfy the competent authority that other materials used are appropriate; and
      (f) surfaces (including surfaces of equipment) in areas where foods are handled and in particular those in contact with food are to be maintained in sound condition and be easy to clean and, where necessary, to disinfect. This will require the use of smooth, washable, corrosion-resistant and non-toxic materials, unless food business operators can satisfy the competent authority that other materials used are appropriate.
   (2) Adequate facilities are to be provided, where necessary, for the cleaning, disinfecting and storage of working utensils and equipment. These facilities are to be constructed of corrosion-resistant materials, be easy to clean and have an adequate supply of hot and cold water.
   (3) Adequate provision is to be made, where necessary, for washing food. Every sink or other such facility provided for the washing of food is to have an adequate supply of hot and/or cold potable water, and be kept clean, and where necessary, disinfected.

Chapter III (Requirements for market stalls or mobile sales vehicles, premises used primarily as a private dwelling house, premises used occasionally for catering purposes and vending machines)

Premises and vending machines are, so far as is reasonably practicable, to be sited, designed, constructed, kept clean and maintained in good repair and condition as to avoid the risk of **contamination**, in particular by animals and pests.

Surfaces in contact with food are to be in a sound condition and be easy to clean and, where necessary, to disinfect.

Where necessary, there are to be facilities to maintain adequate personal hygiene and for the cleaning and, where necessary, disinfecting of working utensils and equipment.

Adequate facilities and/or arrangements for maintaining and monitoring suitable food temperature conditions are to be available.

Foodstuffs are to be placed so as to avoid the risk of contamination.

Chapter IV (Transport)

Conveyances and/or containers used for transporting foodstuffs are to be kept clean and maintained in good repair and condition to protect foodstuffs from contamination, and are, where necessary, to be designed and constructed to permit adequate cleaning and/or disinfection.

Chapter V (Equipment requirements)

All articles, fittings and equipment with which food comes into contact are to:

- (a) be effectively cleaned and, where necessary, disinfected. Cleaning and disinfection are to take place at a frequency sufficient to avoid any risk of contamination;
- (b & c) be so constructed, be of such materials and be kept in such good order, repair and condition as to minimize any risk of contamination and to enable them to be kept clean and, where necessary, to be disinfected; and
- (d) be installed in such a manner as to allow adequate cleaning of the equipment and the surrounding area. Where necessary, equipment is to be fitted with any appropriate control device to guarantee fulfilment of this regulation's objective.

Chapter VI (Food waste)

Food waste, non-edible by-products and other refuse are to be removed from food rooms as quickly as possible, so as to avoid their accumulation.

Waste is to be deposited in closable containers (unless the competent authority agree otherwise). Containers are to be of an appropriate construction, kept in sound condition, be easy to clean and, where necessary, to disinfect.

Adequate provision is to be made for the storage of waste. Refuse stores are to be kept clean and free of animals and pests.

All waste is to be eliminated in a hygienic and environmentally friendly way and is not to constitute a direct or indirect source of contamination.

Chapter VII (Water supply)

An adequate supply of potable water must be provided and used to ensure foodstuffs are not contaminated. Ice is to be made from potable water, if it could contaminate the food. It is to be made, handled and stored to protect it from **contamination**.

Where heat treatment is applied to foodstuffs in hermetically sealed containers it is to be ensured that water used to cool the containers after heat treatment is not a source of contamination for the foodstuff.

Chapter VIII (Personal hygiene)

Persons working in food handling areas are to maintain a high degree of personal cleanliness and wear suitable, clean and, where necessary, protective clothing.

No person suffering from, or being a carrier of a foodborne disease, or afflicted with infected wounds, skin infections, sores or diarrhoea is to be permitted to handle food or enter any food handling area in any capacity if there is a likelihood of direct or indirect contamination. Any person so affected and employed in a food business and who is likely to come into contact with food is to report immediately the illness or symptoms, and if possible their causes, to the food business operator.

Chapter IX (Provisions applicable to foodstuffs)

Contaminated or decomposed raw materials must be rejected unless normal sorting and hygienic preparation will ensure their fitness for human consumption.

All food must be protected against contamination during handling, storage, packaging, display and distribution.

Effective pest control procedures must be implemented.

Food likely to support the reproduction of pathogenic micro-organisms or the formation of toxins is not to be kept at temperatures that might result in a risk to health. Food businesses manufacturing, handling and wrapping processed foodstuffs are to have suitable rooms, large enough to separate the storage of raw materials from processed material, and sufficient separate refrigerated storage.

Where foodstuffs are to be held or served at chilled temperatures they are to be cooled as quickly as possible following the heat-processing stage, or final preparation stage if no heat process is applied, to a temperature which does not result in a risk to health.

The thawing of foodstuffs is to be undertaken in such a way as to minimize the risk of growth of pathogenic micro-organisms or the formation of toxins in the foods.

Hazardous and/or inedible substances, including animal feed, are to be adequately labelled and stored in separate and secure containers.

Chapter X (Provisions applicable to the wrapping and packaging of foodstuffs)

Materials used for wrapping are not to be a source of contamination. They must be stored in such a manner that they are not exposed to a risk of contamination.

Wrapping and packaging operations are to avoid contamination of the products. The integrity and cleanliness of cans and glass jars are to be assured.

Reusable wrapping and packaging material are to be easy to clean and, where necessary, to disinfect.

Chapter XI (Heat treatment)

Heat treatment of food in hermetically sealed containers should conform to an internationally recognized standard (for example, pasteurization, ultra high temperature or sterilization). Every part of the product should achieve a given temperature for a given period of time.

Chapter XII (Training)
Food business operators are to ensure:
(1) that food handlers are supervised and instructed and/or trained in food hygiene matters commensurate with their work activity;
(2) that those responsible for the development and maintenance of the HACCP system or for the operation of relevant guides have received adequate training in the application of the HACCP principles; and
(3) compliance with any requirements of national law concerning training programmes for persons working in certain food sectors.

The Food Hygiene (England) (Scotland) (Wales) (NI) Regulations 2006
Part 1   Preliminary
Reg 1. Each country has its own specific regulations. However, they are almost identical and are all intended to achieve the same objective.

Reg 2. "Authorized officer" means any person authorized in writing by an enforcement authority to act in matters arising under the Hygiene Regulations (examples include environmental health practitioner/officer, technical officer etc.)
"Enforcement authority" means either the Food Standards Agency or the food authority in whose area the food business operator carries out his operations.
"Hygiene Regulations" means these Regulations and Regulation (EC) 852/2004 on the hygiene of foodstuffs, Regulation (EC) 853/2004 for food of animal origin and Regulation (EC) 854/2004 for controls on products of animal origin intended for human consumption as amended by Regulation EC 882/2004 and as read with Directive 2004/41.
"Premises" includes any establishment, any place, vehicle, stall or moveable structure and any ship or aircraft.

Reg 3. Food commonly used for human consumption found on food premises shall be presumed, until the contrary is proved, to be intended for human consumption.

Reg 5. The Food Standards Agency shall enforce the Hygiene Regulations in relation to primary production, slaughterhouses, game handling establishments, fresh meat, cutting plants, egg operations and matters relating to raw cows' milk. The Food Authority shall enforce the Hygiene Regulations in all other food premises.

Part 2   Main provisions
Reg 6. An authorized officer can serve a **hygiene improvement notice** on the food business operator of a food business, for failing to comply with Hygiene Regulations. The **notice** must state the grounds for non-compliance, specify the contraventions and measures necessary to secure compliance, and the time (not less than 14 days) allowed. Failure to comply is an offence.

Reg 7. If a food business operator is convicted of an offence under the above regulations **and** the court is satisfied that the business, any process/treatment, the construction or condition of any premises or the use or condition of any equipment involves a risk of injury to health, they **shall** impose a **hygiene prohibition order**. A **hygiene prohibition order** can apply to the use of a process/treatment, the premises (or part thereof) or any equipment. A copy of the **hygiene prohibition order** must be conspicuously fixed on the premises and contravention of the **order** is an offence. The hygiene prohibition order ceases to have effect when the enforcement authority issues a certificate, which states that there is no longer a health risk. On application by the food business operator, the enforcement authority must determine within 14 days whether the health risk has been removed and if so satisfied, issue the certificate within three days.

The court may also impose a prohibition on the proprietor or manager participating in the management of any food business. (Only a court can lift a **hygiene prohibition order** on a food business operator.) This prohibition applies for at least six months.

Reg 8. If an authorized officer of an enforcement authority is satisfied that there is an imminent risk of injury to health, he/she may issue a **hygiene emergency prohibition notice** requiring the immediate closure of the premises. An application for a **hygiene emergency prohibition order** must then be made to the court within three days of serving the notice, and at least one day before the date of application, the food business operator must be advised of this intention. (Saturdays, Sundays and Bank Holidays are excluded.)

The **hygiene emergency prohibition notice** and **hygiene emergency prohibition order** must be served on the food business operator and conspicuously displayed on the premises. Any contravention is an offence. A hygiene emergency prohibition notice ceases to have effect if no application for an order is made to the court. A **hygiene emergency prohibition notice/order** ceases to have effect when the enforcement authority issues a certificate, stating that there is no longer a health risk. Compensation is payable by the enforcement authority in respect of loss suffered in complying with the **notice**, unless an application is made for a **hygiene emergency prohibition order** within three days (five days in Scotland) **and** the court is satisfied that the health risk condition was fulfilled.

The Food Law Code of Practice provides guidance on the use of hygiene improvement notices and prohibition procedures.

Reg 9. If an authorized officer of an enforcement authority is satisfied that the Hygiene Regulations are being breached in respect of an establishment subject to approval under Regulation 853/2004 (mainly slaughterhouses and cutting premises), he may serve a remedial action notice. This notice may prohibit the use of any equipment or part of the establishment, impose conditions on or prohibit a process or require the rate of operation to be reduced or stopped.

An authorized officer may also serve a detention notice requiring the detention of any animal or food for the purposes of examination or sampling.

Failure to comply with either of the above notices is an offence.

Reg 10. Enables proceedings to be taken against another person when the offence was due to his/her act or default.

Reg 11. It is a defence for the accused to prove that he took all reasonable precautions and exercised all due diligence to avoid commission of the offence by himself or by a person under his control.

If the defence relates to the offence being due to another person, at least 7 clear days before the hearing or within one month of his first court appearance, the accused must notify the prosecutor in writing identifying that other person.

Part 3 Administration and enforcement

Reg 12. Empowers an authorized officer to purchase or take samples of food, food sources,
& 13. contact materials or any article or substance required as evidence. Such samples should, if considered necessary, be submitted for analysis by a public analyst or

examination by a food examiner, who shall provide a certificate specifying the result of the analysis or examination.

Reg 14. Empowers an authorized officer, on production of an authenticated document showing his/her authority, to enter any premises within his/her area at all reasonable hours to carry out his/her duties under the hygiene regulations. In the case of a private dwelling house, entry cannot be demanded unless 24 hours' notice has been given to the occupier. An authorized officer is also empowered to enter business premises outside his/her area for the purpose of ascertaining whether or not there are any contraventions of the hygiene regulations or regulations/orders made thereunder. Warrants may be issued by a justice of the peace authorizing entry by force if necessary, when entry is refused.

Authorized officers can inspect, seize and detain records, including computer records, required as evidence. Improper disclosure of information so obtained is an offence.

Reg 15. It is an offence to obstruct persons executing the provisions of the hygiene regulations, including the failure to assist or provide information (unless it incriminates them) or the furnishing of false information.

Reg 16. Prosecutions must commence before the expiry of three years from the commission of the offence or one year from its discovery by the prosecutor, whichever is the earlier.

Reg 17. A person guilty of an offence under these Regulations shall be liable on:
- (a) summary conviction to a fine not exceeding the statutory maximum (level 5*); or
- (b) conviction on indictment to imprisonment for up to two years and/or an unlimited fine.

The penalty for obstruction on summary conviction shall be a fine not exceeding Level 5* and/or up to three months' imprisonment.

Reg 18. Where an offence by a body corporate has been committed with the consent or connivance or due to the neglect on the part of any director, manager, secretary or similar officer, he shall also be liable to prosecution.

Reg 20. & 21. Enable aggrieved persons to appeal to the magistrates' court or the crown court.

Reg 22. Allows for appeals against hygiene improvement notices and remedial action notices.

Reg 23. Section 9 of the Food Safety Act 1990 (inspection and seizure of suspected food) applies to these regulations as regards an authorized officer of an enforcement authority.

Part 4 Miscellaneous and supplementary provisions

Reg 24. Empowers the Secretary of State to issue codes of recommended practice for the guidance of food authorities as regards the execution and enforcement of the Hygiene Regulations. The Food Standards Agency may issue a food authority a direction requiring their compliance with any code. Food authorities must have regard to any such code.

Reg 25. Officers of enforcement authorities are not personally liable for acts involved with the execution of the Hygiene Regulations provided they honestly believed such action was necessary.

*The Criminal Justice Act, 1991 provided fines for summary offences in magistrates' courts to be placed on a scale of levels 1 to 5, unless otherwise stipulated in a particular act.
Level 1 – £200   Level 2 – £500   Level 3 – £1,000   Level 4 – £2,500   Level 5 – £5,000

Reg 27. When an authorized officer certifies that food has not been produced, processed or distributed in accordance with the Hygiene Regulations, it shall be treated as failing to comply with food safety requirements in accord with Section 9 of the Food Safety Act. All food within a batch shall be treated as certified until it is proved that the rest of the batch was dealt with in compliance with the Hygiene Regulations.

Reg 28. Relates to the service of documents.

Schedule — Temperature control requirements

(1) This Schedule does not apply to slaughterhouses and cutting premises or any food business operations on ships or aircraft.

Chill holding requirements

(2) No person shall keep any food likely to support the growth of pathogenic micro-organisms or the formation of toxins in a food premises at a temperature above 8°C. (Excludes mail order to an ultimate consumer as long as there is no risk to health).

General exemptions from the chill holding requirements

(3) Paragraph 2 shall not apply to:
   (a) food which:
   - has been cooked or reheated;
   - is for service or on display for sale; and
   - needs to be kept at or above 63°C in order to control the growth of pathogenic micro-organisms or the formation of toxins;

   (b) food that, for the duration of its **shelf life**, may be kept at **ambient temperatures** with no risk to health;

   (c) food that is being or has been subjected to a process such as **dehydration** or canning, intended to prevent the growth of pathogenic micro-organisms at ambient temperatures, but this paragraph shall not apply in circumstances where:
   - after or by virtue of that process, the food was contained in a hermetically sealed container; and
   - that container has been opened;

   (d) food that must be ripened or matured at ambient temperatures, but this paragraph shall cease to apply once the process of ripening or maturation is completed; and

   (e) raw food intended for further processing (which includes cooking) before human consumption, but only if that processing, if undertaken correctly, will render that food fit for human consumption.

(4) If food is kept above 8°C, it is a defence for the person charged to prove that:
   (a) a food business responsible for manufacturing, preparing or processing the food has recommended that it is kept:
   - at or below a specified temperature above 8°C; and
   - for a period not exceeding a specified shelf life;

   (b) the specified temperature is on a label on the packaging or is a written instruction;
   (c) the food was not kept above the specified temperature; and
   (d) the specified shelf life had not been exceeded.

HYGIENE for MANAGEMENT *Food safety legislation*

A food business shall not recommend food is kept above 8°C for a specified shelf life, unless there has been a well-founded scientific assessment.

Chill holding tolerance periods
(5) It is a defence for a person charged under paragraph 2 to prove that the food:
   (a) was for service or on display for sale;
   (b) had not previously been kept for service or on display for sale at a temperature above 8°C; and
   (c) had been kept for service or on display for sale for less than four hours.

It is also a defence to prove that the food:
   (a) was being transferred to or from a food business vehicle to premises at which the food was going to be kept at or below 8°C;
   (b) was kept above 8°C for an unavoidable reason such as:
      ♦ to accommodate the practicalities of handling during and after processing or preparation;
      ♦ the defrosting of equipment; or
      ♦ the temporary breakdown of equipment
and was kept at a temperature above 8°C for a limited period consistent with food safety.

Hot holding requirements
(6) No person shall in the course of a food business keep any food that:
   (a) has been cooked or reheated;
   (b) is for service or on display for sale; and
   (c) needs to be kept at or above 63°C in order to control the growth of pathogenic micro-organisms or the formation of toxins
at or in food premises at a temperature below 63°C.

Hot holding defences
(7) It is a defence for any person charged under paragraph 6 to prove that:
   (a) a well-founded scientific assessment indicates there is no risk to health if food held for service or on display is kept at a specified temperature below 63°C for a period not exceeding a specified period of time and this time/temperature had been adhered to.

It is also a defence to prove that the food:
   (a) had been kept for service or display for less than two hours; and
   (b) had not previously been kept for service or on display by that person.

Schedule 6 Restrictions on the sale of raw milk intended for direct human consumption.

Raw milk can only be sold direct to the final consumer by the occupier of a premises at which milk producing cows are kept or a distributor. The raw milk must meet the following standards:
   ♦ Plate count at 30°C (cfu* per ml)   <20,000
   ♦ Coliforms (cfu per ml)   <100

* cfu - colony forming units

The Food Hygiene (Scotland) Regulations 2006
Schedule 4 Temperature Control Requirements
Chill and hot holding offences
Food should be kept in a refrigerator, a cool ventilated place or above 63°C unless:
- it is undergoing preparation for sale;
- it is exposed for sale or has been sold;
- it is being cooled under hygienic conditions as quickly as possible to a safe temperature, immediately following cooking or the final processing stage;
- it is reasonable to store it at a different temperature so as to be conveniently available for sale on the premises to consumers; or
- for the duration of its shelf life, it may be kept at ambient temperatures with no risk to health.

Reheating of food
Food which has been heated and is thereafter reheated before being served for immediate consumption or exposed for sale, shall be raised to a temperature of not less than 82°C. (It is a defence to prove that reheated food could not be raised to 82°C without a deterioration of its qualities.)

Schedule 6 Restrictions on the placing on the market of raw milk or raw cream
It is an offence for any person to place on the market raw milk or raw cream intended for direct human consumption.

It is a defence to prove that the raw milk or raw cream was intended for export, for example, to England, Wales, Northern Ireland or other member state or third country as long as it complies with Regulation 253/2004 and any relevant national rules.

Offences and penalties
Any person guilty of an offence against these regulations shall be liable:
- on summary conviction, to a fine not exceeding the statutory maximum; or
- on conviction on indictment, to a fine or imprisonment for a term not exceeding two years, or both.

EU Regulation No 2073/2005 on Microbiological Criteria for Foodstuffs
Food business operators shall ensure that foodstuffs comply with the microbiological criteria set out in Annex 1 of the Regulation. This will require a sampling and testing plan proportionate to the nature and size of their business as part of their risk-based food safety management system.

"Microbiological criterion" means a criterion defining the acceptability of a product, a batch of foodstuffs or a process, based on the absence, presence or number of micro-organisms, and/or on the quantity of their toxins/metabolites, per unit(s) of mass, volume, area or batch.

"Ready-to-eat food" means food intended by the producer or the manufacturer for direct human consumption without the need for cooking or other processing effective to eliminate or reduce to an acceptable level micro-organisms of concern.

Food Safety Act, 1990

As a result of EU Regulations, especially Regulation EC No 852/2004 and the food hygiene regulations 2006, many of the food safety provisions included within the Food Safety Act, 1990 are no longer applicable to the majority of food businesses. This Act is now primarily concerned with food standards.

The Act consists of four parts and five schedules.

Part I Preliminary

Section 1. Extends the definition of "food" to include water that is bottled for sale or used as an ingredient (water supplied to the premises is excluded).

Section 2. Extends the meaning of sale to include food that is offered as a prize or reward or given away in connection with any entertainment for the public.

Section 3. Food, or ingredients, commonly used for human consumption are presumed, until the contrary is proved, to be intended for sale for human consumption.

Section 5. "Food authorities" include the council of London and metropolitan boroughs, districts and counties.

"Authorized officer" means any person authorized in writing by a food authority to act in matters arising under the Food Safety Act, 1990.

Section 6. Requires every food authority to enforce within their area those parts of the act for which they are responsible.*

Part II Main provisions

Section 7. It is an offence to treat food so as to render it injurious to health, with the intent that the food will be sold in that state. Regard shall be had to the cumulative effect of foods consumed over a long period.

Section 9. An authorized officer of a food authority may seize or detain food (for up to 21 days), which fails to comply with **food safety requirements** or that is likely to cause food poisoning or a foodborne disease. Food that is seized has to be dealt with by a justice of the peace. Any person liable to be prosecuted in respect of such food is entitled to make representations to the justice of the peace. If the food is not condemned, or detained food is cleared, compensation can be claimed. Any expenses incurred in the destruction of condemned food must be paid by the owner of the food.

If such food is part of a batch, then the whole batch shall, until the contrary is proved, be presumed to fail to meet the **food safety requirements**.

The Food Law Code of Practice provides guidance on the seizure and detention of food.

---

*The Food Safety (Enforcement Authority) (England and Wales) Order, 1990 (SI 1990 No. 2462) sets out the division of responsibility for enforcing the act.

Section 13. Empowers the minister to issue an **emergency control order**, prohibiting the carrying out of commercial operations in respect of food, food sources or contact materials that involve an imminent risk of injury to health. Failure to comply is an offence and expenses for "work in default" can be recovered.

Section 14. It is an offence to sell, to the prejudice of the purchaser, any food that is not of the **nature** (different kind or variety) or **substance** (not containing proper ingredients) or **quality** (inferior, for example, stale bread) demanded by the purchaser.

Section 15. It is an offence to sell, display or have in possession for the purpose of sale, food that is falsely described or labelled, which is misleading as to the nature or substance or quality.

Sections 16, 18 & 26. Empower the ministers to make regulations, including those relating to composition, fitness, processing, hygiene, labelling and for securing food safety and the training of **food handlers** (Schedule 1).

Section 17. Empowers the ministers to make regulations to secure compliance with Community obligations.

Section 19. Empowers the Ministers to make regulations requiring the registration of food premises, and for the licensing of premises to secure compliance with food safety requirements or in the interest of public health or protecting the interests of consumers.

Section 20. Enables proceedings to be taken against another person when the offence was due to his/her act or default.

Section 21. It is a defence for a person to prove that he/she took all reasonable precautions and exercised all **due diligence** to avoid the commission of the offence, by himself/herself or by a person under his/her control.
   In the case of persons not involved with the preparation or importation, for example, a retailer, charged with an offence under Sections 8, 14 or 15 it is a defence to prove:
   ♦ the offence was due to another person;
   ♦ that he/she carried out all reasonable checks or it was reasonable to rely on checks carried out by his/her supplier; and
   ♦ he/she did not know or suspect his/her act or omission would amount to an offence.

Section 23. Enables a food authority to provide training courses on food hygiene within or outside their area.

Section 25. Empowers ministers to make orders to facilitate the exercise of functions, i.e. to provide for the detailed enforcement of the act. The section also specifies the circumstances when it will not be an offence for a person to disclose information obtained by means of an order.

Part III Administration and enforcement
Section 27. Requires food authorities to appoint a public analyst.

Sections 29 & 30. Empowers an authorized officer to purchase or take samples of food, food sources, contact materials or any article or substance required as evidence. Such samples should, if considered necessary, be submitted for analysis by a public analyst or examination by a food examiner, who shall provide a certificate specifying the result of the analysis or examination.

Section 31. Empowers ministers to make regulations relating to sampling and analysis. (The Food Safety (Sampling and Qualifications) Regulations, 1990 (SI. No. 2463) lay down procedures for enforcement officers when taking samples for analysis or **microbiological** examination.)

Section 32. Empowers an authorized officer, on production of an authenticated document showing his/her authority, to enter any premises within his/her area at all reasonable hours to carry out his/her duties under the act. In the case of a private dwelling house, entry cannot be demanded unless 24 hours' notice has been given to the occupier. An enforcement officer is also empowered to enter any business premises outside his/her area for the purpose of ascertaining whether or not there are any contraventions of the act or regulations/orders made thereunder. Warrants may be issued by a justice of the peace authorizing entry, by force if necessary, when entry is refused.

Authorized officers can inspect, seize and detain records, including computer records, required as evidence. Improper disclosure of information so obtained is an offence.

Section 33. It is an offence to obstruct persons executing the provisions of the act, including the failure to assist or provide information (unless it incriminates them) or the furnishing of false information.

Section 34. Prosecutions must commence before the expiry of three years from the commission of the offence or one year from its discovery by the prosecutor, whichever is the earlier.

Section 35. The penalty for obstruction on summary conviction shall be a fine not exceeding level 5* and/or up to three months' imprisonment.

The penalty for the remaining offences is:
- on conviction on indictment to an unlimited fine and/or up to two years' imprisonment; or
- on summary conviction to a fine not exceeding the relevant amount and/or imprisonment for up to six months. (In the case of Sections 7, 8 or 14 the relevant amount is £20,000; the amount for the other sections is level 5.)

Part IV Miscellaneous and supplemental

Section 40. Empowers the ministers to issue codes of recommended practice for the guidance of food authorities, as regards the execution and enforcement of the act and regulations/orders made thereunder.

Section 41. Requires food authorities to provide reports, returns and information to the minister.

Section 42. Enables the minister to empower another food authority, or one of his/her officers, to discharge the duty of a food authority that has failed to discharge that duty and the failure affects the general interests of consumers of food. The authority in default will be responsible for the payment of reasonable expenses.

Section 49. Relates to the form and authentication of documents.

Section 50. Relates to the service of documents.

Section 54. Deals with the application of the act to the Crown, i.e. the removal of Crown immunity.

Statutory codes of practice

Section 40 of the Food Safety Act 1990 and Regulation 24 of the Food Hygiene (England) (Scotland) (Wales) (NI) Regulations 2006 permit ministers to issue codes of practice for enforcing authorities regarding the execution and enforcement of food law.

Currently there is one Food Law Code of Practice to which enforcement authorities must have regard when discharging their duties. There is also additional advice provided in a Food Law Practice Guidance to which enforcement officers may wish to adhere.

The Food Safety (Sampling and Qualifications) Regulations 1990 (SI 1990 No. 2463)

These Regulations set out the procedures to be followed by enforcement officers, when taking samples for analysis or microbiological examination and also set out qualification requirements for public analysts and food examiners.

Regulation (EC) No. 853/2004 laying down specific hygiene rules for food of animal origin (Effective from 1 January 2006)

The regulation repeals many of the vertical directives and regulations made thereunder relating to food of animal origin. It applies to both processed and unprocessed products of animal origin.

The requirements do not usually apply to food containing both products of plant origin and processed products of animal origin.

"Processed products" means foodstuffs resulting from the processing of unprocessed products. These products may contain ingredients that are necessary for their manufacture or to give them specific characteristics.

"Unprocessed products" means foodstuffs that have not undergone processing, and includes products that have been divided, parted, severed, sliced, boned, minced, skinned, ground, cut, cleaned, trimmed, husked, milled, chilled, frozen, deep-frozen or thawed.

"Mechanically separated meat" means the product obtained by removing meat from flesh-bearing bones after boning or from poultry carcases, using mechanical means resulting in the loss or modification of the muscle fibre structure.

The regulation supplements Regulation (EC) No. 852/2004 and applies to:

slaughterhouses, for both red meat and poultry, cutting plants, game handling establishments, premises producing minced meat, meat preparations or mechanically separated meat, despatch centres for live bivalve molluscs, vessels used to harvest fishery products, establishments handling fishery products, milk production holdings, establishments for the manufacture of eggs and egg products.

It is concerned with:
- the hygiene of the above premises;
- the transport of live animals to the slaughterhouse (including domestic ungulates (domestic cattle and bison, pigs, sheep, goats and horses/mules), poultry and lagomorphs (rabbits and hares);
- the hygiene requirements for the production and harvesting of live bivalve molluscs, and fishery products including heat treatment, relaying of molluscs, health standards, wrapping and packaging, labelling, storage and transport;
- the health requirements for raw milk production;
- hygiene during milking, collection and transport;
- the microbiological criteria for raw milk;
- the hygiene of dairy products, including temperature requirements, wrapping and packaging and labelling;
- the hygiene requirements for the manufacture of egg products, including analytical specifications and labelling;
- the hygiene of businesses preparing frogs legs or snails;
- the hygiene requirements for the preparation of rendered animal fat and greaves;
- the hygiene requirements for businesses treating stomachs, bladders and intestines;
- the hygiene requirements for the manufacture of gelatin and collagen;
- slaughter hygiene, including cutting and boning;
- emergency slaughter outside the slaughterhouse and slaughter on the farm; the hygiene requirements during storage and transport;
- the hygiene requirements for cutting premises;
- the hygiene requirements for hunting and handling wild game; and
- the hygiene requirements of businesses producing minced meat, meat preparations, mechanically separated meat and meat products.

The regulation does not usually apply to retail businesses. However, it may apply to operations supplying food of animal origin to another establishment (excluding businesses only involved in storage or transport). In this latter case specific temperature requirements outlined in Annex III will apply, for example offal<3ºC and other meat<7ºC.

The regulation does not apply to retail premises supplying other retailers if the supply is marginal, localized or restricted.

Primary production for private domestic use and the direct supply of small quantities of product to the final consumer are also excluded. National rules are required to ensure the hygiene and welfare of such operations.

Food business operators shall:
- be registered with or approved by the competent authority;
- health mark products of animal origin placed on the market;
- ensure certificates or other documents accompany consignments of products of animal origin.

Regulation (EC) No. 854/2004 laying down rules for official controls on products of animal origin

This regulation applies to the activities and persons within the scope of Regulation 853/2004. It includes requirements for:

- the approval of establishments (must comply with Regulations 852/2004 and 853/2004);
- the auditing and inspection of businesses to ensure compliance with the regulations;
- the inspection of red meat animals and poultry (ante-mortem and post-mortem) including animal welfare. Inspection includes microbiological criteria;
- live bivalve molluscs;
- fishery products;
- raw milk and dairy products;
- action to be taken in the event of non-compliance;
- tasks of the official veterinarian and official auxiliaries;
- lagomorphs (rabbits and hares); and
- farmed and wild game.

Regulation (EC) No. 882/2004 on official controls to ensure compliance with food law and animal health

The regulation lays down general rules for the performance of official controls which should be carried out regularly, on a risk basis and with appropriate frequency.

Official controls e.g.. audits, inspections, surveillance, verification, sampling and analysis should take account of:
- identified risks associated with food or food business;
- the food business operator's past record of compliance;
- the reliability of any own checks carried out; and
- any information that might indicate non-compliance

Official controls, except audits, should be without prior warning
The competent authorities shall ensure that:
- official controls on food are effective;
- staff have no conflict of interest;
- they have sufficient qualified and experienced staff;
- they have access to an adequate laboratory;
- they have appropriate facilities and equipment;
- they have the legal powers to carry out official controls;
- they have contingency plans in the event of an emergency;
- food business operators assist staff to accomplish their tasks;
- official controls are impartial, consistent and transparent; and
- staff receiving appropriate training.

Requirements are also included for:
- inspection methods and techniques;
- sampling and analysis;
- third country imports;
- fees or charges;
- registration/approval of food business establishments;
- community and national reference laboratories;
- multi-annual national control plans and reports;
- community controls, including audits of member states; and
- enforcement measures for non-compliance.

The Official Feed and Food Controls (England) Regulations 2006

These regulations enable the competent authorities to meet their obligations regarding Regulation (EC) 882/2004 principally with respect to the monitoring and reporting of enforcement activity and in relation to the import of food of non-animal origin from outside the community.

The competent authorities include the Food Standards Agency, DEFRA (Department of the Environment, Food and Rural Affairs), food authorities (local authorities and port health authorities) and HM Customs and Excise.

The Imported Food Regulations 1997 are repealed.

The Quick-frozen Foodstuffs Regulations, 1990 (SI No. 2615) as amended 1994

These regulations apply to "quick-frozen foodstuff", i.e. food that has undergone "quick-freezing", whereby the zone of maximum crystallization is crossed as rapidly as possible and is labelled for the purpose of sale to indicate it has been subjected to quick-freezing.

Quick-frozen foodstuff intended for supply to the ultimate consumer, must be suitably pre-packaged so as to protect it from microbial and other contamination and dehydration and must remain so packaged until sale. Such food shall, if intended for supply without further processing, be labelled with:
- an indication of the date of minimum durability;
- an indication of maximum advisable storage time;
- an indication of the temperature and the equipment in which, it is advisable to store it;
- a batch reference; and
- a clear message not to refreeze after defrosting.

Quick-frozen food must, subject to permitted exceptions, be maintained at or below −18ºC. (Exceptions relate to local distribution and storage in retail display cabinets consistent with good storage practice.) Air temperature monitoring or recording instrumentation must be fitted by businesses involved with the storage, distribution and sale of quick-frozen foodstuffs. Records must be retained for at least one year.

The Diseases of Animals (Waste Food) Order, 1973 (SI No. 1936)

All waste food, including meat or animal products, must be processed before being used for feeding to animals. It is an offence for any caterer to allow the removal of waste food unless the person holds a current licence issued by the Department of the Environment, Food and Rural Affairs (DEFRA). Conditions regarding the type and cleaning of vehicles used to convey waste are laid down. The order also applies to Scotland.

Office of Fair Trading

The Director General of Fair Trading is empowered, under Part III of the Fair Trading Act, 1973, to seek written assurances from a person carrying on a business, which has persistently operated to the detriment of the interests of consumers. Assurances are usually sought from persons who ignore repeated complaints, warnings and prosecutions.

Where a person refuses to give, or breaks, an assurance, the Director General can apply to a county court for a court order requiring the person to operate within the law, for example, by giving the court an undertaking to protect food from risk of contamination. If the person fails to observe the order, the court can impose very heavy penalties for contempt of court.

The Police and Criminal Evidence Act, 1984 (PACE)

Section 67 of this act requires persons charged with a duty of investigating offences to have regard to codes of practice made under Sections 60 and 66. The act, therefore, applies to environmental health practitioners/officers investigating contraventions of food legislation. The first code of practice came into force on 1 January 1986 and has superseded the judges' rules in relation to questioning of suspects.

Persons suspected of an offence must be cautioned prior to asking questions intended to obtain evidence for use in court. The Criminal Justice and Public Order Act, 1994 requires the caution to be in the following terms: "You do not have to say anything. But it may harm your defence if you do not mention, when questioned, something which you later rely on in court. Anything you do say may be given in evidence." (Minor deviations are allowed.) Persons must also be advised that they are not under arrest and may leave the premises but if they stay they may wish to seek legal advice.

In most cases involving food/hygiene offences, persons suspected of the offence will be invited to attend an interview at the council's offices and advised to discuss the matter with their solicitor, who may wish to be present at the interview. A signed record of the time and content of the caution, together with relevant information regarding the interview, including all those present, must be retained. Tape recorded interviews are commonly held.

Code B of PACE requires enforcement officers to go through a set procedure for obtaining warrants and conducting searches (and seizures), when there are grounds for suspicion that an offence has been committed. Provisions are available for enforcement officers to issue "Search with Consent" notices, which advise the person whose premises are being searched of his/her rights. However, these are only required to be issued in circumstances where the enforcement officer is acting outside of his/her statutory powers. Particular procedures are laid down for dealing with juveniles, the mentally ill or handicapped, and persons requiring an interpreter, and written statements.

SPECIFIC LEGISLATION RELATING TO FOOD POISONING AND FOODBORNE ILLNESS

Public Health (Control of Disease) Act, 1984
Section 11. Medical practitioners must notify cases, or suspect cases, of food poisoning to the proper officer of the local authority immediately. This requirement also applies to medical practitioners in hospitals. Telephoned or faxed notifications are preferred.

Section 18. The occupier of any premises in which there is, or has been, someone suffering from food poisoning, must provide any information within his/her knowledge that may be required to enable the proper officer to prevent the spread, or trace the source, of food poisoning.

Section 20. The proper officer can require any person to discontinue work to prevent the spread of typhoid and paratyphoid, dysentery, scarlet fever, acute inflammation of the throat, gastroenteritis and undulant fever. The local authority must compensate persons suffering any loss by discontinuing their work.

The Public Health (Infectious Diseases) Regulations, 1988 (SI No. 1546)
Regulation 9 and Schedule 4. The local authority can, on receipt of a report from the proper officer, require that any person suffering from, or a carrier of, food poisoning, typhoid, paratyphoid and other salmonella infections, dysentery and staphylococcal infections likely to cause food poisoning, discontinue work in connection with food until the risk of causing infection is removed. Managers of food businesses must assist the local authority if it is considered necessary for a medical officer to carry out a medical examination, or bacteriological tests, of a person suspected of carrying one of the above diseases, in order to prevent the spread of infection.

National Health Service (Amendment) Act, 1986

This act removed Crown immunity from health authority premises.

Section 1 removes Crown immunity from health authorities in England and Wales and health boards in Scotland for the purposes of the Food Safety Act, 1990 and any regulation made thereunder.

Section 2 removes Crown immunity from health authorities and health boards for the purposes of the Health and Safety at Work, etc. Act, 1974 and regulations, orders and other instruments made thereunder.

The act also removes any protection afforded to members and officers of health authorities and health boards by Section 125 of the National Health Service Act, 1977 or *Section 101* of the National Health Service (Scotland) Act, 1978. In effect this means that, where appropriate, legal action can be taken against individuals for contraventions of food or health and safety legislation.

The National Health Service (Food Premises) Regulations, 1987 (SI No. 18)

These Regulations provide that health authorities in England and Wales shall be treated as both owners and occupiers for the purposes of food legislation.

The Health Protection Agency (HPA)

The HPA is a national organization for England and Wales, created on 1 April 2003 to protect people's health by minimizing risks from infectious diseases, poisons, chemicals, biological and radiation hazards. The HPA incorporates several organizations, including the Public Health Laboratory Service (PHLS), the Communicable Disease Surveillance Centre (CDSC), the Central Public Health Laboratory and NHS public health staff responsible for infectious disease control. The HPA is responsible for: advising the government on public health matters; delivering services to protect public health; providing impartial advice and information to professionals and the public; providing a rapid response to health protection emergencies; and improving knowledge of health protection through research, development, education and training.

THE LAW RELATING TO FOOD STANDARDS

Food standards requirements are those that relate to the labelling, composition, quality and fair trading of foods. Food standards legislation is usually enforced by trading standards officers except in London and in metropolitan and unitary authorities where environmental health practitioners/officers have this responsibility.

The Food Labelling Regulations, 1996 (as amended)(SI No. 1499)

These regulations principally implement Council Directive 79/112/EEC. They apply directly in England, Scotland and Wales and require all prepacked food for sale to the ultimate consumer or caterer to be labelled with:
- the name of the food;
- a list of ingredients;
- a quantitative ingredients declaration ("Quid")
- a "best-before" date or a "use-by" date;
- any special storage conditions or conditions of use;
- the name and address of the manufacturer, packer or seller; and
- where necessary, place of origin and instructions for use.

The particulars that are required by the regulations to appear on a label must be easy to understand, be legible, indelible and, when sold to the ultimate consumer, easily visible.

The name of the food may sometimes be a prescribed "legal name", for example, "King Edwards potatoes" or "Galia melon". However, where no legal name exists, a "customary name" or one that is commonly in use and would be recognized by the consumer, may be used. Examples of customary names include "Bakewell tart" and "fish fingers".

Ingredients lists should be preceded with the word "Ingredients:" and should then provide details of all of the components of the food in weight-descending order, according to their use at the time the food was prepared or at the "mixing bowl" stage.

Subject to certain exemptions, food must be date-marked to indicate its shelf life. Two types of date marking are allowed:

- "best-before" or "best-before end" followed by the date up to and including which the food should retain its **optimum** condition if properly stored (for example, it will not be stale). This type of date marking applies to most foods and provides an indication of minimum durability (shelf life);
- "use-by" followed by the date in terms of either the day and the month or the day, month and year after which time the food is likely to be unfit. This form of date marking applies to highly perishable foods (microbiologically) with a short shelf life. Correct storage of such food is essential and any storage conditions that need to be observed to ensure the stated shelf life, for example, store under refrigeration, must be indicated.

The specific requirements of labelling, depend on the shelf life of the food. The following are examples of the various labels allowed.

Expected shelf life	Date marking allowable
Highly perishable and likely to constitute an immediate danger to health.	Use-by 1st April 2009 (year can be omitted).
3 months or less.	Best-before 1st April 2009 (year can be omitted).
3 to 18 months.	Best-before 1st December 2010 or Best-before end of December 2010.
More than 18 months.	Best-before December 2011 or Best-before end of 2011.

The following foods are exempt: fresh fruit and vegetables which have not been peeled or cut into pieces; wine, alcoholic drinks (10% or more alcoholic strength); soft drinks (greater than 5 litres for catering premises); flour confectionery and bread; vinegar; salt; sugar; chewing gum and edible ices.

It is an offence to sell foods bearing an expired "use-by" date and for anyone, other than the person originally responsible for applying the date mark, to change it.

Nutrition labelling is only required for foods where a claim is made about the nutritional properties, for example, if the label declares that a food is "low fat" or "high in fibre".

The Genetically Modified and Novel Foods (Labelling)(England) Regulations, 2000 (SI No. 768)

These regulations require businesses to provide advice to consumers regarding the presence of genetically modified (GM) ingredients. In restaurants, cafes and the like, this advice may be provided verbally by trained staff, as long as a notice is displayed on the premises that informs customers how to obtain such information.

European Regulations 49/2000 and 50/2000 brought in additional requirements for the labelling of GM ingredients by manufacturers. The regulations introduce a threshold of 1% adventitious contamination of ingredients by GM food, below which there is no requirement to declare the presence of the GM food. This can only be done where the manufacturer has taken positive steps to prevent such adventitious contamination from occurring.

The Food (Lot Marking) Regulations, 1996 (SI No. 1502)

These regulations require that a unique identification or "lot" number be applied to prepacked foods. The purpose of this number, is to facilitate easy product recall in the event of a defect occurring in the food. The lot mark does not have to be understood by the consumer but if not obvious, should be prefixed with the letter "L". It is acceptable for the durability date to also act as a lot number.

Food additives are substances that are not usually consumed as a food on their own and are not normally used as an essential ingredient of a food. However, they are intentionally added to give the food certain desirable properties. Examples include colours, sweeteners and preservatives and their use in food is controlled by:

- **the Sweeteners in Food Regulations, 1995 (as amended)(SI No. 3123)**
- **the Colours in Food Regulations, 1995 (as amended)(SI No. 3124)**
- **the Miscellaneous Food Additives Regulations, 1995 (as amended)(SI No. 3187)**

The Materials and Articles in Contact with Foods Regulations, 1987 (SI No. 1523)

These regulations set out the general requirement that all food contact materials and articles should not transfer their constituents to food in quantities which could endanger human health, or make the food otherwise unacceptable to consumers.

The Plastic Materials and Articles in Contact with Foods Regulations, 1998 (SI No. 1376)

These regulations lay down specific requirements for the manufacture and use of food grade plastics.

THE LAW AFFECTING HEALTH AND SAFETY

The Health and Safety at Work, etc. Act, 1974

A major piece of legislation that incorporated a completely new approach to health, safety and welfare. It applies not only to employees but also to members of the general public and the self-employed. The act created the health and safety executive, working under the direction of the Health and Safety Commission and also applies to Scotland. The objectives of the act include:

- to secure the health, safety and welfare of persons at work;
- to protect non-employees and members of the public against risks to health and safety, arising from the activities of persons at work; and
- to control the keeping and use of dangerous substances.

The Act attempts to create an awareness of the need for high standards and encourages management to promote health and safety. Supervision and training must be given to all staff.

A further requirement is the duty placed upon employers to prepare and revise a written safety policy and to bring this document to the attention of their staff. Businesses with less than five employees are exempt. Safety policies should consist of three parts, the first part being a general statement of policy, the second the organization and thirdly the arrangements for implementing the policy and bringing it to the attention of all employees. Unions may appoint safety representatives and, at the request of the representatives, an employer must set up a safety committee.

The Control of Substances Hazardous to Health Regulations, 1999 (SI No. 437)

These regulations introduce a legal framework for the control of substances hazardous to health in all types of businesses, including factories, farms, retail and catering premises, stalls and delivery vehicles. The regulations set out essential measures that employers (and employees) have to take to ensure people are protected from the hazardous substances they may encounter.

Substances **hazardous to health** include substances labelled as dangerous (i.e. toxic, harmful, irritant or corrosive) pathogens and substantial quantities of dust.

Employers must carry out an assessment of all work that is liable to expose any employee to hazardous solids, liquids, dusts, fumes, vapours, gases or micro-organisms. The assessment involves evaluating the risks to health and then deciding on the action necessary to remove or reduce those risks. Equipment must be properly maintained and procedures observed. Employees should be informed, instructed and trained about the risks and the precautions to be taken. In some circumstances it may be necessary to monitor the exposure of employees and to carry out an appropriate form of surveillance of their health.

The Reporting of Injuries, Diseases and Dangerous Occurrences Regulations, 1995 (SI No. 1995)

These regulations require employers to notify the enforcing authority of all injuries resulting from accidents at work which cause incapacity for more than three days. In addition, certain diseases associated with specified work activities must be reported. Notification must be made on an accident report form (F2508) within ten days. Dangerous occurrences and accidents resulting in death or major injury (defined in the regulations) must be reported immediately, for example, by telephone. Written confirmation on form F2508 must follow within ten days.

The Electricity at Work Regulations, 1989 (SI No. 635)

Employers must ensure electrical equipment and systems are in good condition and have been correctly installed. Fixed wiring systems and portable equipment must be checked and records of inspection and maintenance retained. Electricians employed must be competent.

The Management of Health and Safety at Work Regulations, 1999 (SI No. 3242)

These regulations set out broad duties for employers, which include a requirement that the health and safety of employees is assessed, so that any necessary preventive and protective measures can be identified. Emergency procedures must be set up and employees must be given appropriate health and safety information and training.

The Workplace (Health, Safety and Welfare) Regulations, 1992 (SI No. 3004)

Aspects covered include maintenance of workplaces and equipment, ventilation, temperatures, lighting, cleaning, room dimensions, workstations, seating and sanitary accommodation.

The Health and Safety (Display Screen Equipment) Regulations, 1992 (SI No. 2792)

Covers the use of computers including the assessment of the workstation, eyesight tests, training and the provision of information.

The Provision and Use of Work Equipment Regulations, 1998 (SI No. 2306)

General duties are placed on employers to ensure that equipment is suitable for its use, in good repair and, where necessary, guarded. Employees must be given adequate information, instruction and training to use equipment.

The Manual Handling Operations Regulations, 1992 (SI No. 2793)

Include provisions relating to the avoidance of manual handling, assessment of risk, the reduction of risk of injury, training and information as to loads.

The Personal Protective Equipment at Work Regulations, 1992 (SI No. 2966)

These regulations require employers to provide and maintain suitable protective equipment, following risk assessment. Equipment must be replaced when necessary and employees must be provided with appropriate information, instruction and training.

THE ENFORCEMENT OF FOOD LAW IN ENGLAND AND WALES

The Food Standards Agency (FSA) (www.food.gov.uk) was brought into being on 1st April 2000 by the Food Standards Act, 1999 and has the overall responsibility for the enforcement of food law. The FSA is accountable to Parliament but acts as an independent watchdog, with the intention of putting consumers first and being accessible.

The FSA's primary aim is to protect the health of the public in relation to food by:
- reducing foodborne illness;
- helping people eat more healthily;
- promoting honest and informative labelling to help consumers;
- promoting best practice within the food industry;
- improving the enforcement of food law; and
- earning people's trust by what they do and how they do it.

Food hygiene legislation is enforced by authorized officers (environmental health practitioners/officers) employed by local authorities or port health authorities, and food standards legislation is enforced by trading standards officers and technical officers.

Some legislation is enforced by both local authorities and by officers employed by central government. In the case of meat legislation, the Meat Hygiene Service (MHS), a division of the Food Standards Agency, regulates conditions in slaughterhouses, cutting plants and cold stores licensed under fresh meat legislation. The Dairy Hygiene Inspectorate (DHI), a division of DEFRA, is responsible for the inspection of dairy farms.

Food incidents and hazards

A food incident occurs when a food authority or the FSA becomes aware that food or its labelling fails to meet food law requirements. If the incident has the potential to cause harm to the customer it becomes a food hazard. Localized food hazards which are not considered serious are dealt with by the food authority. Serious localized food hazards, for example, involving *E. coli* O157, *C. botulinum* or *Salmonella* Typhi or affecting vulnerable groups, or likely to involve large numbers or deaths, and non-localized hazards when food is distributed beyond the boundaries of the food authority, should be notified to the FSA and other relevant agencies at the earliest opportunity by the quickest available means. The Food Law Code of Practice provides details of the way in which food hazards should be assessed and how they should be dealt with.

Food alerts

A 'food alert' is a communication from the FSA to a food authority concerning a food hazard. The authority may or may not be required to rake action. The FSA also issues information on product recalls. The largest recall made to date involved the illegal carcinogenic dye Sudan 1 which was found in hundreds of food products in the UK in 2005. The Food Law Code of Practice provides food authorities with guidance for responding to food alerts.

SCOTTISH LEGISLATION

Scotland has a different legal system to England, being founded on principles originating from Roman law. In the event of any legal query, it is always best to consult someone conversant with the specifics of both the legal system and the subject matter in question.

Scottish legal terms

Licensing board

The statutory body made up of appointed councillors, which meets every three months to grant licences to premises serving alcohol.

Licensing committee

Part of a district council's committee structure, which meets every committee cycle to grant licences to premises under the Civic Government (Scotland) Act. A major part of their time is taken in processing applications for licences not connected with the food trade, for example, taxi drivers.

Procurator fiscal (sometimes abbreviated to either procurator or fiscal)

A full-time head of a department servicing each sheriff court. Officers of this department are responsible for the independent assessment of the validity of all cases that would be taken to a sheriff court or above. Members of this department are responsible for the presentation of prosecution evidence to the court. The staff who present evidence to a sheriff court are solicitors.

Sheriff court

This court is the equivalent of the English stipendiary magistrates' court. It is the usual location for all food cases to be heard.

Sheriff

A full-time paid official presiding alone over a sheriff court. They are usually QCs with considerable experience. The sheriff principal is the senior sheriff for a Scottish region.

High court

If a sheriff feels that a case merits a greater sentence than he/she is empowered to impose, he/she can refer it to this court for sentence. The court moves around the country on circuit. It is presided over by a single Scottish Law Lord and has 15 jurors.

Court of Session

The highest Scottish Court – presided over by three Scottish Law Lords.

Licensing (Scotland) Act, 1976 as amended by Law Reform (Miscellaneous Provisions) (Scotland) Act, 1990

This act requires that every new licensed premises (i.e. premises serving alcohol) must be in possession of a food hygiene certificate, indicating compliance (or likely compliance in the case of a planned development) with the relevant food hygiene legislation. This certificate must be issued by the environmental health department before a licence can be issued by the licensing board. Furthermore, the local authority is included as a statutory objector.

Civic Government (Scotland) Act, 1982
This act requires the licensing of street traders, market stalls, places of public entertainment and late night caterers. These licences are issued by the licensing committee of the relevant local council.

THE LAW RELATING TO FOOD POISONING AND FOODBORNE ILLNESS IN SCOTLAND

Public Health (Notification of Infectious Diseases) (Scotland) Regulation 1988
Regulation 3

Requires medical practitioners to notify cases, or suspect cases, of food poisoning to the Chief Administrative Medical Officer of the NHS Board, immediately.

Notification of suspects or outbreaks
Ice-cream (Scotland) Regulations, 1948 Regulation 12
Slaughterhouse Hygiene (Scotland) Regulations, 1978 Regulation 54

These regulations require food handlers with suspect, or confirmed, food poisoning to tell their supervisor, who must notify the Chief Administrative Medical Officer (CAMO).

Examination of individuals
Ice-cream (Scotland) Regulations, 1948 Regulation 13

The above legislation empowers the CAMO to examine individuals working with food, when it is believed that they are associated with an outbreak of infectious disease.

Health Services & Public Health Act, 1968 Section 72
Public Health (Infectious Disease) (Scotland) Regulations, 1975 Regulation 9

The above legislation empowers a Sheriff, on written evidence, to order the medical examination of person(s) believed to be sufferers or carriers of an infectious disease. A person specified as a carrier should be treated as a continuing danger to others for a period specified (which can be up to three months).

Discontinuation of an individual's employment
Public Health (Scotland) Acts, 1897 to 1907 Section 7
Ice-cream (Scotland) Regulations, 1948 Regulation 12
Poultry Meat (Hygiene) (Scotland) Regulations, 1976 Schedule 3 Part 1

The above legislation prohibits persons suffering from infectious disease from continuing work.

Health Services & Public Health Act, 1968 Section 71

The designated medical officer may make a written request to an individual to discontinue his/her work, with a view to preventing the spread of food poisoning.

Social Security (Unemployment, Sickness & Invalidity Benefit) Regulations, 1975

Reg. 3. Where a person is excluded from work on certification from the designated medical officer and is under medical observation. By reason of being a carrier or having been in contact with a case of infectious disease, he/she may be deemed incapable of work for the purposes of benefit.

Compensation for loss of employment
Health Services & Public Health Act, 1968 Section 71

A person who has suffered loss in complying with a request to discontinue work shall be compensated by the local authority.

Appendix I Food processing

These days, there is not only a greater consumer demand for quality foods but also for foods that are convenient to prepare and serve. Many consumers are particularly concerned that the food is as "natural" as possible and free from food additives and contaminants. This is not always easy to achieve. Virtually all foods that we consume are subjected to some form of preservation, even if it is only chilling, packaging in a modified gas atmosphere or freezing. There is a demand for "fresh food" but usually to present such food to the consumer, who might be miles away from the production point, requires the application of very sophisticated food technology. This may range from plant breeding to give the desired properties, for example, seedless grapes or potato varieties suitable for crisp making, to packaging in containers that will control the ripening rates of fruits.

The technology of food processing draws upon the principles of chemistry, physics and biology, along with those of engineering, to ensure safe food manufacture.

The general techniques of preservation used in food processing have been dealt with in Chapter 9. This section is primarily concerned with specific processing techniques and foods.

HEAT PRESERVED FOODS

These foods are made commercially sterile, principally by the application of a predetermined amount of lethal heat. Various types of hermetically sealed containers are used such as cans, retortable plastic trays and pouches, to prevent post-process contamination. The use of the term "canning" tends to be generic to all these types of packs. Unlike most other forms of preservation, the food inside the can remains an ideal medium for bacterial growth. It is therefore imperative that:

- the heat process is adequate to destroy all pathogenic and spoilage organisms, which are capable of growth within the anaerobic conditions of the can;
- the closure of the can precludes the entry of micro-organisms; and
- the post-process handling of the can prevents damage and subsequent contamination.

The most heat-resistant pathogenic organism is *Clostridium botulinum*. This bacterium will not grow below a pH of 4.5 (or an a_w of 0.94). Consequently, when determining the heat process, regard must be had for the pH of the can contents. Foods with a pH of less than 4.5 are known as acid foods and those with a pH of more than 4.5 are termed low-acid foods.

Fruits have a pH of less than 4.5 and consequently only receive a relatively low pasteurizing heat process. Vegetables and meats have pH's much higher than 4.5 and are given a process known as a "botulinum cook" to render them commercially sterile.

Before canned food is heat processed, it is normally prepared in some way. The actual preparation will vary depending on the food type but the following flow chart shows the usual processes involved.

Exhausting

Exhausting is only carried out for cold fill ingredients. The exhausting process is usually done by blowing steam into the headspace (space above can contents at top of can). This has the effect of producing concave ends on the can after processing and cooling, due to the

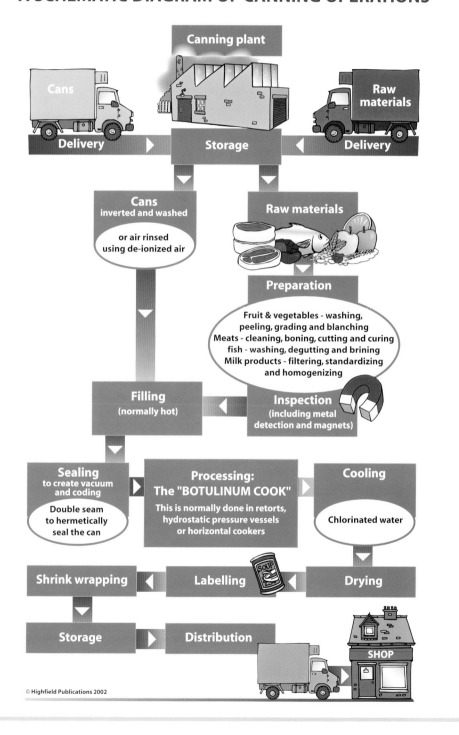

formation of a vacuum when the steam condenses. The removal of oxygen from the can also reduces the risk of oxidative corrosion of the inner can surface and helps to maintain the nutritive quality of the food.

Coding and sealing

After filling, the cans are coded, usually by video jet, and then sealed before being sent for retorting.

Container seams and seals

The construction and size of the container seams are very important to ensure integrity and safety of the food. The diagram below shows the construction of a can seam.

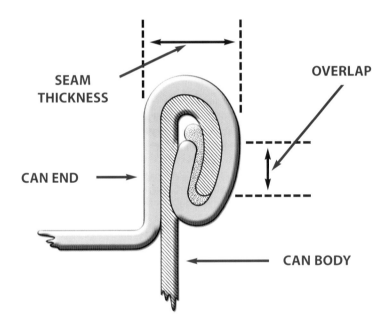

Processing

The actual process time and temperature must be determined for each can size of low-acid food and will depend on such things as weight, composition, headspace, viscosity, presence of preservatives and intended storage conditions. Scheduled processes for all low-acid products have been established by the Campden and Chorleywood Food Research Association.

The minimum safe thermal process for a low-acid canned food, is one which would reduce the chance of survival of one spore of Clostridium botulinum to less than one in 10^{12} (considered to be the equivalent of 121°C for three minutes, i.e. an F_O value of 3.0).

F_O

F_O is the measure of the lethal effect of processing. It is the sum of the lethal effects of a heat process equated to minutes at 121°C. An F_O of 6.0 is therefore equivalent to heating at 121°C for six minutes. The higher the temperature, the shorter the time of heat processing necessary to achieve the same degree of microbial destruction.

The can contents are heated by either convection or conduction. The headspace is critical if the pack heats by convection, for example, mixed packs of solid and liquid, such as potatoes in brine, as there will be dimensional increases and movement of contents.

For solid fill packs, such as corned beef, heat by conduction and headspace is not an issue.

To destroy the more heat-resistant food spoilage organisms, such as *Bacillus stearothermophilus*, a process in excess of an F_O of 3.0 must be given.

SOME TYPICAL EXAMPLES OF F_O VALUES*

Product	F_O
Canned salmon	7.0
Canned strawberries	1.0 (high-acid product)
Beans in tomato sauce	5.0
Whole chicken in broth	4.0
Sterilized milk	5.0
Dog food	14.0

Various types of retorts depend on different heating media including steam, hot water, steam-air mixtures, and "raining" hot water systems using recycled water. The use of microwaves has now been reported in Europe, but so far not in the UK. The validation and control of these heating systems are essential to the assured achievement of sterility.

Cooling

After the heating process, the containers must be quickly cooled with chlorinated water. It is essential that cooling water is brominated or chlorinated before use to ensure its freedom from bacteria. A **contact time** of 20 minutes must be allowed to ensure complete **disinfection**. The level of chlorine in the water should be kept as low as possible (two to five parts per million) to avoid the corrosion of cans. A residual level of 0.5 ppm is normally the target. Pasivation (prevention of electrolysis) and rust inhibitors may be added to the water. Regular samples, both chemical and bacteriological, should be taken from the cooling water. To avoid post-process contamination, handling of the hot cans must be prohibited. Cans should be cooled to between 32°C and 43°C to prevent "stack burn".

After processing, the cans dry naturally, are labelled and then shrink wrapped. The cooled and labelled cans must be completely dry before storage so as to prevent any rusting. Cans that are made of sheet steel with only a thin layer of tin on each side, may suffer damage which results in rusting of the iron.

Underprocessing

Underprocessing may result in the production of gas, due to the action of the surviving micro-organisms, which causes the can to swell. The contents may also undergo some other undesirable changes without gas production. In the latter case, spoilage is only detected after opening, these cans are known as flat sours.

* Source: Introduction to Food Technology, Open Learning Module, Humberside University, 1989. ISBN 0 903057 06 9.

Sources of contamination in canneries
Contamination may be either bacteriological, chemical or physical (see chapter 6).

Bacteriological contamination

Contamination or spoilage of food in cans may occur because of:
- defects in the can such as unsatisfactory welding, pinholes in the plate or a damaged flange;
- a change in raw materials with an increased loading of **thermophilic** organisms;
- poor hygiene prior to filling, which may result in a build-up of thermophiles in the blancher;
- underprocessing or uneven processing;
- post-process contamination, for example, because of a defective end-seam or infected cooling water;
- storage temperatures above 45°C, allowing the multiplication of thermophiles; or
- damage to the can in distribution, storage or retail.

If the examination of blown cans reveals the presence of one or two heat resistant thermophiles, then underprocessing is the likely cause. If a general flora of **mesophiles** is isolated, this usually indicates leaker spoilage. Apart from the **moulds** Byssochlamys fulva and B. nivea, which are occasionally responsible for the spoilage of canned fruit, **yeasts** and moulds are relatively easy to destroy by heat. They are usually present because of post-process contamination. Similar faults may also occur with plastic containers.

Post-process contamination
The greatest care must be taken after heat treatment, especially whilst the can is still warm and wet, and before the sealing compound has hardened. Bacteria are capable of being sucked into visually satisfactory cans through microscopic holes in the seams. If these bacteria are spoilage or pathogenic organisms, problems will occur.

Post-process contamination may arise because of:
- contaminated or unchlorinated cooling water;
- contaminated runways;
- operatives handling warm and wet cans; or
- damage to cans, especially while still warm.

It is recommended that cooling water contains less than 100 organisms per ml and that post-process conveying systems have less than 100 organisms per 500 square millimetres.

Blown cans
Cans may be blown, for example, because of carbon dioxide produced by micro-organisms or because of hydrogen production. The vacuum created during processing normally ensures that the can ends are concave, but the production of gas eventually destroys the vacuum and the can becomes blown. The first stage of this positive pressure is known as a flipper. This is a can with a slightly distended end, which can be pressed back to the normal position with very light pressure. When the end is banged it flips out. The next stage is known as a springer. This is a can with one end distended, which when pushed back to the normal position, causes the other end to spring out. A hard swell is the final stage of blowing when both ends of the can are permanently distended.

COMMON THERMOPHILIC SPOILAGE BACTERIA OF CANNED FOODS

BACTERIUM	Foods commonly spoiled	REMARKS
Aerobic spore formers *Bacillus stearothermophilus* *Bacillus coagulans*	Low-acid foods, such as vegetables. Low and medium acid foods, especially foods such as tomato juice.	Can referred to as a flat sour. Acid produced but no gas production and no swelling of the can.
Anaerobic spore formers *Clostridium thermosaccharolyticum*	Meat, fish and vegetables.	Carbon dioxide and hydrogen produced, contents often have a cheesy odour.
Desulfotomaculum nigrificans	Meat, fish and vegetables.	Hydrogen sulphide gas produced but is soluble and may be absorbed, so preventing swelling. Rotten egg smell. Contents may go black as the hydrogen sulphide reacts with the iron of the container.
Clostridium sporogenes	Fish and meat.	Gas production, contents foul smelling.

Hydrogen swells

This condition results from the action of the contents on the internal surface of the can. It is a problem with acid foods such as canned tomatoes and fruits. Not only is there a build-up of hydrogen causing the can to pass through the flipper, springer and hard swell stages, but also lacquer stripping and a metallic **taint** of the contents occur, along with a high tin content of the food (maximum limit is 200mg/kg). Hydrogen swells are indistinguishable from a can damaged by microbial action and the cans must be discarded.

Certain cans containing foods rich in sulphur, for example, corned beef, sardines and legumes may show a dark staining of the inner surface of the can. This is due to the formation of hydrogen sulphide during processing, which reacts with the tin lining to give tin sulphide. Whilst looking unpleasant, it will normally present no danger to the consumer unless the can has become perforated.

Over filling

Due to the expansion of the contents during heating, the can becomes strained and distorted. These cans must be regarded with suspicion and can only be confirmed by physical or bacteriological checks of the contents.

Damage

Damage is caused by a variety of reasons including perforation by nails, sharp crates, dockers' hooks, knives used to open cardboard cartons in shops, and denting due to mishandling cases. Considerable bacteriological significance should be placed on damaged cans: leakage introduces an infection from the atmosphere and possibly subsequent bacterial spoilage and/or mould growth. Cans with badly damaged seams should be rejected. External wetting, for example, in flooding or fire-fighting, may re-introduce the risk of post-process contamination through seams, especially on handling.

Incorrect storage

Apart from thermophilic spoilage, high temperature storage may result in internal corrosion. Furthermore, if conditions are too humid, external rusting and eventual pinholing of the can may occur.

Pasteurized canned foods

As discussed earlier, most low-acid canned foods are processed to a minimum standard of the "botulinum cook". However, some foods if fully processed, would be inedible. One example of this is canned cured ham. This product only receives a pasteurization process (a centre temperature of about 70°C). This means that other precautions are necessary to ensure that it does not become a source of *Clostridium botulinum* food poisoning. It is the presence of curing salts (sodium or potassium nitrite and sodium chloride), which help to inhibit the growth of botulinum organisms and make canned cured ham a safe food. The combined inhibitory effect of curing salts and sodium chloride on spore germination is referred to as the perigo effect. A sufficient combination of heat process temperature and curing salt level may render the product ambient stable. Lesser treatments may necessitate the use of subsequent chilled storage. It is essential to first evaluate the stability of new products, in order to establish if chilled storage is needed.

Retortable pouches

The development of thermally processed foods in flexible packaging was led by the United States during the 1970's. The aim was to serve the Apollo astronauts with high quality, nutritious food, which did not require a "toothpaste tube" style system. The commercial development of the pouches and much of the filling/sealing equipment has been led by Chinese and Japanese manufacturers. It is only in the last few years that UK processors have exploited the potential of flexible packaging.

The pouch itself is a tri-layer polymeric construction with the most common laminate layers being:

Outer layer: Polyester (affords tensile strength to the pouch).
Middle layer: Aluminium foil (acts as an effective oxygen, light and water barrier).
Inner Layer: Polypropylene/polyolefin (acts as an inert barrier in contact with the food).

In common with traditional metal cans, flexible pouches are double-sealed. However, this sealing is achieved by applying heat and pressure to the pouch in two stages. The first stage creates a seal 3mm in width, whilst the second stage creates two additional seals, 5mm either side of the first one. The filling of these pouches can have a direct influence on the integrity of these seals. Splashing of the product needs to be avoided during the process.

Once filled, the pouch is processed in a horizontal orientation that offers the highest surface area to volume ratio, typically the pouch presenting a profile only 19mm thick. This thickness, provides the processor with the opportunity to use a shorter sterilization time than with the same product weight in a traditional can, resulting in greater retention of nutrients, flavour and taste whilst still achieving a suitable F_0. However, the shorter time/temperature regime can present a problem if the pack is poorly loaded into the formers or is otherwise distended. An increase in pack thickness of just 2mm can reduce the F_0 value from 3.0 to an unsafe 1.6, leading to the possibility of *Clostridium botulinum* spore survival. Furthermore the fill weight must be tightly controlled to prevent over-filling, which would alter the surface area/volume ratio and also reduce the F_0 value of the product.

Aseptic production technology

This differs from canning as the sterilization of the product is carried out as a separate operation from package sterilization, prior to filling aseptically. Various types of heat exchangers are used for these continuous flow processes. Ohmic heating may be used, whereby heating is induced by passing a high-voltage electric current directly through the food material. It is then packed under aseptic conditions into sterilized packaging containers, for example, multi-layered plastic film bags. The finished product has an extended shelf life and is stable at ambient temperatures. Ohmic heating may be combined with other forms of heating, such as plate heat exchange systems.

DAIRY PRODUCTS

The processing and testing of milk

In 1887, Anthony Hailwood produced sterilized milk and by 1920 pasteurized milk had been developed. From this time, there has been a gradual improvement in the quality of milk and hygienic methods of production. The most important factor in securing improvements, has been the introduction of refrigeration on the farm and the use of insulated tankers to transport milk to the dairy. Moreover, better use of refrigeration and reducing contamination during packaging and bottling have extended the shelf life of milk even more.

Most milk sold in this country is heat-treated by one of three methods:
- pasteurization;
- sterilization; or
- ultra heat-treated (U.H.T.).

Pasteurization ensures the destruction of pathogens and most vegetative organisms. However, heat-resistant organisms and spores survive, which accounts for milk souring after a few days. Sterilization and UHT treatment destroy all organisms and this ensures a prolonged shelf life, even without refrigeration. It is essential that the pathogenic contamination of heat-treated milk is prevented, as the milk, devoid of its natural flora, is more likely to support the rapid multiplication of the pathogens.

The production of pasteurized milk

The heating process for producing pasteurized milk in the UK was based on the destruction of two potential pathogens, namely, *Mycobacterium tuberculosis/bovis* and *Brucella abortus*. Whilst destroying these organisms and many other pathogens and spoilage organisms, pasteurization is a relatively mild heat treatment and has a minimal effect on the flavour and nutritional value of milk. However, even with pasteurization, up to 10% of many vitamins will be lost.

Pasteurized milk is usually produced by a heat treatment involving a high temperature for a short time (HTST process, which involves heating to at least 71.7°C for 15 seconds), or a process using different time/temperature combinations to obtain an equivalent effect (e.g. the Holder Process involves 63°C for 30 minutes).

Pasteurized milk must:
- show a negative reaction to the phosphatase test and a positive reaction to the peroxidase test (high temperature pasteurization produces a negative reaction to both tests); and
- meet certain microbiological standards in random sampling checks carried out at the treatment establishment, including the absence of pathogens in 25g and only one out of five samples with a coliform (per ml) count of less than 5.

Milk arriving at the dairy is weighed and checked for its quality, for example, added water, the presence of any blood and its microbiological load. It is then placed in holding tanks where it will be mixed with milk from many other farms, so ensuring a consistent product. It is then filtered to remove gross contamination and pumped to a balance tank, which maintains an even flow rate through the next part of the HTST plant where the heat exchange takes place.

The cold milk entering the plant is preheated by the hot milk leaving the processing plant. It is finally heated to the required temperature of 72°C by using hot water at around 75°C. At no time do the water and milk come into direct contact. The heat exchange takes place through thin stainless steel plates. Regular checks of the unit, particularly the seals, must be made to ensure its integrity. The heated milk flows through a holding tube for 15 seconds, before passing a flow diversion valve. If the milk is below 71.7°C, a bell rings, the milk is diverted back to the balance tank and goes through the whole process again. If the temperature is at or above 71.7°C, the milk passes out of the heat exchange, heating up incoming milk.

To preserve the cream line and reduce caramelization, dairies operate as close as possible to 72°C for 15 seconds. After heat treatment, the milk is cooled rapidly to below 10°C. This prevents the multiplication of heat-resistant organisms, maintains the cream line and improves the keeping quality. As a temperature of 72°C stimulates the germination of any *Bacillus cereus* spores that may be present, cold storage is particularly important to prevent the multiplication of the vegetative bacteria produced. After cooling, the milk is bottled or filled into cartons and stored under refrigeration. Strict stock rotation is essential to avoid complaints of souring, especially during the warm summer months. Regular testing ensures that the milk has a reasonable shelf life and has been given a satisfactory heat treatment.

The methylene blue test

This dye reduction test is a chemical method of assessing the microbiological quality of the milk. Methylene blue is added to the milk which is incubated at 37°C for 30 minutes. If the milk/methylene blue is decolourized in this time, it indicates that the level of micro-organisms present is unacceptably high and the milk is rejected. The test does not differentiate the type of organism causing the contamination.

The phosphatase test

This test is used to determine whether or not the milk has been pasteurized satisfactorily and to ensure that raw milk has not been added after processing. Such contamination is likely to occur if the seals on the heat exchange units are defective. The test relies on the fact that the enzyme phosphatase, which is present in raw milk, is destroyed by effective pasteurization.

The chemical di-sodium para nitrophenyl, which turns yellow if phosphatase is present, is added to the milk and the depth of colour is examined against a standard sample using a comparator. A figure of more than 10g represents a failure of the test.

The production of sterilized milk

The prolonged life of sterilized milk is obtained at the expense of flavour changes. This is due to the caramelization of lactose, and results in a loss of nutritional value (approximately 50% of vitamins are destroyed). Sterilized milk can be produced by heating milk at around 110°C for 30 minutes in a hermetically sealed container, the seal of which must remain intact during such heat treatment, or by use of the continuous flow process.

The turbidity test

This test is to measure the degree of protein denaturation that has taken place. When heated to a sterilization level the proteins albumen and globulin are denatured. If ammonium sulphate is added, the protein casein is precipitated and can be filtered out. The filtrate will be clear and if this is further heated to boiling point, it will turn cloudy (turbid) if any protein has not been already precipitated. Thus, a turbid solution indicates that the original milk had not been adequately sterilized.

Ultra heat treated milk

UHT milk is produced by a continuous flow process with the application of a very high temperature for a short time (not less than 135°C for not less than a second). This is so all residual spoilage micro-organisms and their spores are destroyed, but the chemical, physical and organoleptic changes to the milk are minimal. Immediately after heat treatment, the milk is filled into opaque containers, or containers made opaque by the packaging. It will store for several months without refrigeration.

If the heat treatment involves direct contact with steam, it must be from potable water, must not adversely affect the milk, for example, by leaving foreign matter, and must not change the water content of the treated milk.

Sterilized (and UHT) milk must:
- not contain any unacceptable levels of any pharmacologically active substance;
- not contain any added water;
- after 15 days in a closed container at 30°C (or seven days at 55°C) be organoleptically normal, not show any sign of deterioration and have a plate count at 30°C ≤ 100 per ml.

Since 1997, problems of spoilage have resulted from spores of *Bacillus sporothermodurans* which have survived UHT processing.

Thermized milk

Thermized milk is obtained by heating raw milk for at least 15 seconds, between 57°C and 68°C, so as to show a positive reaction to the phosphatase test.

Raw milk tests at the dairy

Raw milk must come from healthy animals on a registered production holding. Such animals must not show any disease symptoms, udder wounds or have recently been treated with dangerous substances. Herds must be tuberculosis and brucellosis free.

Raw milk must not contain added water or unacceptable antibiotic, hormonal, pesticide or detergent residues or other harmful substances. Raw cow's milk must have a plate count at 30°C of ≤100,000 per ml and a somatic cell count of ≤400,000 per ml.

If raw milk is not heat treated at the dairy within four hours of acceptance, it must be cooled to 6°C. If not treated within 36 hours and the plate count at 30°C exceeds 300,000 per ml, it must not be used for the production of heat-treated milk.

Milk is examined for the presence of taint, dirt, blood and antibiotics and several tests are carried out before processing. Unsatisfactory milk can then be rejected.

Brucella abortus ring test

An **antigen** is added to milk at 37°C and if the milk is positive for the organism, a blue ring rises to the top with the cream.

Freezing point test

This test is carried out to ascertain the addition of water to milk. The milk is frozen and if this occurs at between 0.53°C and 0.565°C it has not been adulterated. Freezing at higher temperatures indicates the addition of water.

The Gerber test (fat content)

The Gerber test is used to determine the fat content of milk. It is an electronic method, which involves using a milkotester. The result is obtained in 15 minutes.

The hydrometer test

The hydrometer test is used to check the density (total solids) of milk. If the fat content is known, the solids-not-fat can then be calculated by subtracting the fat per cent from the total solids per cent (approximately 87 % of milk is water).

The resazurin test

Resazurin is a violet dye, which is decolourized by souring organisms. The resazurin is added to the milk and then left in a water bath at 37°C for ten minutes. The milk will first go pink and eventually colourless, depending on the number of organisms present. After ten minutes the colour is checked using a Lovibond comparator. A reading of 3.5 or less indicates the milk is unsatisfactory.

Problems encountered with milk

Milkstone

When milk is heated, phosphate and albumen are deposited as milkstone. This normally occurs above 72°C and is not a great problem in pasteurization. Milkstone harbours bacteria and causes **cleaning** problems. When it does occur, the plant should be given an **acid clean**, for example, with phosphoric acid.

Milk taint

Either odour or taste taint may be caused by:
- treatment of cows with particular medicines;
- certain types of animal feed such as turnips or kale;
- certain illnesses such as **mastitis**, which may produce bitterness;
- storage of uncovered milk near paint, manure or strong-smelling substances;

- cows near the end of lactation producing a rancid taste;
- sunlight or contact with copper, which may produce an oily taint; or
- phenol from cleaning chemicals at the farm, a rare but serious problem.

Taints of bacteriological origin increase with age; those due to feeding are strong, fresh from the udder. Taints may be retained by milk products.

Blood

May be introduced into milk because of animals suffering from clinical mastitis or from sore udders. Herds infected with mastitis produce a reduced yield and the compositional quality of the milk falls. The milk is tested for the presence of blood at the dairy.

Ropey milk

The milk is slimy, which is caused by *Bacillus aerogenes*. It may be associated with faulty cleaning or disinfection. This is now rare due to improved cooling and cold storage.

Leucocytes

Occasionally, complaints are received about rusty droplets in bottles of homogenized milk. These deposits consist of milk fat, casein and large numbers of leucocytes, commonly referred to as milk slime. Leucocytes are a type of cell found in the udder. The number present in milk increases if cows are suffering from mastitis. The presence of leucocytes in heat-treated milk is purely an aesthetic problem.

Souring of milk

Lactic acid bacteria break down lactose to produce lactic acid. Bacteria involved include *Staphylococcus* and *Streptococcus lactis*. *Bacillus proteus* forms whey.

Cream

The butterfat in milk is present as small globules, which gradually rise to the surface, as they are less dense than milk. This cream line can be seen in bottles of pasteurized milk that have not been homogenized. To speed up the process, a mechanical separator that acts like a centrifuge is used. The stainless steel separator consists of an outer casing, which encloses a stack of conical plates. Pre-warmed milk is fed into the separator, which rotates at around 6,000 revolutions per minute. The lighter cream collects in the centre of the unit and is drawn off, heat treated and cooled. The heavier skimmed milk passes through the plates and is also drawn off, flash heated to around 80°C and then cooled to around 5°C. A valve may be used to control the fat content of the cream.

The cream may contain the same pathogens as milk. It is normally pasteurized in an HTST plant at not less than 72°C (or the equivalent), for not less than 15 seconds and then cooled to less than 5°C. It is normally aged before filling into suitable cartons. Homogenization is sometimes used to thicken the cream.

Sterilized cream

Homogenized cream is canned or bottled and then sterilized, at a temperature of not less than 108°C, for not less than 45 minutes (or the equivalent) and then cooled. Sterilized cream will keep for prolonged periods but obviously has a different flavour from pasteurized. The legislation relating to milk applies to cream. In addition, specific regulations lay down the appropriate compositional standards for particular types of cream; double cream must contain at least 48% milk fat; single cream must contain at least 18% milk fat.

Imitation or non-dairy cream

This is a substance, which is produced by emulsifying edible oils or fats with water. Imitation cream is mainly used by bakeries for filling, in place of dairy cream. It is often prepared from palm kernel oils with egg yolk, water, sugar and milk powder.

Yogurt

To manufacture yogurt, skimmed milk, sugar and water are mixed together and heated to approximately 95°C, for around 30 minutes. The mix is then cooled to around 43°C, filled into incubation tanks, and a starter culture, for example, *Lactobacillus bulgaricus* or *Streptococcus thermophilus*, is added. These bacteria produce lactic acid, the fermentation process is allowed to continue until a pH of approximately 4.0 is achieved and the yogurt has thickened considerably. The mix is then cooled to below 5°C. Unless plain yogurt is required, fruit is added and the mix is then filled into cartons and placed in cold storage.

Yogurt spoilage

Bacterial spoilage is inhibited by the acidity of the yogurt and low storage temperatures. However, unless the mix is heat treated after fermentation, the starter culture will continue to produce acid and after approximately ten days, the yogurt becomes unpalatable. Yeast and moulds are more likely to cause problems. Fermentation of sugar by yeast causes the production of carbon dioxide, which results in a strained convex cap. Mould may be observed as a white or green clump on the surface.

Butter

Cream is held at 5°C in large, insulated, stainless steel tanks until required. It may be necessary to add lactic acid-producing bacteria (starters) and warm the cream to ensure the correct acidity. The temperature and pH are dictated by the flavour required. The cream is then pumped into a large, stainless steel churn which operates half full and at a temperature of around 7°C. The churn revolves for 40 minutes which allows the fat globules to coalesce. The fat granules formed are the size of a grain of wheat.

The buttermilk is then drawn off and the butter sprayed with chilled water to remove any traces of buttermilk. Excess water is drained off and salt may be added, depending on the flavour required. The churn is rotated slowly to disperse moisture and salt and to ensure a smooth finished product. This stage is known as "working the butter". When the butter is removed, the moisture content must, by law, not exceed 16%. After packing, the butter should be stored away from chemicals or other food with strong odours. Modern manufacture is by continuous buttermakers, which operate on the same principle but produce five tonnes per hour.

Defects in butter may be due to poor quality cream, production errors or post pasteurization contamination. The butter may be sour, rancid or cheesy, normally because of bacterial action. It is essential, therefore, to maintain high standards of factory hygiene and develop suitable quality control checks, including bacteriological examination of raw materials and finished product.

Growth of bacteria, yeast and moulds may occur rapidly if butter, especially unsalted, is stored above 16°C. The keeping quality depends on the bacterial load, type of organisms and storage temperature. Chilled or refrigerated storage is advisable. Growth of salmonellae has been recorded in butter kept at room temperatures (25°C).

Margarine

Margarine is produced from a blend of fat-free milk and vegetable, animal and marine oils or fats.

Cheese

Cheese may be made from:
- whole milk, for example, Cheddar;
- skimmed milk, for example, Dutch; or
- milk to which cream has been added, for example, Stilton.

There are three main types of cheese:
- *Hard cheeses*

These have relatively low-moisture content and possess excellent keeping qualities. Examples include Cheddar, Lancashire and Cheshire.

- *Soft cheeses*

Those which are allowed to drain naturally as the curd is not pressed, such as, Camembert, Limburger and Cambridge.

- *Blue-veined varieties*

These cheeses have a high-moisture content and soft curd. They are ripened by normal organisms and by the growth of specific moulds.

Cheese manufacture

Pasteurized milk is run into long stall troughs, and milk cultures of *Streptococcus lactis* are added to ensure the correct acidity. Rennet is then added and the milk-clots form a curd and a liquid whey, which is drained off. The curd is cut into small cubes and then heated to expel moisture and assist in whey drainage. Higher temperatures are used to produce the drier, hard cheeses.

The cubes coalesce and the resultant mass is cut into slabs, which are piled up to increase whey drainage. The dry, firm curd is then passed through a milling machine, which tears it into small pieces, and salt is added. After salting, the cheese is filled into moulds or vacuum packs, pressed and then ripened at 10°C and 80 to 90% relative humidity. Ripening can take from three months to a year. Most types of cheese, once cut, should be stored in a refrigerator, provided that they are well-wrapped.

CHEESE FAULTS

Defects in cheese may occur because of:
- the use of unsatisfactory milk;
- the presence of foreign substances;
- the growth of abnormal bacteria; or
- production errors.

Bacterial faults include sliminess, bitterness, rancidity and pigmentation. Scrupulous attention must be paid to the quality of raw materials and to hygiene during production. Food poisoning incidents, including salmonella and staphylococcal, have been traced back to contaminated cheese. In 1984, imported Camembert and Brie from France resulted in several cases of *E. coli* food poisoning. Listeriosis has often resulted from the consumption of contaminated soft cheeses. Food poisoning and foodborne disease incidents are more commonly associated with cheese made from unpasteurized milk.

Mycotoxin-producing moulds have been isolated from a number of cheeses including Cheddar.

Condensed milk

There are two types of condensed milk:
- sweetened condensed milk; and
- unsweetened evaporated milk.

Sweetened condensed milk

The milk is preheated, filtered and separated (separation is not carried out if full-cream condensed milk is to be produced). Further heating is carried out to destroy bacteria, moulds and enzymes. Sugar is added at a concentration that ensures preservation, by producing a high osmotic pressure. The sweetened milk is pumped into an evaporator, where approximately 60% of the water is boiled off under vacuum at around 55°C. The milk is cooled carefully, to prevent the formation of large lactose (milk sugar) crystals, and canned.

Thickening of condensed milk may be caused by micrococci and contamination by gas-producing yeast may cause the can to blow.

Unsweetened evaporated milk

The production of this milk is very similar to sweetened condensed milk, apart from the fact that no sugar is added. After evaporation, the milk is homogenized to prevent the separation of fat and then cooled and filled into cans. The filled cans are heated to 115°C for 20 minutes to ensure their safety and shelf life.

Underprocessing of evaporated milk may allow the survival of *Bacillus subtilis*, *Bacillus coagulans* or *Bacillus cereus*, which give the can contents a coagulated appearance.

Dried milk

The standardized raw milk is preheated and filtered before the removal of water. There are several methods available for the production of dried milk.

Roller drying

The milk is spread, as a very thin film, on an internally-heated, revolving, horizontal steel cylinder. The resultant dried milk is automatically scraped off.

Spray drying

After an initial boiling under vacuum, the concentrated milk is pumped into the top of a hot-air tower as a very fine spray. The resulting dried milk powder collects at the base of the tower and is cooled. The burnt particles are removed and it is then packed in air-tight containers.

Accelerated freeze drying

To improve reconstitution, this method may be used. The milk is frozen and then heated under vacuum, so that the water is removed by **sublimation**. Salmonellae have been isolated from dried milk powder made from unpasteurized milk. There is also a possible risk if *Staphylococcus aureus* is present and allowed to multiply prior to drying, as heat-resistant **toxins** may be produced.

Ice cream

Ice cream may be made with milk fat, sugar, eggs and cornflour, or vegetable fat, separated milk powder, sugar, gelatin, emulsifiers and water. Manufacture consists of mixing, pasteurizing, homogenizing, freezing and packaging. Alternatively, the mix may be sterilized and frozen, or **dehydrated** and sold as a complete cold mix, which is reconstituted by the addition of water.

Pasteurized ice cream

Pasteurization may be carried out by heating to not less than:
- 65.6C for at least 30 minutes; or
- 71.1C for at least ten minutes; or
- 79.4C for at least 15 seconds.

After heating, the mixture is homogenized and cooled to below 7.2°C, within one and a half hours and then frozen and packaged. The HTST plant must be provided with a flow diversion valve to redirect ice cream, which has not been heated to the correct temperature. Indicating and recording thermometers should be provided and records kept for at least one month.

Sterilized ice cream

If the mixture is sterilized, it must be heated to not less than 148.9°C for at least two seconds and then cooled to below 7.2°C within one and a half hours, or alternatively canned under sterile conditions. If the can is opened, the ice cream must be cooled to below 7.2°C and kept at this temperature until frozen.

Soft ice cream

Soft ice cream is freshly frozen ice cream, which is normally sold to the public direct from the freezer. When using a complete cold mix, it is essential that after reconstitution, the mix is frozen in the soft ice cream machine within one hour. (For cleaning a soft ice cream machine, see chapter 14.)

Ice cream spoilage

Ice cream may become either sour or rancid. Sour ice cream is caused by prolonged storage before freezing or the use of sour milk. Rancid ice cream is caused by the use of rancid fat.

The methylene blue test

This test is used to determine the bacteriological standard of ice cream. The ice cream is mixed with methylene blue and Ringer solution and then incubated in a water-bath at 20°C for 17 hours. It is then incubated at 37°C and inverted every 30 minutes until decolourized.

INTERPRETATION OF RESULTS

Grade	Time to decolourize (hrs)
1	4.5
2	2.5 to 4
3	0.5 to 2
4	0

Most samples should be grade 1 or 2 and, although not legally binding, consistent grade 3 or 4 samples indicate poor hygiene at some stage of production or handling.

Liquid egg

The objective in pasteurizing liquid egg is to destroy pathogenic organisms, especially salmonellae, without affecting the physical and functional properties of the raw egg. Thus, the time and temperature used are critical to achieve this state of affairs. Liquid whole egg can be pasteurized at 60°C for three and a half minutes, or a temperature of not less than 64.4°C for at least two and a half minutes, or another time and temperature that has the same lethal effect on vegetative, pathogenic organisms. Following the pasteurization process, the treated egg must satisfy the alpha-amylase test. This is another enzyme test (similar to the phosphatase test for pasteurized milk). Absence of the enzyme signifies that the liquid egg has been adequately pasteurized.

Pasteurized shell egg

Eggs can now be pasteurized, in the shell, by using warm water or humidity-controlled air. Eggs pass through a warm water bath of aerated air at a temperature and time which results in a 5 log reduction (100,000 to 1) of any salmonella that may be present. For example, 57°C for 30 minutes. At this temperature, the albumen is not denatured. The eggs are then chilled in a cool water bath. They can be sealed with a food grade wax to prevent future contamination and then stored at 7°C.

Shellfish by Vic George

Shellfish may often be contaminated with food poisoning organisms, especially bivalves such as cockles, mussels, oysters and clams, as they obtain their nutrients by filter feeding in coastal waters that may contain sewage effluent. A stringent heat processing routine for commercially cooked shellfish, mainly cockles, should eliminate the risk, however those shellfish that are eaten raw (oysters), or lightly cooked mussels, need to contain minimal amounts of contamination, if illness is not to result.

Significant numbers of shellfish (mainly oyster) associated incidents of viral gastroenteritis still occur, however, bacterial food poisoning is less common and usually results from post-process contamination.

Bivalve molluscs are controlled from the growing areas through to sale under the provisions of Regulation (EC) No 853/2004, Regulation (EC) No 854/2004, Regulation 852/2004 and the Food Hygiene (England) (Scotland) (Wales) (NI) Regulations 2006. The Food Law Code of Practice gives guidance to enforcement authorities on how these regulations should be enforced.

The Regulations place a duty on local authorities to take regular (usually monthly) microbiological samples from bivalve harvesting areas, the results of the samples enable the Food Standards Agency to classify areas, depending on the E. coli content of the shellfish.

CLASSIFICATION OF SHELLFISH AREAS BASED ON *E. COLI*

CAT A*	<230 *E. coli*/100g Nil Salmonella/25g	May go for direct consumption
CAT B+	<4,600 *E. coli*/100g	Must be depurated
CAT C+	<46,000 *E. coli*/100g	Relay (2 months) to meet A or B category
Unclassified	>46,000 *E. coli*/100g	Prohibited

*Draft EU Regulations for Microbiological Criteria for Foodstuffs 2005
+Annex III Section VII, Chapter V of the Regulation (EC) 853/2004

The most common form of depuration, usually referred to as "purification", involves immersing the shellfish in a tank of seawater under strictly controlled conditions. The shellfish should resume their normal filter feeding within the depuration tank and thereby expel pathogens into the water, which will then be destroyed as the water is passed by ultra violet light. The UK operate a standard purification time of 42 hours, although this varies in other countries, from 24 to 48 hours.

The requirement to healthmark each batch of molluscs from a purification plant with the premises approval number and date of dispatch provides traceability of product. Separate monitoring is undertaken for algal toxins. Diarrhetic, paralytic and amnesic shellfish poisoning toxins are becoming more widespread and are causing increasing concern to the enforcement agencies and the shellfish trade.

A food authority may make a temporary prohibition order, prohibiting the collection of any live shellfish from a specified area, if it is satisfied that the consumption of those shellfish would be a risk to public health. Although used in cases of chemical contamination (oil spillages) and sewage contamination, local authorities are increasingly having to use these powers in respect of algal toxins. Many algal toxins are heat stable and the normal cooking process will not inactivate them. The temporary prohibition orders last for up to 28 days after which, if the problem still exists, they must be renewed.

Appendix II

EMPLOYEE MEDICAL QUESTIONNAIRE

NAME .... OCCUPATION ....

ADDRESS .... DEPARTMENT ....

....

....

....

(1) Have you ever had or been a carrier of:
    A foodborne disease   **YES/NO**
    Typhoid or paratyphoid   **YES/NO**
    Tuberculosis   **YES/NO**
    Parasitic infections   **YES/NO**

(2) Has any close family contact suffered from any of the above?   **YES/NO**

(3) Have you suffered from any of the following:
    Serious diarrhoea or vomiting   **YES/NO**
    Skin trouble   **YES/NO**
    Boils, styes or septic fingers   **YES/NO**
    Discharge from the ears, eyes, gums/mouth   **YES/NO**

(4) Please give details of any other medical problems that may affect your employment as a food handler, for example, recurring gastrointestinal disorder.

....

....

(5) Have you been abroad within the last two years?   **YES/NO**

WHERE? ....

(6) Should it be necessary, will you agree to provide such specimens that may be required by the company, to ensure that you are not a carrier of any organism that may infect food?   **YES/NO**

I declare that all the foregoing statements are true and complete to the best of my knowledge and belief.

Signed ....

Appendix III Food safety courses

FOOD SAFETY COURSES AND QUALIFICATIONS

Food businesses operators must ensure that food handlers are supervised and instructed and/or trained in food hygiene matters commensurate with their work activity. This means that staff must have the necessary competencies to produce safe food. The flexibility of EU legislation enables staff to achieve competencies in a variety of ways, including accredited training, in-house training, on the job training or e-learning.

However, in order to obtain a nationally recognised qualification many food handlers attend an accredited course provided by one of the following awarding bodies:

Organizations providing hygiene courses for food handlers
- The Royal Institute of Public Health
  28 Portland Place, London W1B 1DE. Tel: 020 7291 8350 www.riph.org.uk
- The Royal Society for the Promotion of Health
  RSH House, 38A St. George's Drive, London SW1V 4BH. Tel: 020 7630 0121
  www.rsph.org
- Chartered Institute of Environmental Health
  Chadwick Court, 15 Hatfields, London SE1 8DJ. Tel: 020 7928 6006 www.cieh.org
- The Society of Food Hygiene and Technology
  The Granary, Middleton House Farm, Tamworth Road, Middleton, Staffs B78 2BD. Tel: 01827 872500
- The Royal Environmental Health Institute of Scotland
  3 Manor Place, Edinburgh EH3 7DH Scotland. Tel: 0131 225 5444
  www.royal-environmental-health.org.uk

A list of trainers and courses can be found on www.foodsafetytrainers.co.uk.

In 2006, New National Occupational Standards (NOS) in Food Safety were produced for Catering, Retailing and Manufacturing, by their respective Sector Skills Councils. The Awarding bodies used the NOS to develop new food safety qualifications specific to the three food sectors. The new qualifications will appear on the National Qualifications Framework once the courses have been accredited by the Qualifications and Curriculum Authority (QCA).

The following revised levels apply:
- **Level 1 (up to 3 hours)** - Will probably be part of the induction programme. Aimed at those handlers who are new or have relatively low-risk jobs.
- **Level 2 (6 to 9 hours)** - Intended for high-risk food handlers and replaced the Foundation course.
- **Level 3 (20 to 24 hours)** - Intended for supervisors, especially in high-risk food businesses. This replaced the Intermediate course.
- **Level 4 (around 40 hours)** - Intended for managers and those food handlers who have significant responsibilities for food safety, for example, QA staff in factories, trainers or auditors. This replaced the Advanced course.

The syllabuses for the new courses remains very similar to those of the old courses, although there is increased emphasis on allergens, hazards, controls, monitoring and recording (HACCP) and items of food safety of practical relevance to their particular sector. Less emphasis should be placed on unnecessary information, for example, level 2 catering should not require knowledge of preservation, names of Acts and Regulations, details of specific food poisoning organisms or hygiene practices relating solely to manufacture, for example, CIP. The level 3 and level 4 syllabuses have a greater emphasis on supervisory and management responsibilities together with a requirement for a greater knowledge of HACCP principles.

Courses offered by the Royal Institute of Public Health (RIPH)

- Level 2 Award in Food Safety in Catering OR Retail OR Manufacturing.
- Level 3 Award in Supervising Food Safety in Catering OR Retail OR Manufacturing.
- Level 4 Award in Managing Food Safety in Catering OR Manufacturing.

RIPH Foundation Certificate in Food Hygiene (not accredited)

Foundation Certificate in HACCP Principles
Intermediate Certificate in Applied HACCP Principles
Advanced Certificate in Applied HACCP Principles

Courses offered by the Royal Society for the Promotion of Health (RSPH)

- Level 2 Award in Food Safety in Catering OR Retail OR Manufacturing.
- Level 3 Award in Supervising Food Safety in Catering OR Retail OR Manufacturing.
- Level 4 Award in Managing Food Safety in Catering OR Manufacturing.

Certificate in Food Hygiene Awareness
Fundamentals of Food Hygiene } Not accredited
Fundamentals of Food Hygiene for the Food Industry

Fundamentals of HACCP (not accredited)
Foundation Certificate in HACCP

Courses offered by the Chartered Institute of Environmental Health (CIEH)

- Level 2 Award in Food Safety in Catering OR Retail OR Manufacturing.
- Level 3 Award in Supervising Food Safety in Catering OR Retail OR Manufacturing.
- Level 4 Award in Managing Food Safety in Catering OR Manufacturing.

Intermediate Certificate in Hazard Analysis Principles and Practice
Level 3 Award in Implementing Food Safety Management Procedures
HACCP in Practice Certificate (not accredited)

HYGIENE for MANAGEMENT *Appendix III Food safety courses*

Courses offered by the Society of Food Hygiene and Technology (SOFHT)

- **The SOFHT Food Hygiene Awareness Pack**
Intended for in-house delivery of a 90-minute training session.

- **The SOFHT Hygiene Training Scheme (six hours)**

- **Foundation Certificate in HACCP Principles**

Courses offered by the Royal Environmental Health Institute of Scotland (REHIS)

- **Introduction to Food Hygiene Course (two hours)**
- **Elementary Food Hygiene Course (six hours)**
- **Intermediate Food Hygiene Course (17 hours)**
- **Diploma in Advanced Food Hygiene Course (minimum 36 hours)**

Principles of HACCP Course (nine hours)

Intermediate HACCP Practice Course (17 hours)

The times provided are the minimum recommended and it may be beneficial to extend the time allowed for the level 2 and 3 courses.

Glossary

Acid	A chemical, with a pH of less than 7, which forms hydrogen ions in solution, the hydrogen being replaceable by a metal to form a salt. Used for removing hard water scale. Generally corrosive and turns litmus paper red.
Acute disease	A disease that develops rapidly and produces symptoms quickly after infection. Patients soon recover, or die. Does not imply severity: an acute disease may be mild.
Acute poison	A poison, that takes effect rapidly. In the case of rodenticides, it usually involves a single dose.
Additives	Preservatives, antioxidants, colourings, emulsifiers, stabilizers, artificial sweeteners and flavourings added to food to improve quality, taste, shelf life, function or appearance.
Aerobic	Using oxygen.
Aerobic colony count	The count of viable aerobic bacteria per gram of product (based on the number of colonies grown on nutrient agar plate).
Aerosol	Airborne contamination.
Agar	A seaweed extract used to form a gel as the basis for culture plates, on which bacteria can be grown.
Air-on temperature	The temperature of the warmer air measured at the return air vent in the multi-deck refrigerated cabinet.
Air-off temperature	The temperature of the cold air, leaving the evaporator coil, where it enters the multi-deck refrigerated cabinet.
Algae	Simple plants capable of photosynthesis, most commonly found in aquatic environments or damp soil, for example, seaweed and spirogyra (forms bright green slimy masses in ponds).
Alkali	A chemical, with a pH greater than 7, which reacts with an acid to form a salt and water only.
Allergy	An identifiable immunological response to food or food additives, which may involve the respiratory system, the gastrointestinal tract, the skin or the central nervous system. In severe cases this may result in anaphylactic shock.
Ambient temperature	The temperature of the surroundings. Usually refers to the room temperature.
Anaerobic	Using little or no oxygen.
Anorexia	Loss of appetite.
Antibiotic	A chemical used to destroy, or inhibit the growth of, pathogenic bacteria within human or animal bodies. Often extracted and refined from moulds.
Antibody	A protein produced by white blood cells to destroy antigens and protect the body against disease. Antibodies are produced when we are vaccinated and provide immunity from specific infections.
Antigen	A foreign protein that can trigger an immune response resulting in the production of an antibody, as part of the body's defence against infection. Antigens include viruses, bacteria and their toxins.
Antiseptic	A substance that prevents the growth of bacteria and moulds, specifically on or in the human body.
Antiserum	Blood serum containing specific antibodies. Used for typing species of bacteria and the treatment of some diseases.
Aseptic	Free from micro-organisms.
Aspic	A clear savoury meat jelly used as a decorative glaze on prepared foods. An ideal growth medium for bacteria.
Asymptomatic	Without symptoms.

HYGIENE for MANAGEMENT *Glossary*

Audit	A systematic and independent examination to determine whether activities and related results comply with planned arrangements and whether these arrangements are implemented effectively and are suitable to achieve objectives (Regulation (EC) No 854/2004).
Authorized officer	Any person who is authorized by a food authority, in writing, to act in matters arising under the Food Safety, Act 1990.
Autolysis	The self-destruction of cells, including bacteria, as a result of the action of their own enzymes.
A_W (water activity)	A measure of the water in food available to micro-organisms. Most bacteria prefer values of 0.995 to 0.980. $$A_W = \frac{\text{Water vapour pressure of food}}{\text{Water vapour pressure of pure water}}$$
Bacteraemia	The presence of bacteria in the blood, usually without multiplication.
Bacteria (singular bacterium)	Single-celled micro-organisms with rigid cell walls that multiply by dividing into two. Some bacteria cause illness and others cause spoilage.
Bactericide	A substance that destroys bacteria.
Bacteriostat	A substance that inhibits the growth and multiplication of bacteria.
Bacteriology	The study of bacteria, especially those that cause disease.
Bacteriophage	A parasitic virus of bacteria.
Binary fission	Asexual method of reproduction by the division of the nucleus into two daughter nuclei, followed by similar division of the cell body. The method of reproduction used by bacteria.
Biodegradable	Chemicals and materials that can be broken down by bacteria or other biological means (usually during sewage treatment).
Biosecurity (of poultry and farm animals)	Taking all necessary measures to prevent the contamination of feed and water and the infection of animals from all sources and vehicles of pathogens, including pests, birds, animals, dust, soil, feed, people, waste, the environment, premises and equipment. It may also include the exclusion of toxic chemicals.
Blanching	The immersion of, for example, vegetables in hot water for up to a minute to destroy enzymes and reduce spoilage, during the storage of frozen food. It also helps to preserve colour and flavour.
Botulinum cook	The thermal process applied to low-acid canned food, which reduces the chance of survival of one spore of *Clostridium botulinum* to less than one in 10^{12}. (Equivalent to 121°C for three minutes)
Carcinogen	A substance capable of causing cancer. Carcinogens may be chemical, physical or biological.
Carrier	A person who harbours, and may transmit, pathogenic organisms without showing signs of illness.
Case	See clinical case.
Case control study	An epidemiological study, in which the characteristics of cases of disease (for example, food histories) are compared with a matched control group of people without the disease, for example, to determine a food vehicle.
Challenge testing (microbiological)	The inoculation of a product with specific pathogens to determine, for example, the critical limits for cooking, by subjecting the product to various time/temperature combinations.
Causal factors (food poisoning)	How the food vehicle became contaminated and the stage of food preparation that allowed bacterial multiplication or survival.
Causative agent	The organism, toxin or poison associated with the illness, which is recovered from sufferers and/or food and/or the environment under investigation, for example, salmonella, scombrotoxin or mercury.
Chronic disease	A disease, that usually develops slowly, and symptoms last for a prolonged period.

HYGIENE for MANAGEMENT Glossary

Chronic poison	A substance, that is used at low concentration and relies on repeated intake by the target pest to ensure elimination.
Cleaning	The process of removing soil, food residues, dirt, grease and other objectionable matter.
Clean surface	A surface that is free from residual film or soil, has no objectionable odour, is not greasy to touch and will not discolour a white paper tissue that is wiped over it.
Clinical case	A person with symptoms who has become ill.
Cohort study	An epidemiological study of a population to compare the incidence of disease in persons exposed and not exposed to possible causal factors. For example, comparing the attack rates of persons with food poisoning, based on those who ate a specific food and those that did not.
Colony	A cluster of bacteria grown on a culture plate in sufficient numbers as to be visible to the naked eye.
Commensal	A usually harmless micro-organism, but one that may cause illness in immunocompromized persons.
Commercially sterile food	Food that has been subjected to treatment, that has destroyed all pathogens and organisms capable of causing problems, such as spoilage, under normal storage conditions, for example, low-acid canned food.
Competent authority	The central authority of a member state competent to ensure compliance with relevant regulations, or any other authority to which that central authority has delegated that competence.
Compliance	Measures that satisfy a legal requirement.
Compressor (refrigerator)	A mechanical pump which moves up and down in a cylinder containing the refrigerant gas and pumps it round the system, starting with the condenser.
Condenser (refrigerator)	A unit that looks like a car radiator with a fan in front of it. The fan draws air from the room across the surface of the condenser, which cools the gaseous refrigerant, delivered from the compressor, and returns it to a liquid state.
Contact time	The period of time required by a disinfectant to achieve disinfection, for example, of a surface.
Contamination	The presence or introduction of a hazard. (EC Regulation No. 852/2004)
Cook-chill	A system of food preparation in which food is cooked in advance and rapidly cooled for chilled storage to be reheated several days later. Strict control of chilled storage temperature is required to ensure the safety of the food.
Core temperature	The temperature found at the centre of the thickest part of a piece of food.
Coved	Rounded finish to the junctions between walls and floors, or between two walls, to make cleaning easier.
Critical control point (CCP)	A step in the process where control can be applied and is essential to prevent or eliminate a food safety hazard or reduce it to an acceptable level.
Cross-contamination	The transfer of bacteria from contaminated foods (usually raw) to ready-to-eat foods by direct contact, drip or indirect contact using a vehicle such as hands or a cloth.
Cryogenics	A system of refrigeration using injection of liquefied gas into the storage chamber.
Cuisine sous vide	A system of cooking (pasteurizing) raw or par-cooked food in a sealed pouch under vacuum. After pasteurization, the pouch is cooled and stored below 3ºC. Products have a shelf life of up to 21 days (eight in the UK) and are regenerated using heat, immediately before consumption.
Culture	A liquid or solid medium used to grow a population of a particular type of micro-organism as a result of the inoculation and incubation of the medium.

HYGIENE for MANAGEMENT *Glossary*

Danger zone of bacterial growth	The temperature range within which the multiplication of some pathogenic, mesophilic bacteria is possible, i.e. from 5°C to 63°C. Most rapid growth occurs between 20°C and 50°C.
Dehydration (of food)	The reduction of the available water in food to prevent the growth of micro-organisms.
Dehydration (of people)	The loss of water and salt from the body, for example, as a result of serious diarrhoea. 60% of a man's body weight and 50% of a woman's body weight is water. This percentage is necessary for the healthy functioning of cells.
Denaturation	Disruption.
Deoxyribonucleic acid (DNA)	The genetic material found in all cells of every organism.
Depuration	The process of purifying shellfish, by re-laying them in clean water for around 42 hours.
Detergent	A chemical or mixture of chemicals made of soap or synthetic substitutes; it facilitates the removal of grease and food particles from dishes and utensils and promotes cleanliness, so that all surfaces are readily accessible to the action of disinfectants.
Disinfectant	A chemical used for disinfection.
Disinfection	The reduction of micro-organisms to a level that will not lead to harmful contamination or spoilage of food. Chemical agents and/or physical methods, which are used, should not adversely affect the food. The term disinfection normally refers to the treatment of premises, surfaces and equipment, but may also be applied to the treatment of skin.
Dormant	The description of bacteria that are not growing or multiplying, but are waiting to multiply when favourable conditions return.
D value	The time, in minutes, required at a given temperature, to reduce the number of viable cells or spores of a given micro-organism by 90%.
Due diligence	The legal defence in the Food Safety Act, 1990, and the Food Hygiene (England) (Scotland) (Wales) (NI) Regulations 2006 that a person charged with an offence had taken all reasonable precautions and exercised all due diligence to avoid the commission of the offence by himself or a person under his control.
Electronic or UV fly killers	Equipment to control flying insects consisting of a UV light, that attracts the insect, which is then destroyed on a high voltage grid or captured on a sticky board.
Emulsification	The production of a mixture of two immiscible liquids, one being dispersed in the other in the form of fine droplets.
Encephalitis	Inflammation of the brain.
Endemic	A disease, that is prevalent in a particular area.
Endocarditis	Inflammation of the internal lining of the heart, especially the valves.
Endotoxin	Toxins present in the outer membrane of the cell of many gram-negative bacteria. They are released on the death of the bacteria.
Enteritis	Inflammation of the intestine.
Enterotoxigenic	Able to produce a toxin adversely affecting the intestines.
Enterotoxins	Exotoxins that affect the gastrointestinal tract.
Environmental control/management	The denial of access to pests, by design maintenance and proofing of buildings and the denial of food and harbourage by good housekeeping.
Enzyme	A protein that regulates the rate of chemical reaction in the body.
Epidemic	A sudden spread of a large number of cases of disease within a community, when there are usually much smaller numbers present.
Epidemiological surveillance	The study of the distribution and trends in incidence of disease in a population through the systematic collection, consolidation and evaluation of morbidity and mortality reports and other relevant data, including incidence, sources, causes, mode of spread, distribution and control. Surveillance includes the

HYGIENE for MANAGEMENT Glossary

	dissemination of information to all those who need to know, so that appropriate action can be taken.
Evaporator (refrigerator)	Consists of a long tube bent many times and passed through hundreds of aluminium fins. Air from the cabinet is drawn over the evaporator and heat is transferred to the liquid refrigerant, delivered from the condenser, which boils and reverts to a gas.
Evisceration	Removal of the internal organs and intestines.
Exotoxin	Highly toxic proteins, usually produced during the multiplication or sporulation of some gram-positive bacteria. They are often produced in food.
Extraneous matter	Any contaminant of raw material or food. Usually referred to as "foreign bodies", in the finished product.
Extrinsic contaminants	Contaminants, that are not part of the food.
Faecal material	Excreta. Stools. The indigestible residues voided from the alimentary canal after digestion and absorption of food and water. One third of the dry weight of human faeces is bacteria, mainly *Escherichia coli* and *Streptococcus faecalis*.
Fermentation	The process involving the growth of beneficial micro-organism and the production of acid in food such as yogurt, cheese, salami and sauerkraut.
First aid materials	Suitable and sufficient bandages and dressings, including waterproof dressings and antiseptic. (All dressings to be individually wrapped.)
Fomites (singular fomes)	Inanimate objects such as clothes, books and bedding that can harbour pathogens and act as vehicles of infection.
Food	Food includes drink, ice and chewing gum but excludes live animals or birds, fodder or animal feeding stuffs, or controlled drugs.
Foodborne disease	Illness resulting from the consumption of food or water contaminated by pathogenic micro-organisms and/or their toxins. Characterized by having a low infective dose and no requirement for the multiplication of the organism within the food to cause illness.
Food business	Means any undertaking, whether for profit or not, and whether public or private, carrying out of any of the activities related to any stage of production, processing and sale of food.
Food-contact surface	A surface that comes into contact with food, for example, a cutting board or a knife.
Food handler	Any person in a food business who handles food, whether open or packaged (food includes drink and ice).
Food handling	Any operation in the production, preparation, processing, packaging, storage, transport, distribution and sale of food.
Food hygiene	The measures and conditions necessary to control hazards and to ensure fitness for human consumption of a foodstuff taking into account its intended use. (EC Regulation No. 852/2004)
Food poisoning	An acute illness of sudden onset caused by the recent consumption of contaminated or poisonous food.
Food vehicle	The food eaten, that contained the pathogen (causative agent), which gave rise to, for example, food poisoning.
Freezer burn	The loss of moisture from the surface of unwrapped frozen food. Results in a parchment-like appearance or spongy pale brown or yellow patches.
Fumigation	The application of a toxic chemical in the form of a gas, vapour or volatile liquid in a closed container or to a food stack under gas-proof sheets.
Fungi	Plants unable to synthesize their own food and are usually parasitic or saprophytic. Includes single-celled microscopic yeasts, moulds, mildews and toadstools. Yeasts are used to produce alcohol, moulds cause food spoilage and ringworm is a fungal disease of animals.

HYGIENE for MANAGEMENT Glossary

Fungicide	A substance that kills fungi and mould.
Galvanized metal	Iron or steel, that has been coated with zinc for protection against corrosion.
Gastroenteritis	An inflammation of the stomach and intestinal tract, which normally results in diarrhoea.
Germicide	An agent used for killing micro-organisms.
Germination	The process resulting in the formation of a new mature vegetative bacterium from a spore, following heat activation and slow cooling.
Grease trap	A device fitted into a drainage system to prevent fat and grease entering the sewer.
HACCP (hazard analysis critical control point)	A food safety management system which identifies, evaluates and controls hazards which are significant for food safety.
Hazard	A biological, chemical or physical agent in, or condition of, food with the potential to cause harm (an adverse health effect) to the consumer. (NB most biological hazards are microbiological).
Hazard analysis (codex alimentarius)	The process of collecting and evaluating information on hazards and conditions leading to their presence to decide which are significant for food safety and therefore should be addressed in the HACCP plan.
Haemoglobin	The oxygen-carrying pigment of red blood cells. The oxygen and haemoglobin join together to form oxyhaemoglobin.
Halophilic organisms	Organisms that can grow in high concentrations of salt.
Haloduric organisms	Organisms that can survive in high levels of salt but do not grow.
Hand-contact surface	A surface touched by the hand, for example, door handle, drawer handle, tap, nail brush, soap dispenser lever, toilet seat and knife handle.
Heat labile	Unstable at high temperatures.
Hermetically sealed	A package for a foodstuff with an airtight seal, to protect it from contamination, for example, cans, plastic pouches and cartons. Food may be pasteurized or sterilized after sealing in the pack, or beforehand and then packed in a sterile environment.
Health declaration	A medical questionnaire completed, for example, when applying for a job or on return to work after an illness, to determine fitness for food handling duties.
High-risk foods	Ready-to-eat foods, which, under favourable conditions, support the multiplication of pathogenic bacteria and are intended for consumption without treatment, which would destroy such organisms.
Host	An organism providing food and shelter for a parasite that lives in or on the host.
Hygiene	The science of preserving health and involves all measures necessary to ensure the safety and wholesomeness of food.
Immune system	The cells and proteins that fight invading pathogens and protect the body from infection.
Immunity	The ability to resist an invading organism so that the body does not develop the disease.
Immuno-compromized	An individual who is unable to produce a normal immune response to an infection.
Incubation period	The period between infection and the first signs of illness.
Indicator organism	Generally a non-pathogenic organism, not usually present in the food, the presence of which suggests poor hygiene and that the food may be contaminated with pathogenic organisms.
Infective dose	The number of a particular micro-organism required under normal circumstances to produce clinical signs of a disease.

HYGIENE for MANAGEMENT Glossary

Infestation	The presence of rats, mice, insects or mites in numbers or under conditions that involve an immediate or potential risk of contamination, loss or damage to food. The term usually implies the existence of a breeding population but may be used to denote the presence of individuals.
Inorganic	Substances that do not contain carbon, such as salt or metal.
Insecticides	Chemical substances used to kill insects.
Inspection	The systematic gathering and recording of data from observations, examinations, and discussions with food handlers and managers, the analysis and interpretation of data collected and the preparation of an understandable and actionable report.
Integrated pest management	The cost-effective implementation of prevention and eradication strategies based on the biology of pests, intended to ensure a pest free operation.
Intrinsic contaminants	Already present in the food, for example, stalks and leaves in vegetables or bones in fish.
Isolate	A single species of a micro-organism, originating from a particular sample or environment, growing in a pure culture.
Lag phase	The first phase of bacterial growth when there is no multiplication and the bacteria acclimatize to their new environment.
Larva (plural larvae)	The immature stage of an insect that undergoes complete metamorphosis (Egg >larva > pupa > adult).
Lassitude	A feeling of weakness or tiredness.
Load lines	Lines marked on refrigerated units above, or in front of, which product will be out of refrigeration or obstruct airflow, resulting in a failure to keep food at the correct temperature.
Low-acid foods	Food with a pH above 4.5.
Low-dose pathogen	A pathogen capable of causing illness with very few organisms, for example, less than ten. Foodborne diseases are usually caused by low-dose pathogens.
Low-risk foods	Ambient stable foods that do not normally support the multiplication of pathogens. However, they may occasionally be responsible for foodborne illness if they are ready-to-eat foods contaminated with low-dose pathogens, for example, lettuce.
Lux	A measure of light levels.
Malaise	A vague feeling of being ill.
Mastitis	Inflammation of the mammary gland.
Median	Situated towards the middle. The middle number or average of the two middle numbers in an ordered sequence of numbers, for example, 8 is the median of 2, 8, 32 and 1, 6, 10, 14.
Mesophiles	Organisms that have a growth range of 10°C to 56°C, with an optimum of 20°C to 46°C.
Metabolism	A term used to describe all of the chemical processes that occur in the body.
Microaerophilic organisms	Organisms that prefer to grow in an environment of approximately 5% oxygen, for example, campylobacter.
Microbiology	The science of studying micro-organisms.
Micro-organisms	Any minute living organisms including bacteria, viruses, yeasts, moulds, protozoa and prions.
Modified atmosphere packaging	The replacement of air in a package by one or more gases followed by sealing to prevent re-entry of air.
Mildew	A type of fungus similar to mould.
Morphology	The study of the structure and form of micro-organisms.
Moulds	Microscopic plants (fungi) that may appear as woolly patches on food.
Mycotoxins	Poisonous chemicals (toxins) produced by some moulds, for example, *Aspergillus flavus*.

HYGIENE for MANAGEMENT *Glossary*

Neonate	A newborn child up to four weeks old.
Neurotoxin	A poison/toxin produced by pathogens, for example, *Cl. botulinum*, which affects the nervous system.
Onset period	The period between consumption of the food and the first signs of illness. (Where incubation of micro-organisms within the body does not take place.)
Ootheca	The egg case of a cockroach.
Open food	Unwrapped food that may be exposed to contamination.
Optimum	Best.
Organic	Relating to or derived from plants or animals and having a carbon basis.
Organoleptic	Involves the use of the five senses: sight, smell, hear, taste and touch.
Osmophiles	Organisms that can grow in high concentrations of sugar.
Osmoduric	Organisms that can survive in high concentrations of sugar without multiplying.
Outbreak	An incident in which two or more people, thought to have a common exposure, experience a similar illness or proven infection (at least one of them having been ill).
Outbreak location	The place where the food vehicle was prepared or served.
Parasite	An organism that lives and feeds in or on another living creature, known as the host, in a way that benefits the parasite and disadvantages the host. In some cases, the host eventually dies.
Pasteurization	A heat treatment of food at a relatively low temperature that destroys vegetative pathogens and most spoilage organisms, so prolonging the shelf life. Toxins and spores generally survive and rapid cooling and refrigerated storage is usually essential.
Pathogen	Disease-producing organism.
Personal hygiene	Measures taken by food handlers to protect food from contamination from themselves.
Pests (food)	Animals which live in or on our food. They contaminate food and are destructive noxious or troublesome.
Pesticide	A chemical used to kill pests.
pH	An index used as a measure of acidity or alkalinity.
Phage type	A variety of bacteria within a particular species, distinguished on the basis of their susceptibilities to a range of bacteriophages.
Plate count	The number of colonies of bacteria growing on an agar plate.
Potable	Safe to drink and acceptable for use in food preparation.
Preservation of food	The treatment of food to prevent or delay spoilage and inhibit growth of pathogenic organisms, which would render the food unfit.
Primary production	Those stages in the food chain up to and including, for example, harvesting, slaughtering, milking and fishing.
Prions	Proteinaceous infectious particles.
Probe	The part of the thermometer that is inserted into food or between packs to obtain temperature readings.
Product code	The date marking on product packaging to show its safe shelf life within which time it should be consumed.
Protective clothing	The clothing worn by food handlers to prevent contamination of food, including hairnets, hats, coats, apron, gloves and boots.
Proteolytic	Having the ability to break down proteins.
Protozoa (singular protozoan)	Single-celled organisms, which form the basis of the food chain. Live in moist habitats such as oceans, rivers, soil and decaying matter. Some are pathogenic, for example, *Entamoeba histolytica*.
Psychrophiles	Organisms that have a growth range of –8°C to 25°C with an optimum below 20°C.

HYGIENE for MANAGEMENT *Glossary*

Psychrotrophs	Organisms that have a growth range of −5°C to 40°C with an optimum above 20°C.
Pulmonary	Pertaining to the lungs.
Pupa (plural pupae)	The third stage of development of an insect that undergoes complete metamorphosis.
Quats	A popular name for quaternary ammonium compounds (disinfectant).
Radicidation	The use of irradiation to destroy non-spore-forming pathogens, pests and parasites.
Radurization	The use of irradiation to destroy spoilage organisms, prolong shelf life and reduce micro-organisms to a safe level.
Ready-to-eat foods	Food, that are intended for consumption without any treatment that is intended to destroy any pathogens that may be present. They include all high-risk foods and such foods as fruit, salad, vegetables and bread.
Rehydration	The addition of liquid, usually water or milk, to dry foods.
Residual insecticide	A long-lasting insecticide applied in such a way that it remains active for a considerable period of time.
Respiration	The loss of water by evaporation, for example, from fruit.
Risk	The likelihood of a hazard occurring in food.
Risk assessment	The process of identifying hazards, assessing risks and severity and evaluating their significance.
Route	The path along which bacteria are transferred from a source to ready-to-eat foods.
Safe food	Food that is nutritious, compositionally sound, free from contaminants at levels that could cause harm or illness and is labelled with the correct safety instructions for use and storage.
Sanitizer	A chemical agent used for cleansing and disinfecting surfaces and equipment.
Saprophyte	An organism that lives on dead organic matter.
Septicaemia	Blood poisoning. Rapid multiplication of bacteria and toxin production within the blood.
Serotyping	A test used to distinguish between different sub-types of the same species of bacteria, for example, *Salmonella* Typhi and *Salmonella* Enteritidis.
Shelf life	The period within which food is safe and of the best quality.
Shelf-stable	Foods that do not normally suffer microbiological spoilage at room temperature.
Sneeze guards	Screens fitted to food display units, intended to protect food from customers sneezing or coughing.
Source (of pathogens)	A source may be considered as the origin of the pathogen (causative agent), for example, the cow on the farm or the vehicle that brought the pathogen into the premises, for example, the milk.
Spoilage	Food deterioration resulting in off-flavours, odours and change in appearance, indicating the products are unsuitable for sale or consumption.
Sporadic case	A single case of disease apparently unrelated to other cases or carriers.
Spore (bacterial)	A resistant resting-phase of bacteria, which protects them against adverse conditions.
Spore (fungal)	The reproductive body of a fungus.
Sporulation	The production of spores, for example, in adverse conditions.
Sterile	Free from all living organisms.
Sterilization	A process that destroys all living organisms.
Stock rotation	The practice of ensuring the oldest stock is used first and that all stock is used within its shelf life.
Sublimation	The process of ice, under vacuum, changing directly into water vapour without going through a liquid phase.
Surveillance	See epidemiological surveillance.

HYGIENE for MANAGEMENT Glossary

Symptomatic	With symptoms.
Synergism	A phenomenon whereby two chemicals in combination have a greater effect than the sum of their individual efforts.
Systemic disease	A disease affecting the whole of the body, as opposed to being localized.
Taint	Contamination of food from undesirable flavours or odours, for example, butter absorbs paint fumes, and chocolate stored next to detergent washing powders will taste soapy.
Tap proportioner	A device fitted to a tap, which delivers the right amount of detergent or sanitizer to the water.
Taxonomy	The naming of organisms (nomenclature), the grouping of organisms (classification) and the identification of organisms.
Terrazzo flooring	Coloured marble chips set in portland cement in a mosaic fashion. (May be attacked by acids and alkalis.)
Thermophilic organisms	Organisms that prefer to multiply above 45°C.
Thermoduric organisms	Organisms that can survive but do not multiply at temperatures above 45°C.
Total viable count	The total number of living cells detectable in a sample. The number of cells is assessed from the number of colonies, which develop on incubation of a suitable medium, that has been inoculated with the sample of bacteria.
Toxins (bacterial)	Poisons produced by pathogens, either in the food or in the body after consumption of contaminated food.
Typing (bacteria)	Any method used to distinguish between closely related strains of micro-organisms, for example, members of the same species.
Use-by-date	The date mark required on high-risk perishable prepacked food, which must be stored under refrigeration. The food should be consumed on or before the use-by-date. It is an offence to sell food after its use-by date.
Vegetative bacteria	Viable bacteria capable of multiplication.
Vehicles (for pathogens)	Things used by bacteria to transfer them from sources to ready-to-eat foods, including hands, cloths and equipment, hand-contact surfaces and food-contact surfaces.
Virology	The study of viruses.
Virulence	The ability of a pathogen to cause disease, for example, the number of deaths, the proportion of people exposed to infection who became ill and how rapidly the infection spreads through the body.
Viruses	Microscopic pathogens that multiply in the living cells of their host.
Water activity a_w	A measure of the available water in a food.
Water hardness	Water containing dissolved calcium and magnesium salts. Temporary hardness results from the presence of bicarbonates and these form a scale when water is heated. Permanent hardness involves sulphates and is unaffected by heat. Soft water usually has up to 60ppm of salts, whereas hard water is over 120ppm. Hardness interferes with the action of soap causing a scum.
Water softener	A unit to remove water hardness salts to prevent a scale build-up in water heating equipment.
Wholesome food	Sound food, fit for human consumption.
Xerophilic organisms	Organisms capable of growth under dry conditions at lower levels of air.
Yeast	A unicellular fungus, which reproduces by budding and grows rapidly on certain foodstuffs, especially those containing sugar. Yeasts are the chief agents of fermentation. (Sugar converted to alcohol and carbon dioxide.)
Zoonoses	Diseases, that can be transmitted naturally from animal to man and vice versa.
Z value	The change in temperature in degrees celsius required for a ten-fold change in D value.

Abbreviations

Commonly used abbreviations in Hygiene for Management

ACC	Aerobic colony count
ATP	Adenosine triphosphate
A_W	Water activity
BMI	Body mass index
BMR	Basal metabolic rate
BPCA	British Pest Control Association
BSE	Bovine spongiform encephalopathy
CAP	Controlled atmosphere packaging
CCDC	Consultant in communicable disease control
CCP	Critical control point
CCTV	Closed circuit television
CDSC	Communicable Disease Surveillance Centre
CFU	Colony forming unit
CHIPS	Chemicals, hazards, information and packaging for supply
CIEH	Chartered Institute of Environmental Health
CIP	Clean-in-place
CJD	Creutzfeldt-Jacob disease
COSHH	Control of substances hazardous to health
CPU	Central production unit
DEFRA	Department of the Environment, Food and Rural Affairs
DH	Department of Health
DNA	Deoxyribo nucleic acid
EDTA	Ethylene diamine teracetic acid
EFKs	Electronic fly killers
EHO	Environmental health officer
EHP	Environmental health practitioner
ELISA	Enzyme linked immunosorbant assay
EU	European Union
FSA	Food Standards Agency
FSA (Ireland)	Food Safety Authority of Ireland
GMP	Good manufacturing practice
HACCP	Hazard analysis critical control point
HDL	High density lipoprotein
HOCl	Hypochlorous acid
HPA	Health Protection Agency
HTST	High temperature short time

HYGIENE for MANAGEMENT Abbreviations

HUS	Haemolytic uraemic syndrome
IPM	Integrated pest management
LACORS	Local Authorities Coordinators of Regulatory Services
LDL	Low density lipoprotein
MAP	Modified atmosphere packaging
MEL	Maximum exposure limits
NAMAS	National measurement accreditation service
NVQs	National vocational qualifications
PCR	Polymerase chain reaction
PPE	Personal protective equipment
PVC	Polyvinyl chloride
QA	Quality assurance
QACs	Quaternary ammonium compounds
QC	Quality control
REHIS	Royal Environmental Health Institute of Scotland
RIPH	Royal Institute of Public Health
RNA	Ribo nucleic acid
RSPH	Royal Society for the Promotion of Health
SI	Statutory instrument
SOFHT	Society of Food Hygiene and Technology
SPI	Stored product insect
TVC	Total viable count
UHT	Ultra heat treatment
UK	United Kingdom
USA	United States of America
UV	Ultraviolet
ULV	Ultra low volume
VCJD	Variant Creutzfeldt-Jacob disease
VTEC	Vero cytotoxin-producing *E. coli*
WHO	World Health Organization

Useful websites

Awarding Bodies	See Appendix III
British Nutrition Foundation	www.nutrition.org.uk
Centers for Disease Control (USA)	www.cdc.gov
Codex Alimentarius Commission	www.codexalimentarius.net
Dept of Environment, Food & Rural Affairs	www.defra.gov.uk
Department of Health (DH)	www.doh.gov.uk
European Law	www.europa.eu.int/eur-lex
European Food Safety Authority	www.efsa.eu.int
Food & Drink Federation	www.fdf.org.uk
Foodlaw - Reading	www.foodlaw.rdg.ac.uk
Foodlink	www.foodlink.org.uk
Food Safety Authority of Ireland	www.fsai.ie
Food Safety Information (USA)	www.foodsafety.gov
Food Safety Trainers	www.foodsafetytrainers.co.uk
Foodservice (USA)	www.foodservice.com
Food Standards Agency	www.food.gov.uk
Her Majesty's Stationery Office	www.hmso.gov.uk
Hospitality Institute of Technology & Management (USA)	www.hi-tm.com
Health Protection Agency	www.hpa.org.uk
Health Protection Scotland	www.hps.scot.nhs.uk
Institute of Food Science & Technology	www.ifst.org
LACORS	www.lacors.com
United States Dept of Agriculture (USDA)	www.usda.gov
World Health Organization	www.who.int

Bibliography

Eley R. (Ed.) (1995). Microbial Food Poisoning. London: Chapman & Hall.

Garbutt J. (1997). Essentials of Food Microbiology. London: Arnold.

Heldman R. Dennis and Hartel W. Richard (1997). Principles of Food Processing. New York: Chapman and Hall.

Hobbs B. C. and Roberts D. (1995). Food Poisoning and Food Hygiene. London: Edward Arnold.

Jay James M. (1996). Modern Food Microbiology. New York: Chapman and Hall.

MacAuslan E. M.R. (2004). Effective Food Hygiene Training. Doncaster: Highfield.co.uk ltd.

North R. (1999). Some Observations on Food Hygiene Inspections. London: Chadwick House Group Ltd.

Sheard M. and Church I. (1992). Sous-Vide Cook-Chill. Leeds: Leeds Metropolitan University.

Smith T. (Ed) (1995). Complete Family Health Encyclopedia. London: Dorling Kindersley.

Sprenger R.A. (2006). The Foundation HACCP Handbook. Doncaster: Highfield.co.uk ltd.

Taylor E and J. (1990). Master Catering Theory. London: Macmillan Press Ltd.

Wallace C. (2006). Intermediate HACCP. Doncaster: Highfield.co.uk ltd

Index

A

Aberdeen typhoid outbreak, 74
Abrasives, 218
Accelerated freeze drying, 150
Access, for pests, 253
Accidental spillage, 250
Acetic acid, 153, 230
Acid descaler, 252
Acid detergent, 224
Acid foods, 27, 120, 363
Acidification, 153
Acids, 225, 380
Acts of Parliament, 328
Acute disease, 380
Acute poison, 380
Additives, 107, 151, 358, 380
Adenosine triphosphate (ATP), 232
Adenosine triphosphate measurement, 309
Adenoviruses, 57
Adolescents, diet of, 104
Adults, diet of, 104
Advanced Certificate in Applied HACCP Principles (RIPH), 378
Advisory Committee on the Microbiological Safety of Foods, 73, 138
Aerobes, 28
Aerobic, 380
Aeromonas spp., 38
Aerosol, 380
Aflatoxicosis, 47
Aflatoxin, 47
Aflatoxin M1, 46
AFRC Institute of Food Research, 125, 130
Agar, 380
Agglutination, 24
Aide-memoire, 303
Air-blast freezing, 147
Airbricks, 116
Air-conditioning systems, 193
Air drying, 243
Air-off temperature, 121, 380
Air-on temperature, 121, 380
Air velocity, 111, 119
Albumen, 67, 73, 123, 367

Alcohols, 228
Aldehydes, 228
Alfalfa, 10
Algae, 380
Alkali, 380
Alkaline detergent, 224
Allergen, 11
Allergic reaction, 107
Allergy, 11, 272, 380
Alpha-amylase test for liquid egg, 374
Alphachloralose, 265, 280
Aluminium, 44, 207, 225, 234
Ambient temperature, 340, 380
American Cockroach, 275
Amines, 152
Amino acids, 66, 67, 95, 96, 108
Amoebic dysentery, 53
Amphoteric surfactants, 218
Amylase, 108
Anaerobes, 28
Anaerobic, 358, 380
Angles, internal, 209
Ångström units, 157
Animals, 75
Anionic surfactants, 218
Anisakis, 145
Anorexia, 380
Antibiotics, 11, 153, 367, 380
Antibody, 310, 380
Anticoagulants, 264
Antigen, 308, 380
Antimony, 44
Anti-motility agents, 11
Antioxidant, 107, 153
Antiseptic, 380
Antiserum, 380
Ants, 277
Apple juice, 61, 75
Apricot kernels, 45
Aprons, 168
Aquatic biotoxins, 46
Aseptic, 380
Aseptic production technology, 365
Aspergillus flavus, 28, 47

Aspergillus parasiticus, 47
Asperigillus ustus, 46
Aspic, 55, 380
Astroviruses, 57
Asymptomatic, 380
Atmospheres, controlled, 154
Auditing, 299, 381
Australian Spider Beetle, 273
Authorized officer, 12, 28, 49, 314, 316, 337, 338, 339, 340, 343, 345, 354, 381
Authorized persons, 280
Autolysis, 381
Automatic defrost, 122
A_W (water activity), 27, 145, 149, 358, 381, 389

B

Bacillary dysentery, 53
Bacilli, 21
Bacillus cereus, 21, 22, 37, 42, 124, 134, 137, 366, 372
Bacillus coagulans, 363, 372
Bacillus licheniformis, 37
Bacillus nivea, 362
Bacillus sporothermodurans, 74, 367
Bacillus stearothermophilus, 361, 363
Bacillus subtilis, 37, 143, 372
Bacteraemia, 381
Bacteria, 8, 10, 28, 381
  definition of, 20
  colonies of, 26
  multiplication of, 25-29
  vegetative, 389
Bacterial food poisoning, 30-38
  case of, 13
  prevention of, 39-42
Bacterial growth, 358
Bactericidal soaps, 162, 164, 165
Bactericide, 381
Bacteriological counts, 312
Bacteriological food sampling
  by enforcement officers, 311
  disadvantages of, 307
Bacteriological monitoring, 306, 307
Bacteriology, 381
Bacteriophages, 24, 381
Bacteriostat, 381

Bait boxes, 259
Bait shy, 265
Baiting
  pulsed, 264
  surplus, 264
Baits, 265, 267
  safety precautions in the use of, 266
Baked products, 142
Balanced diet, 96, 99
Balloon guards, 193
Basal metabolic rate (BMR), 91
Bean sprouts, 10, 74
Bearings, 210
Beef, rare, 123
Beetroot, 142
Belgium pâté, 54
Benzene, 43
Benzoic acid, 152
Best-before date, 351
Beverage vending machine, 246
Biguanides, 227
Bile salts, 108
Bimetallic coil thermometers, 120
Binary fission, 25, 381
Biochemical reactions, 24, 68
  determination, 308
Biodegradability, 218, 381
Biological principles, 63
Biosecurity (of poultry and farm animals), 381
Birds, 75
Birdwire systems, 279
Biscuit Beetle, 273
Bitty cream, 37
Bivalve molluscs, 375
Black Rat, 262, 263
Blackmail, 288
Blanching, 146, 150, 230, 381
Blast chillers, 119, 125, 126
Blast-freezing, 139
Blatta orientalis, 275
Blattella germanica, 275
Bleach, 251
Blood, 369
  pressure, 95
Blowflies, 268

HYGIENE for MANAGEMENT Index

Blown cans, 362
Bluebottle, 268
Body defences, 108
Body mass index (BMI), 100
Boils, 166
Bolts, contamination from, 82
Bone, 88
Booklice, 271, 273
Boric acid, 277
Botrytis, 142
Botulinum cook, 358, 364, 381
Botulinum, type E, 155
Bovine spongiform encephalopathy (BSE), 60
Bread, spoilage of, 142
Breakfast, 102
Brining, 152
Bristles, 240
Brodifacoum, 264
Bromadiolone, 264
Brown House Moth, 274
Brown Rat, 260, 262
Browning, 151, 153
Brucella abortus, 55, 74
 ring test, 368
Brucella melitensis, 55
Brushes, 240
Buckets, 240
Burgers, 124
Butter, 370
Butterfat, 369
Buttermilk, 370
Byelaws, 328
Byssochlamys fulva, 362

C
Cadmium, 44
Calciferol, 265
Calcium, 98
 propionate, 153
Caliciviruses, 57
Calibration, 120
Calliphora spp., 268
Calories, 90
Campylobacter, 14, 16, 17, 19, 22, 23, 72, 75
Campylobacter coli, 53

Campylobacter enteritis, 12, 53
Campylobacter jejuni, 53
Canned foods, 113
 pasteurized, 364
 thermophilic spoilage bacteria of, 363
Canning, 358-364
Canopies, 193
Cans
 blown, 362
 damage of, 363
 exhausting of, 358
 overfilling of, 363
 sealing of, 360
Capsule, 20
Caramelization, 366
Carbohydrates, 90, 93, 109
Carcinogen, 381
Cardboard, 81
Carrier, 11, 169, 170, 336, 376, 381
Case control study, 381
Catering Industry Guide, 176, 194
Cationic surfactants, 219
Causal factors, 48, 381
Causative agent, 48, 381
Caustic potash, 225
Caustic soda, 224, 225, 249
CE marking, 205
Ceilings, 188, 334
Cell membranes, 20, 66
Cell wall, 20
Central production unit (CPU), 132
Centre temperature, 123
Cereals, 113
Certificate in Food Hygiene Awareness (RSPH), 378
Challenge testing, 381
Chartered Institute of Environmental Health (CIEH), 377
Checklist, 176
Cheese 54, 371
 Cheddar, 37
Chelates, 224
Chemical
 contamination, 89
 stores, 249
 supplier, 222

HYGIENE for MANAGEMENT Index

Chemical and metallic poisoning, 42-44
Chemicals (Hazards, Information and Packaging for Supply) Regulations, 1994 (CHIPS), 248
Chicken, thawing of, 69
Chill holding requirements, 340, 342
Chilled display cabinet, 120
Chilled foods, the safety of, 138
Chilled storage and distribution, 138
Chilling, 137
Chloramphenico, 43
Chlorhexidine, 165
Chlorinated water, 361
Chlorine gas, 224
Chlorine release agents, 226
Chlorophenols, 226
Cholesterol, 95
Chopping boards, 213
Chronic disease, 381
Chronic poison, 382
Chronic rodenticides, 263
Ciguatera poisoning, 46
Cirrhosis, 47
Civic Government (Scotland) Act, 1982, 357
Cladosporium herbarum (Blackspot), 142
Cleaner, heavy duty, 251
Cleaning, 7, 118, 223, 382
    activities, 84
    aids, 238
    and HACCP, 217
    and the law, 216
    benefits of, 216
    control and monitoring of, 221
    definition of, 222
    effective methods of, 237-241
    energy in, 222
    foam or gel, 233
    hazards from, 217
    manual, 232
    methods of, 232-236
    post actions, 242
    preliminary actions, 242
    problems from, 218
    slicing machine, 245
    soft ice-cream machine, 245
    technology of, 222
    work surface, 245, 382
Cleaning and Disinfection, 216-252
Cleaning and disinfection facilities, 194
Cleaning chemicals, 89
    choosing of, 237
    dilution of, 252
    time to function, 238
Cleaning costs, 223
Cleaning procedures, 242-247
Cleaning products, 251
Cleaning schedules, 220
Cleaning terms, 218-220
Cleaning-in-place (CIP), 235
Cling film, 114
Clinical case, 382
Cloakrooms, 197
Closing meeting, 317
Clostridium, 21
Clostridium botulinum, 21, 22, 35, 41, 115, 134, 137, 358, 360, 364
Clostridium perfringens, 14, 23, 25, 33, 41, 126, 140, 314
Clostridium sporogenes, 363
Clostridium thermosaccharolyticum, 363
Closure of food premises, 255, 338
Cloth, 76, 78
Clothes moth, 274
Coaching, 172
Cobalt, 60, 154
Cocci, 21
Cockle fishing, 46
Cockroaches, 275-277
    control of, 276, 277
    traps, 277
Codes, 116
Codes of practice (statutory), 346
Codex Alimentarius, 287, 290, 333
Coding, 360
Cohort study, 382
Cold stores, 127
Coleslaw, 54, 74
Coliforms, 313, 341, 366
Colonies, 23, 382
Colour coding, 214, 238
Commensals, 10, 382
Commercially sterile food, 382
Common House Fly, 268

HYGIENE for MANAGEMENT Index

Communicable Disease Surveillance Centre, 12
Communication skills, 182
Community guides to good hygiene practice, 333
Compactor, 198
Compatibility, 219
Competency, of the food handler, 172, 180
Competent authority, 382
Competition, 28
Compressor (refrigerator), 382
Computer-based programmes, 178
Condenser (refrigerator), 382
Connections, gas and electrical, 212
Construction of Food Premises, 186-203
Consultant environmental health practitioner/officer, 321
Contact dust and gels, 265
Contact time, 382
Containers, 81
Contamination, 7, 8, 70, 119, 214, 328, 332, 333, 334, 335, 336, 382
   by micro-organisms, 70, 362
   chemical, 70, 217
   control, 77, 86
   microbiological, 70, 217, 362
   physical, 70, 217
   sources of, 48, 71, 76, 388
Contemporaneous notes, 50
Contractors, 220
   liaison with, 259
   selection of, 258
Control and Monitoring of Food Standards and Operations, 282-325
Control of Pesticides Regulations 1986, 258
Control of Substances Hazardous to Health (COSHH) Regulations, 1999, 248, 267, 354
Control measures, 284, 290, 296-298
Control point, 284
Control systems, 86
Controlled atmospheres, 154
Convalescent carriers, 11
Conviction, 339, 345
Conveyors, 212
   belts, 83
Cook-chill systems, stages of, 133

Cook-chill, 132-136, 382
   benefits of, 134
   management of, 136
   microbiological guidelines of, 136
   safety, 134
Cooked food, 68
Cook-freeze, 139
Cooking, 123
Cool spots, 34, 35
Cooling, 125, 361
   slow, 126
   water, 361
Cooling towers, 193
Copper, 44, 98, 207
Core temperature, 382
Corrective action, 284, 292, 296-298
COSHH, 248, 267, 354
Council Directive relating to machinery 89/392/EEC, 205
County court, 329
Courses, classroom based, 171
Court of Appeal, 329
Court of Session, 356
Coving, 382
Craft knives, 70
Cream line, 366
Cream, 112, 369
   imitation, 370
   non-dairy, 370
Cream cleaners, 218
Creutzfeldt-Jacob disease, 60
Criminal Justice Act 1967, 50
Criminal Justice Act, 1991, 339
Criminal Justice and Public Order Act, 1994, 350
Critical control point (CCP), 284, 290, 332, 382
   monitoring of, 292
Critical limit, 284, 291, 296-298
Cross-contamination, 73, 76, 176, 304, 382
Crown court, 329
Crown immunity, 351
Cryogenic freezing, 147
Cryogenics, 382
Cryptosporidiosis, 16, 19, 61
Cryptosporidium, 16, 17, 19

Cuisine sous-vide, 136, 382
Culling, of birds, 280
Culture, 382
Culture media, 310
Curing, 152, 364
Customer complaints, 79
Cuts, 166
Cyclobutanones, 156
Cyclospora cayentanensis, 61
Cyclosporiasis, 61
Cysticerus bovis, 61, 145
Cysts, 62
Cytoplasm, 20

D
D value, 148, 383
Dairy products, 365
Damage, caused by pests, 254
Damaged stock, 115
Danger zone, 39, 63, 67, 118, 383
Data, 304
Data loggers, 122
Dead spaces, 209
Deadly nightshade, 45
De-boxing, 186
Decision tree, 284, 290, 291
Decline phase, 26
Decomposition, 142
Defrosting, 118
Degreaser, 251
Dehydration, 100, 109, 114, 149, 150, 340, 383
Deliveries of food, 110, 296
Denaturation of proteins, 66, 67, 383
Dental sepsis, 169
Deoxyribo nucleic acid (DNA), 67, 68, 383
Department of the Environment, Food and Rural Affairs, 114
Depuration, 59, 375, 383
Dermatitis, 272
Desiccation, 21
Design and Construction of Food Premises, 186-203
Design of Equipment, 204-215
Design, general principles, 186
Desulfotomaculum nigrificans, 363

Detergents, 223, 224, 251, 383
  dishwash, 252
  neutral, 252
Deterioration, of equipment, 214
Deviation, 285
Diabetes, 107
Diarrhetic shellfish poisoning, 46
Dieldrin, 43
Diet and health, the link between, 102
Diffusers, 83
Digestion, 108
Dioxins, 42, 89
Diploma in Advanced Food Hygiene (REHIS), 379
Diptera, 268
Directives, 328
Dirty cans and bottles, removal of, 88
Disease of Animals (Waste Food) Order 1973, 349
Dishcloths, 243
Dishwash rinse aid, 252
Dishwashing, 233
  mechanical, 243
Disinfectants, 21, 252, 334, 383
  amphoteric, 227
  choosing of, 228
  inactivation of, 231
  QAC based, 251
  resistance to, 231
Disinfecting, of fruit and vegetables, 229
Disinfection, 7, 225-232, 334, 335, 361, 369, 383
  chemical, 226
  frequency of, 230
  heat, 226
  of airspace, 230
  of cleaning materials and equipment, 229
  of food and hand-contact surfaces, 229
  of hands, 229
  steam, 226
  where, 229
Dispenser
  mushroom, 237
  swan-neck, 237
Dispersion, 223

Display cabinets, 120
DNA, 20
　detection, 308
Documentation for HACCP, 293
Doors, 189, 334
Dormant, 383
Dose, infective, 385
Dosing aids and applicators, 237
Double-sink washing, 242
Double-wash procedure, 162
Drain Flies, 268
Drainage, 74, 192, 212
Drains, 192, 256
Drawing pins, 83
Drinks vending machines, 140
Drosophilia spp., 268
Dry clean, 233
Dry-food stores, 116
Drying
　accelerated freeze, 372
　methods of, 150
　roller, 372
　spray, 372
Ducting, 193
Due diligence, 79, 86, 121, 170, 175, 258, 259, 260, 283, 294, 323, 338, 344, 383
Duodenum, 108, 109
Dust, 75
Dysentery, 53
Dyslexia, 180

E

Ears, 166
Echinococcus granulosus, 62
Eczema, 169
Eggs, 71, 73, 111
　liquid, 374
E-learning, 179
Electrical supplies, to food premises, 191
Electricity at Work Regulations, 1989, 354
Electromagnetic spectrum, 157
Electron microscope, 58
Electronic fly killer, 85, 269, 383
Elementary Food Hygiene Course (REHIS), 379
ELISA (enzyme linked immunosorbant assay), 24

Emergency control order, 344
Emollients, 164, 165
Employee medical questionnaire, 376
Emulsification, 370, 383
Emulsion, 219
Encephalitis, 383
End kitchens, 133
Endemic, 383
Endocarditis, 383
Endotoxin, 23, 31, 383
End-product testing, 284
Energy, 64, 90, 91
Energy-providing nutrients, 90
Enforcement authority, 337
Enforcement officers, 158
Enrichment culture, 308
Entamoeba histolytica, 53
Enteric fever, 52
Enteritis, 383
Enterobacter spp., 160
Enterobacteriaceae family, 312, 313
Enterococcus faecalis, 38
Enterotoxigenic, 383
Enterotoxin, 23, 33, 383
Environment Agency, 250
Environmental
　considerations, 250-252
　control/management, 383
Environmental health department, 113, 115, 130, 183, 184
Environmental health practitioners/officers, 49, 182, 186, 328, 337, 355, 390
Enzymes, 67, 372, 383
　autolytic, 142
Ephestia elutella, 274
Epidemic, 383
Epidemiological surveillance, 383
Epidermis, 160
Equipment
　cleaning of, 214
　exterior, 210
　fixing and siting of, 210
　for cleaning, 238
　requirements, 335
　selecting the right type, 238
Erepsin, 108
Ergotism, 28

HYGIENE for MANAGEMENT Index

Escherichia coli, 24, 38, 42, 55, 160, 314, 375
Escherichia coli O157, 7, 10, 14, 16, 17, 18, 22, 55, 75, 314
Ethanol, 228
Ethnic groups, diet of, 106
European Commission, 60
European Committee for Standardization, 206
European six pack regulations, 247
European Union (EU), 311, 328
Eutrophication of rivers, 224
Evaporated milk, 372
Evaporator coil, 118, 384
Evisceration, 384
Examination, 180
Exclusion of food handlers, 169
Exhausting of cans, 358
Exotoxin, 21, 36, 37, 384
Expert witness, 324
Extraneous matter, 384
Extrinsic foreign bodies, 79, 384

F

F_0, 360, 365
Facultative anaerobes, 28
Faecal matter, 384
Faecal streptococci, 123, 310
Fan motors, 193
Fannia canicularis, 268
Fans, 118
Farm animals, 56
Fasciola hepatica, 62
Fasteners, 209
Fats, 90, 93
Fatty acids, 93, 94
Feral pigeons, 278
Fermentation, 153, 370, 384
Fibres, 81, 92
Filling hoppers, 83
Fimbriae (or pili), 20
Final consumer, 330
Fipronil, 277
Firebrats, 272
First aid, 249, 384
First-in, first-out, 116
Fish and shellfish, 73, 374
Fish and shellfish trade, law affecting, 374

Flagella, 20
Flaking paint, 82
Flat sours, 361, 363
Flies and flying insects, 268-270
Flipper, 362
Floors, of food premises, 190, 336
Flour, 113
Flow diagram/chart, 285, 288
Flow diversion valve, 366
Fluorescent tubes, 194
Fluoride, 99
Fly screens, 269
Flying insect control, 269
Flying insects, chemical control of, 270
Foaming activity, 219
Folate, 97
Fomites (singular fomes), 384
Food, 330, 343, 384
    appearance of, 306
    cooking of, 66
    cooling of, 65, 336
    disinfecting, 230
    hot holding, 139, 341, 342
    packaging of, 120
    reheating of, 342
    samples of, 311
    smell of, 306
    sound of, 306
    taste of, 306
    texture of, 306
    vehicle, 48,
    washing, 230, 334
    waste, 335
    wholesomeness of, 7, 389
Food additives, 107, 353
Food alerts, 355
Food and Drug Administration (US), 124
Food authorities, 51, 331, 337, 343
Foodborne Diseases, 16, **52-62**, 336, 343, 372, 376, 384
Foodborne disease in Northern Ireland, 18
Foodborne disease in Scotland, 16, 17
    law relating to, 357
Foodborne illness, law relating to, 350
Food business, 330, 384
Food business operator, 330
Food complaints, investigation of, 321

Food-contact surfaces, 229, 384
Food Contamination and its Prevention, 70-89
Food (Control of Irradiation) Regulations 1990, 157
Food control, organizations involved in, 326
Food employees, 8
Food handlers, 84, 376, 384
    exclusion of, 170
    fitness to work, 169
    motivation of, 173
Food hygiene, 7-19, 332, 377, 384
Food Hygiene (England) (Scotland) (Wales) (NI) Regulations 2006, 255, 323, 337, 346, 374
Food incidents and hazards, 355
Food Labelling (Amendment) (Irradiated Food) Regulations, 1990, 157
Food Labelling Regulations, 1996, 351
Food Law Code of Practice, 323, 338, 346, 356
Food law, enforcement of, 355
Food (Lot Marking) Regulations, 1996, 353
Food MicroModel, 311
Food Poisoning, 7-19, **30 -51**, 328, 343, 384
    bacterial, 10, 30
    chain, 39
    costs of, 7
    incidence of, 12-19
    law relating to, 350
    notifications of, 12
    statistics, 12
    types of, 30
Food poisoning in Ireland, 17
Food poisoning in Scotland, 16, 17
    law relating to, 357
Food poisoning outbreaks, investigation of, 48-51
Food premises, 333
    construction and design of, 186-203
    official control of, 315
    painting of, 189
Food Processing, 358-375
Food rooms, 334
Food safety, 63
    management system, 285, 332
    policies, 283
    requirements, 331, 340, 343
    training, 171-185, 337

Food Safety Act 1990, 12, 151, 311, 323, 331, 343
Food Safety & Hygiene Legislation, 328-357
Food Safety Authority of the Republic of Ireland, 17
Food Safety (Enforcement Authority) England and Wales Order, 1990, 343
Food Safety (Sampling and Qualifications) Regulations, 1990, 50, 346
Food Service Sanitation Manual (USA), 70
Food spoilage, 142-145
Food Spoilage and Preservation, 142-157
Food Standards Act, 1999, 355
Food Standards Agency (FSA), 13, 14, 31, 42, 47, 54, 57, 83, 176, 339, 331, 337, 355, 375
Food standards, law relating to, 351
Foodstuffs, 336
Food law enforcement, 355
Foreign bodies, 79, 209
    detection of, 86
    extrinsic, 79, 384
    intrinsic, 79, 386
Free fatty acids, 143
Free radicals, 68, 155
Freezer breakdown, 130
Freezer burn, 129, 147, 384
Freezing
    of food, 129
    rates of, 146
Freezing point test, 368
Freezing systems, 147
Frozen food, 127
Frozen poultry, thawing and cooking, 131
Fruit, 74, 112
Fruit Flies, 268
Fumigation, 271, 384
Fungi, 28, 384
Fungicide, 385
Furocoumarins, 45
***Fusarium* spp.**, 28

G
Galvanized equipment, 44, 385
Gambierdiscus toxicus, 46
Gamma rays, 154, 157

Garden Ant, 277
Gas connections, 212
Gas supplies, to food premises, 191
Gastroenteritis, 12, 374, 385
Gel cleaning, 233
Gelatin, 124
General Food Regulations 2004, 331
Generation time, 25
Genes, 67
Genetically Modified and Novel Foods (Labelling) (England) Regulations, 2000, 352
Gerber test, 368
German Cockroach, 275
Germicide, 385
Germination, 21, 385
Germs, 10
Giardia lamblia, 16, 17, 19, 60
Giardiasis, 60
Glass, 83
Glasswashers, 195
Globulin, 367
Gloves, 165, 168
Glucose, 93, 109
Glycoalkaloids, 45
Goats milk, 52, 74
Grain Weevil, 273
Gram stain, 23
Gram-negative, 23
Gram-positive, 23
Granolithic, 191
Grease, 82
    polymerized, 25
    traps, 192, 385
Greenbottle, 268
Greenhouse effect, 115
Guards, 211
Gulls, 278

H

HACCP 8, 56, 63, 64, 79, 80, 86, 158, 283-299, 332, 385
    advantages of, 285
    and catering, 295
    control chart, 295, 296-298
    EU legislation, 328
    plan, 293, 294
    prerequisites for, 286
    reasons for failure, 286
    seven principles of, 287
    stages in implementing, 287-294
    team, 287
Haemoglobin, 104, 382
Haemolytic uraemic syndrome (HUS), 56
Hafnia spp., 45
Hair, 166
Hairnet, 166
Haloduric organisms, 152, 385
Halophiles, 152, 385
Hand
    disinfection, 165
    drying, 163
    hygiene, 159
    sprayers, 237
Hand-contact surfaces, 229, 385
Handwashing
    facilities, 194
    reasons for failure of, 164
    requirements for, 161
Harbourage, of pests, 256
Hard swell, 362
Hardwood, 213
Haricot beans, 45
Hazard analysis and critical control point (HACCP), 283-299, 332, 385
Hazard analysis (Codex), 285, 385
Hazardous chemicals, handling of, 249
Hazardous products, storage, handling and dispensing of, 249
Hazards, 70, 186, 285, 289, 296-298
    biological, 289
    chemical, 289
    physical, 289
Hazards and control measures, 80-89
Hazchem symbols, 248
Head covering, 166
Headspace, 361
Health and Safety, law affecting, 353
Health and Safety at Work etc. Act, 1974, 247, 255, 267, 283, 353
Health and Safety (Display Screen Equipment) Regulations, 1992, 248, 354
Health and safety of cleaning, 247
Health and Safety Commission, 353
Health declaration, 385

Health Protection Agency, 351
Health Services and Public Health Act, 1968, 357
Healthy carriers, 11
Healthy diet, 100
Heart disease, 95
Heat disinfection, 226
Heat labile, 35, 385
Heat resistant organisms, 149
Heat treatment, 336
Hepatitis, 59
Herbal tea, 45
High court, 356
High-risk foods, 8, 126, 176, 377, 385
Histamine, 45
Histidine, 45
Hofmannophila pseudospretella, 274
Home Authority Principle, 326
Homogenization, 369
Honey, 36, 43
Horse meat, 61
Host, 385
Hot food, 119
    cooling of, 125, 336
Hot holding requirements, 139, 341, 342
Hot water supplies, 191
House Fly, 268
House Mouse, 261, 262
Housekeeping, of food premises, 257, 278
HTST (high temperature short time) process, 366
Human body, 109
Humidity, 111, 116, 149
Hydramethylnon, 277
Hydrazine derivatives, 45
Hydrogen bond, 67
Hydrogen cyanide, 45
Hydrogen peroxide, 228
Hydrogen sulphide, 144
Hydrogen sulphide gas, 363
Hydrogen swells, 363
Hydrogen-ion concentration (pH), 27
Hydrometer test, 368
Hygiene, 7, 332, 385
    committees, 181
    cost of poor hygiene, 7
    courses, 377-379

    emergency prohibition notice, 338
    emergency prohibition order, 338
    improvement notice, 337
    prohibition order, 337, 338
    qualifications, 377
    regulations, 337
    training, 171-185, 337
Hyphae, 28
Hypochlorite dips, 54
Hypochlorites, 226, 230
Hypochlorous acid (HOCl), 226

I

Ice, 191
Ice crystals, 129
Ice machines, 191
Iceberg lettuce, 53
Ice cream, 112, 373
Ice cream machine, cleaning of, 245
Identification of bacteria, 23
Ileum, 108, 109
Imitation cream, 370
Immune system, 385
Immunity, 385
Immunoassay, 24, 308
Immuno-compromized, 385
Impedance testing, 309
Incubation, 312, 385
Indian meal moth, 274
Indicator organisms, 312, 385
Indictable offences, 329, 339, 345
Induction training, 179
Industrial waste oil, 43
Industry guides to good hygiene practice, 158, 175, 178, 333
Infant botulism, 36
Infants, 103
Infectious intestinal diseases, 14
Infective dose, 385
Infestation, 386
Information, 329
Infrared thermometers, 121
Inorganic, 386
Insecticides, 43, 270, 386
    residual, 388
Insects, 75, 267
    control of, 267, 270

Inspection,
    belt, 81
    chambers, 192
Inspection of food premises, 297-303, 386
    by authorized officers, 316-320
Instrumentation, 210
Integrated pest management (IPM), 254, 386
Intermediate Certificate in Applied HACCP
    Principles (RIPH), 378
Intermediate Food Hygiene Course (REHIS), 379
Intermittent excretors, 11
Intrinsic foreign bodies, 79, 386
Introduction to Food Hygiene Course (REHIS),
    379
Investigation of food complaints, 321
Iodine, 99, 227
Iodophors, 227
Iron, 44, 98, 113
Irradiation of food, 154
Irrigation, 74
Isolate, 386
Iso-propanol, 228

J
Jejunum, 108, 109
Jewellery, 168
Joints, 208
Joints,
    cooling of, 125
Judge, 329
Justice of the peace, (JP), 329, 343

K
Kentucky mop, 240
Kick plates, 256
Kitchen
    design, 199
    plan, 201
Klebsiella spp., 45

L
Laboratory isolations, 13
Lactation, diet in, 104
Lactic acid, 153, 370
Lactic acid bacteria, 153
Lactobacillus bulgaricus, 370
Lactose, 93

Lag phase, 26, 386
Laminates, 114
Larvae, 268, 386
Lasius niger, 277
Lassitude, 386
Latent heat, 146
Latex gloves, 165
Lavatories, 333
Law governing the sale of food, 328-350
Law Reform (Miscellaneous Provisions),
    (Scotland) Act, 1990, 356
Lead, 44
Learning methods, 173
Legal terms, 329
Legionella bacteria, 193
Lesser house fly, 268
Lettuce, 75
Leucocytes, 369
Licensing Board, 356
Licensing Committee, 356
Licensing (Scotland) Act, 1976, 356
Lighting of food premises, 193
Limescale remover, 252
Linear workflow, 186
Linoleic acids, 93
Linolenic acids, 93
Lipase, 143
Liposcelis bostrychophilis, 273
Liquid egg, 374
Liquid nitrogen, 147
Liquid soap, 161
Listeria, 16, 17, 19, 161
Listeria monocytogenes, 14, 54, 134, 135
Listeriosis, 54
Load-lines, 120, 386
Local acts, 328
Local Authorities Coordinators of Regulatory
    Services (LACORS), 311, 317, 320, 326
Local Government Miscellaneous Provisions
    Act, 1976, 197
Lockers, 168
Logarithmic phase, 26
Lovibond comparator, 368
Low-acid foods, 360, 386
Low-dose pathogen, 10, 386
Low-risk foods, 9, 386
Luciferin luciferase, 232

Lucilia spp., 268
Lux, 386
Lysozyme, 145

M

MacConkey agar, 23
Machinery, new, 205
Machinery Regulations, 207, 209, 210
Machines, single disc scrubbing, 241
Macronutrients, 90
Maggots, 268
Magistrates' court, 329
Magnesium, 98
Magpies, 54
Maintenance, 215, 280
Maintenance operatives, 84
Malaise, 386
Management, 48
    failures, 15, 174
    functions of cleaning, 220
    knowledge, 174
    responsibility, 174
Management of Health and Safety at Work Regulations, 1999, 247, 354
Manual Handling Operations Regulations, 1992, 247, 355
Margarine, 371
Market stalls, 335
Mastitis, 74, 368, 386
Materials and Articles in Contact with Food Regulations, 1987, 353
Maximum permissible levels, 81
Mayonnaise, 73
Meat pies, 111
Mechanical aids, 241
Mechanical dishwashing, 243
Mechanical ventilation, 333
Median, 386
Medical
    clearance, 169
    questionnaires, 11
    screening, 159
Melons, 9, 43, 74
Mercury, 43
Mesophiles, 27, 362, 386
Metabolism, 386
Metal, galvanized, 384

Metal detection, 87
Metallic taint, 113
Methoprene, 277
Methylene blue test, 366, 373
Mice, barbecued, 263
Microaerophilic, 53, 386
Microbial death, 68
Microbiological
    analysis, 292
    criterion, 342
    guidelines, 314
    quality, 315
    testing, 308-310
Microbiology, 20-29, 386
MicroModel, 311
Micronutrients, 90
Micro-organisms, 20-29, 70, 160, 358, 361, 386
Microscopical examination, 23
Microwave ovens, 65, 124, 213
Mildew, 386
Milk, 74, 112
    condensed, 372
    dried, 372
    pasteurized, 365
    processing of, 365
    raw, 55, 56, 61, 74
    ropey, 369
    souring of, 369
    sterilized, 367
    sweetened condensed, 372
    taint, 368
    testing of, 365
    thermized, 367
    unsweetened evaporated, 372
Milkstone, 368
Mill Moth, 274
Minerals, 90, 96, 98, 99
Mist netting, 281
Mites, 272
Mobile
    equipment, 115
    racks, 115
    sales vehicles, 335
    silos, 115
Modelling, 301
Modified atmosphere packing, 115, 154, 386

Moisture, 27
Molybdenum, 207
Monilia sitophilia, 143
Monitoring, 282, 285, 292, 296-298
Monitoring temperatures, 120
Monomorium pharaonis, 277
Monosodium glutamate, 43
Monounsaturated fatty acids, 94
Mops, 240
Morganella spp., 45
Morphology, 24, 386
Moulds, 28, 46, 70, 142, 145, 152, 362, 386
Mouse traps, 261
Mouth, 166
Mucor spp., 143
Multimedia projectors, 183
Mus domesticus, 261
Musca domestica, 268
Mushroom poisoning, 45
Mussels, 46
Mycelium, 28
Mycobacterium bovis, 55
Mycobacterium tuberculosis, 55
Mycotoxins, 10, 28 46-48, 387
Mycotoxin-producing moulds, 372
Myoglobin, 123

N
Nailbrush, 162
Narcotizing, 280
National guides to good practice, 333
National Health Service (Amendment) Act, 1986, 351
National Health Service Food Premises Regulations, 1987, 351
National physical laboratory, 307
National Vocational Qualifications (NVQs), 179
Neonate, 387
Neophobic, 262
Neurotoxin, 387
Neutral detergents, 224
Niacin, 97
Nisin, 153
Nitrates, 43
Nitrites, 43, 149, 151

Nitrosamines, 152
Non-ionic surfactants, 219
Norovirus, 16, 17, 19, 57
Norway Rat, 260, 262
Nose, 166
Notices, 83, 196
Nucleic acids, 20, 66, 67
Nucleotides, 67
Nutrient agar, 23
Nutrients, 26, 108
Nutritional needs, 103
Nutritional Safety and Food Quality, 90-109
Nuts, 82
Nylon filament, 240

O
Oakwood Deli, 325
Obesity, 91
Observations, 303
Obstruction, 339, 345
Oesophagus, 108
Office of Fair Trading, 349
Official Feed and Food Controls (England) Regs. 2006, 349
Ohmic heating, 149, 365
Oils, 82, 94
Older people, diet of 105
Olive oil, 43
Olive-pomace oil, 42
Onset period, 387
Oocysts, 61
Ootheca, 276, 387
Open food, 387
Oral evidence, 325
Orders, 328
Organic, 387
Organisms,
  haloduric, 152, 385
  halophilic, 385
  heat-resistance of, 149
  osmoduric, 152, 387
  pathogenic, 358
  spoilage, 358
  thermoduric, 389
  thermophilic, 362, 389

vegetative, 365
viral, 16
xerophilic, 149, 389
Organoleptic assessment of food, 306, 387
Oriental Cockroach, 275
Osmoduric organisms, 152, 387
Osmophiles, 152, 387
Osmosis, 151
Outbreak, 48, 387
   location of, 387
Outdoor clothing, 168
Oven cleaner, 251
Over filling, 363
Overhead projectors, 183
Oxidative rancidity, 143, 155
Oxygen
   presence of, 28
   restriction of, 154
Oysters, 53, 58
Ozone, 230

P

Packaging, 73, 81, 113, 332, 336
Pallets, 116
Pancreas, 108, 109
Paralytic shellfish poisoning, 46
Parasites affecting food and/or man, 60, 387
Paratyphoid fever, 16, 52
Pasivation, 361
Pasteurization, 148, 364, 387
Pasteurized ham, cans of, 118
Pasteurized milk, 365
Pasteurized shell eggs, 374
Pasties, 111
Pathogens, 10, 12, 117, 313, 358, 387
Patients, 11
Patulin, 46
Peanut butter, 47
Peanuts, 11, 28, 47
Pellofreeze system, 147
Penicillium, 28, 46, 142
Pepsin, 108
Peptide bond, 67
Peptides, 108
Peptone water, 23
Per-acetic acid, 228, 234
Perfume, 168

Perigo effect, 364
Perimeter area, 199
Periplaneta americana, 275
Perishable foods, 117
Peristalsis, 108
Peroxidase test, 366
Peroxy compounds, 228
Personal Hygiene, 158-170, 333, 335, 336, 387
Personal Protective Equipment at Work Regulations, 1992, 248, 355
Pest Control, 85, **253-281**
   book, 266
   environmental factors, 253
   in-house, 259
   reasons for, 254
   role of management, 260
   strategies, 258-260
Pesticides, 367, 387
Pests, 85, 253, 333, 387
   environmental control, 253
   physical control, 254
Pets, 75
pH, 27, 149, 153, 358, 387
Phage typing, 24, 387
Pharaoh's Ant, 277
Phaseolus vulgaris, 45
Phenols, 43, 89
Pheromone lure, 271
PHLS Salmonella Sub-committee, 170
PHLS survey, 73
Phosphatase test, 366
Phosphoric acid, 224
Phosphorus, 98
Physical contamination, 79
Phytochemicals, 102
Phytoestrogens, 45
Pickling, 153
Pillsbury Company, 284
Pipelines, 212
Pipework, 211
Pistachio nuts, 47
Plankton, 46
Plasters, 166
Plastic Materials and Articles in Contact with Foods Regulations, 1998, 353
Plastic strips, 279

Plastics, 81, 114
Plate counts, 308, 368, 387
Plate freezing, 147
Platforms, 189
Plodia interpunctella, 274
Point systems, 279
Poisonous plants and fish, 45, 46
Police and Criminal Evidence Act 1984, 50, 349
Polyester, 238
Polymerase chain reaction (PCR), 309
Polymerized grease, 25
Polypropylene, 207, 213, 238, 240
Polythene, 81
Polyunsaturated fatty acids, 94
Polyvinylchloride, 114
Port Health Authority, 331
Positive list principle, 151
Post appointment, 159
Post-inspection discussions, 305
Post-process contamination, 84, 361, 362
Potable water, 335, 387
Potassium, 98
Potassium hydroxide, 225
Potassium sorbate, 153
Pouches, 137
PowerPoint, 183
Precautionary principle, 331
Precautions, all reasonable, 86
Pregnancy, diet in 105
Premises, 337
Preparation surfaces, 212
Prerequisite programmes, 286
Preservation of food, 145-157, 358, 387
    by chemical methods, 151-153
    by high temperatures, 148, 149
    by low temperatures, 145-147
    by physical methods, 154-157
Pressure systems, 215
Pre-training assessment, 179
Prevention of Damage by Pests Act, 1949, 255
Primary production, 330
Prions, 60, 387
Processing, 332
Procurator fiscal, 356
Product code, 387
Production, primary, 330, 387

Proofing, 269
Proper officer, 12
Proportioning pumps, 237
Proprionates, 143
Prosecutions, 345
Protective clothing, 167, 336, 387
Proteins, 90, 95, 108
Proteolytic, 387
Protozoa, 53, 61, 387
Provisions and Use of Work Equipment Regulations, 1998, 247, 354
Pseudomonas, 142, 162
Psocids, 271
Psoriasis, 169
Psychrophiles, 27, 145, 387
Psychrotrophs, 27, 388
Ptinus tectus, 273
Public analyst, 327, 344
Public Health Control of Disease Act, 1984, 350
Public Health Infectious Diseases Regulations, 1988, 350
Public Health (Notification of Infectious Diseases) Scotland Regs 1988, 357
Public sanitary accommodation, 197
Puffer fish, 45
Pulmonary, 388
Pupa (plural pupae), 388
Purulent gingivitis, 169
Pyrethroids, 270
Pyrrolizidine alkaloids, 45

Q

Quaternary ammonium compounds (QACs), 227, 388
Quick-freezing, 146
Quick-frozen Foodstuffs Regulations, 1990, 349

R

Radiant heat, 120
    from equipment, 194
Radicidation, 388
Radurization, 388
Rancidity, 143
Rapid cleanliness testing, 232
Raspberries, 59, 74, 75

Rattus norvegicus, 260, 262
Rattus rattus, 262-263
Raw food, 71
Raw materials, 80, 110, 305
Raw meat and poultry, 71, 111
Raw milk, 341
Raw milk test, 367
Reaction, allergic, 165
Ready-to-eat foods, 10, 312, 388
Recording thermometers, 120
Records, 180, 294
Rectum, 109
Red kidney beans, 45
Red whelk poisoning, 45
Refreezing, of food, 129
Refrigerated vehicles, 126
Refrigeration units, effective selection of, 122
Refrigerators, 117-123
Refuse, 198
    compaction, 198
    containers, 269
Regeneration, 133
Registration, of premises, 333
Regulation (EC) for Microbiological Criteria
    for Foodstuffs 2005, 342
Regulation (EC) No. 852/2004 on the hygiene
    to foodstuffs, 331
Regulation (EC) No. 853/2004 laying down
specific hygiene rules for food of animal
    origin, 346
Regulation (EC) No. 854/2004 laying down
    rules for official controls on products of
    animals, 347
Regulation (EC) No. 882/2004 on official
    controls to ensure compliance with food
    law and animal health, 348
Regulations, 328
Reheating and service, 138, 342
Rehydration, 151, 388
Religious and ethnics groups, dietary
    restriction practised by, 107
Rennet, 371
Repellent gel, 279
Report, 305
Reporting of Injuries, Diseases and
    Dangerous Occurrences Regulations,
    1995, 354

Resazurin test, 368
Resident micro-organisms, 160
Residual insecticide, 270
Resistance
    of bacteria, 231
    of rodents, 264
Respiration, 112, 388
Retail, 330
Retail premises, 128
Retinol, 97
Retortable pouches, 364
Returnable bottles, 88
Review (of HACCP), 293, 333
Rhizopus, 142
Rhubarb leaves, 45
Rilsan, 238
Risk assessment, 15, 170, 171, 247, 332, 388
Risk, 186, 331, 388
RNA, 68
Rodent
    control, 263-267
    damage, 261
    smears, 261
    sticky boards, 266
Rodenticides, 264, 265
Rodents, 260-267
Roller drying, 150
Rope, in bread, 143
Rotaviruses, 57
Roughness, of surface finish, 207
Routes of contamination, 76, 388
Royal Environmental Health Institute of
    Scotland (REHIS), 377
Royal Institute of Public Health (RIPH), 377
Royal Society for the Promotion of Health
    (RSPH), 377
Rubber, 81
Rubber gloves, 168
Runaway heating, 131
Rust, 82

S
Sabotage, 288
Safe cooking temperatures, 124
Safe food, 7, 282, 388
Safety committee, 354

HYGIENE for MANAGEMENT Index

Safety data sheet, 249
Safety policies, 353
Saliva, 109
Salmonella, classification of, 30
Salmonella Ealing, 32
Salmonella Eastbourne, 32
Salmonella Enteritidis, 72, 325
Salmonella Enteritidis PT4, 32, 33
Salmonella food poisoning, routes of, 31
Salmonella in Ireland, 18
Salmonella, infection, 40
Salmonella Kedougou, 9
Salmonella Napoli, 32
Salmonella Oranienburg, 32
Salmonella Paratyphi, 52
Salmonella Saint-paul, 32
Salmonella Typhi, 52
Salmonella Typhimurium DT104, 33
Salmonellae, 7, 10, 11, 13, 14, 16, 19, 20, 22, 30, 74, 75, 124, 313
Salt, 100, 151
Sampling, 342, 345, 346
Sampling programme, 312
Sanitary convenience, 196, 333
Sanitary disposal, 197
Sanitizer, 251, 388
Saponification, 219
Saprophyte, 388
Saturated fatty acids, 93
Sausage rolls, 111
Scanners, 89
Scaring, of birds, 279
Schoolchildren, diet of 103
Scientific Principles for Food Safety, 63-69
Scombrotoxic fish poisoning, 45
Scottish Centre for Infection and Environmental Health, 16
Scottish legislation, 356
Scouring powder, 219
Scrubbing machine, 241
Scum, 219
Sealed, hermetically, 385
Seams, of containers, 360
Sebaceous glands, 160
Selenium, 99

Septic spots, 166
Septicaemia, 388
Sequestrants, 219
Serological typing, 24
Serotyping, 388
Services, of food premises, 191
Severity, 285
Sewers, 192
Shelf life, 111, 113, 119, 120, 127, 340, 352, 365, 388
Shelf-stable, 388
Shellfish, 58, 59, 374
   purification of, 375
Sheriff, 356
Sheriff court, 356
Shigella flexneri, 53
Shigella sonnei, 53
Shigellae, 11, 16, 17, 19
Shooting, of birds, 280
Silverfish, 272
Single-wash procedure, 163
Sinks, 195, 241
Site selection, 186
Siting of equipment, 200
Sitophilus granarius, 273
Size of kitchens, 201
Skin infections, 166
Skin lotions, 166
Slaughterhouses, 111
Slimmers, diet of, 105
Slotted trays, 115
Slow cooking, 124
Small round structured viruses (SRSVs), 57
Smoking, 154, 167
Sneeze guards, 388
Soak clean, 233
Society of Food Hygiene and Technology (SOFHT), 377
Socket mop, 240
Sodium, 98
   benzoate, 152
   carbonate, 225
   chloride, 364
   dichloro-isocyanurate, 227
   hydroxide, 224, 225
   hypochlorite, 226

HYGIENE for MANAGEMENT Index

metasilicate, 225
nitrate, 43, 152
nitrite 43
propionate, 153
Soft cheeses, 54, 371
Soft drinks and fruit juices, 144
Soil, 76, 225
Soils, carbonized/polmerized, 234
Solar heat gain, 116, 189
Solders, 208
Solids-not-fat, 368
Solvents, 219
Somatic cell, 368
Sorbic acid, 153
Sources of contamination, 48, 71, 76, 388
Souring of milk, 369
Sous vide cook-chill, 136-138, 382
Soy sauce, 43
Sparrows, 278
Specification, of products, 305
Specifications, 81
Specimens, 13
Spices, 305
Spitting, 167
Spoilage, 127, 388
    bacteria, 143-144, 145
    of food in cans, 362
    of poultry, 144
Sporadic case, 388
Spores, 21, 28, 33, 123, 126, 364, 388
Sporulation, 388
Spotters, 88
Spray balls, 236
Spray drying, 150
Springer, 362
Spirochaetes, 21
Stack burn, 361
Staff
    appraisals, 180
    motivation, 184
    sanitary convenience, 196
    selection, 158
Stainless steel, 207
Stairs, of food premises, 189
Staleness, of bread, 143
Stanley Royd Hospital, Wakefield, 32
Staphyloccus aureus, 21, 22, 24, 27, 36, 41, 71, 75, 124, 160, 166, 313, 314, 373
Staphylococci, 10, 11, 21, 24
Staphylococcus epidermis, 160
Staphylococcus saprophyticus, 160
Staples, 81
Starch, 93
Starlings, 278
Starter culture, 370
Stationary phase, 26
Statutory codes of practice, 346
Statutory Instruments, 328, 330
Steam disinfection, 226
Steam Fly, 275
Stegobium paniceum, 273
Sterile, commercially, 148, 358, 382
Sterilization, 148, 364, 388
Sterilized cream, 369
Sterlized milk, 367
Sterilizing sinks, 195
Sticky flypaper, 270
Sticky traps, 276
Stipendiary magistrate, 330
Stock control, 222
Stock rotation, 116, 366, 388
Stomach, 109
Storage and disposal of waste, 199
Storage and Temperature Control of Food, 110-141
Storage
    correct methods of, 257
    environmental considerations, 250
    of clean equipment, 214
    systems, 115
    times, 129
Stored products insects, 271-274
Streptococci, 21
Streptococcus lactis, 370
Streptococcus thermophilus, 370
Streptococcus viridans, 38
Streptococcus zooepidemicus, 55
String, 81
Sublimation, 150, 373, 388
Sub-typing, 24
Sucrase, 109
Sucrose, 93
Sugars, 93, 152
    extrinsic, 93

intrinsic, 93
milk, 93
Sulphur dioxide, 153
Summary conviction, 339, 345
Summons, 330
Sun drying, 150
Suppliers, 305
Supply of Machinery (Safety) Regulations, 1992, 205
Surface cleaner, hard, 251
Surface finish, 207
Surface tension, 223
Surface
food-contact, 334, 384
hand-contact, 385
Surfactancy, 223
Surfactants, amphoteric 218, 219
Surveillance, 48, 388
Surveys, of rodents, 260
Suspension, 223
Sweetened condensed milk, 372
Symptomatic, 389
Symptomless carriers, 11
Synergism, 220, 389
Systemic disease, 389

T

Taenia saginata, 61
Taenia solium, 170
Taints, 369, 389
Tampering, control measures against, 84
Tanks, 241
Tannin, 246
Tap proportioner, 389
Target level, 285, 291
Taxonomy, 389
Temperature control requirements, 340
Temperatures, 27, 340, 342, 365
above freezing, 145
ambient, 380
below freezing, 145
core, 382
data loggers, 122
effects of fluctuating, 129
workroom, 193
Temporary Prohibition Orders, 375
Terrapins, 75

Terrazzo flooring, 389
Testing, 180
microbiological challenge, 151, 381
Tetramine, 45
Thamnidium elegans (whiskers), 142
Thawing, of frozen food, 130, 336
Thermistor, 120
Thermocouple, 120
Thermometers, 120
Thermoduric organisms, 389
Thermophiles, 27, 148, 362, 389
Threadworm, 170
Tin, 44, 113, 363
Tineola bisselliella, 274
Tolerance, 285, 291
Total viable count (TVC), 312, 389
Toxins, 10, 21, 389
Toxoplasma gondii, 61
Trace element, 90
Tracking powder for rodents, 263
Trading standards officer, 349, 354
Training and Education of Food Handlers, 171-185, 337
Training
committee, 177
factors affecting the trainee, 182
factors affecting the trainer, 182
importance of, 171
legal requirements for, 175, 337
methods, 181-185
programme, 176-181, 183
records, 175-181
reinforcement of, 184
sessions, 181
Training needs analysis, 177
Transparencies, 183
Transport, 335
Traps, 265, 280
Traywasher, 234
Trichina cysts, 145, 254
Trichina spiralis, 61, 156
Trichinosis, 61
Triclosan, 214
Trypsin, 108, 153
Tuberculin tested, 74
Tuberculosis, 55
Tunnel drier, 150

Turbidity test, 367
Typhoid, 16, 52
Typhoid outbreak, Aberdeen, 74
Typing, bacterial, 389

U
Ultimate consumer, 330
Ultra heat treated milk (UHT), 75, 149, 367
Ultraviolet light, 269
Underprocessing, 361
Unhygienic practices, 168
Unloading of high-risk food, 126
Unpacking, 81
Unpasteurized milk, 9
USA Department of Agriculture, 124
Use-by date, 351, 352, 389
UV
    radiation, 225
    treatment, 73, 375

V
Vaccination, 72
Vacherin Mont d'or cheese, 54
Vacuum cleaners, 241
Vacuum packing, 114, 137, 154
Validation, 285, 293
Vegetables, 74, 112
Vegetarians, diet of, 106
Vegetative cell, 21, 389
Vehicles
    of bacterial contamination, 76, 389
    of food poisoning, 8, 76
    refrigerated, 126
Vending machines, 140, 335
Ventilation, 192
Ventilation ducts, 76
Verification, 285, 293
Vero cytotoxins, 55
Vessels, 212
Vibrio parahaemolyticus, 25, 37, 41
Vibrio vulnificus, 38
Vibrios, 21
Vinegar, 142
Viral gastroenteritis, 57
Virology, 389
Virulence, 389
Viruses, 10, 389

Visitors, 71
Visual aids equipment, 183
Vitamin C, 146
Vitamins, 90, 96, 97
    fat soluble, 97
    water soluble, 97
VTEC, 55

W
Wall surfaces, 188, 334
Walls of food premises, 188
Warehouse moth, 274
Warfarin, 264
Warm air driers, 163
Warrants, 339, 345
Washhand basins, 194, 333
Washdown systems, 234
Washing facilities in food premises, 196
Washing machine, automated, 234
Washing soda, 225
Washing-up liquid, 252
Wasps, 269
Wastage, of food, 254
Waste, 198
    food, 198
    pipes, 195
Water, 92
    activity, 27, 389
    authority, 250
    hardness, 389
    softener, 389
    supply, 74, 191, 335
Waterproof dressing, 166
Weedkiller, 43
Weil's disease, 254
Wetting power, 223
Wholesome food, 7, 389
Wicks, 265
Wild birds, 75
Wildlife and Countryside Act 1981, 280
Windows, 189, 334
Wishaw, 56
Wood, 207
    splinters, 82
Workflow, 187
Workforce assessment, 177, 179, 181
Workplace (Health and Safety and Welfare)

HYGIENE for MANAGEMENT Index

Regulations, 1992, 196, 248, 354
Worksurface cleaning of, 245
Wrapping, 113, 332, 336

X

Xerophilic organisms, 149, 389
X-ray inspection systems, 87
X-rays, 154

Y

Yeasts, 29, 142, 152, 160, 362, 370, 389
Yersinia enterocolitica, 16, 17, 38, 42, 134
Yogurt, 370
Young children, 103

Z

Z value, 389
Zinc, 44, 98
   phosphide, 265
Zone of maximum crystallization, 146
Zoonoses, 389

HYGIENE for MANAGEMENT